Developments in Mathematics

Volume 49

More information about this series at http://www.springer.com/series/5834

Michael D. Hirschhorn

The Power of q

A Personal Journey

Springer

Michael D. Hirschhorn
School of Mathematics and Statistics
University of New South Wales
Sydney, NSW
Australia

ISSN 1389-2177 ISSN 2197-795X (electronic)
Developments in Mathematics
ISBN 978-3-319-57761-6 ISBN 978-3-319-57762-3 (eBook)
DOI 10.1007/978-3-319-57762-3

Library of Congress Control Number: 2017939603

Mathematics Subject Classification (2010): 11P83, 11P81, 05A17, 11P84, 11E25, 33D15

Printed on acid-free paper

This Springer imprint is published by Springer Nature
The registered company is Springer International Publishing AG
The registered company address is: Gewerbestrasse 11, 6330 Cham, Switzerland

Dedicated to my wife, Terri, on the 50th Anniversary of our first meeting, July 19, 1966.

The original version of the book has been revised: For detailed information please see Erratum. The Erratum to the book is available at http://dx.doi.org/10.1007/978-3-319-57762-3_44

Foreword

This is a unique book written by a brilliant and unique mathematician. I have known Mike Hirschhorn for more than four decades and I was a co-advisor for his Ph.D. thesis.

The world of q is a broad one. It is primarily focused on aspects and implications of q-hypergeometric functions, theta functions and q-infinite products. All these are composed in various ways from factors of the form $1 - q^n$. The subject has serious applications in various branches of number theory including representations by quadratic forms (especially sums of squares) and the theory of integer partitions.

The Preface commences with the comment “When I am asked what sort of mathematics I do, I say ‘High School algebra, but taken somewhat further’”. This beginning, with an appealing modesty, gently prepares the reader for what will turn out to be a startling display of virtuosity. Mike Hirschhorn has perfected a method that had its start in the works of Jacobi and has been carried on in the works of Schroeter, Rogers, Ramanujan, Watson and others. The book contains a number of novel proofs of famous results, such as the Jacobi triple product identity, the Rogers–Ramanujan identities and the formula described as Ramanujan’s most beautiful identity:

$$\sum_{n\geq 0} p(5n+4)q^n = 5 \prod_{n\geq 1} \frac{(1-q^{5n})^5}{(1-q^n)^6},$$

where $p(n)$ is the number of partitions of n.

However, the majority of the book is devoted to results originally found by the author such as:

$$\left(\sum_{-\infty}^{\infty} a^n q^{n^2}\right)^3 = \left(1+6\sum_{n\geq 0}\left(\frac{q^{6n+2}}{1-q^{6n+2}} - \frac{q^{6n+4}}{1-q^{6n+4}}\right)\right)\sum_{-\infty}^{\infty} a^{3n}q^{3n^2}$$

$$+3q\prod_{n\geq 1}\frac{(1-q^{6n})^3}{1-q^{2n}}\sum_{-\infty}^{\infty}(a^{3n+1}+a^{-3n-1})q^{3n^2+2n},$$

or the delightful:

The number $n \equiv 4 \pmod 8$ *is the sum of four distinct even squares if and only if it is the sum of four distinct odd squares. Moreover, the number of partitions into four distinct even squares is never less than half, and never more than, the number of partitions of n into four distinct odd squares.*

I began this Foreword noting the uniqueness of Mike Hirschhorn's mathematics. I truly meant this as much more than a vague compliment. Mike's special style, technique and choice of topic will become vivid as you read this book. I have many times examined a new paper by him and commented: "Ah! Vintage Hirschhorn!"

I recommend that you read his Preface, Introduction and the next two chapters. Then, dip in anywhere to find "vintage Hirschhorn". Enjoy!

George E. Andrews
Evan Pugh University Professor in Mathematics
The Pennsylvania State University

Preface

When I am asked what sort of mathematics I do, I say "High School algebra, but taken somewhat further". This is the study and application of q-series, technically known as basic hypergeometric series. The theory of q-series has its roots in the work of Euler, Cauchy, Abel and, in particular, Jacobi. It was given great impetus by Ramanujan, and has been contributed to by many others, too numerous to list here.

We learn in High School that the geometric series

$$1+x+x^2+\cdots=\sum_{n\geq 0} x^n$$

sums to

$$\frac{1}{1-x}$$

provided that $|x|<1$, that is $-1<x<1$.

At some stage in the 20th Century, it became popular to use the symbol q instead of x (hence "q-series"), so

$$1+q+q^2+\cdots=\frac{1}{1-q}$$

provided $|q|<1$. But it is noteworthy that Ramanujan used the symbol x (and did not use $\sum$ notation!).

We shall see that in the hands of such luminaries as Euler, Jacobi and Ramanujan, the simple fact stated above became a fundamental tool in the development of q-series and its application to problems in the theory of numbers.

In 1621, Bachet made the claim that every number (non-negative integer) is the sum of four squares, but he had no proof. This problem remained open through the next 150 years, defeated Euler and was finally settled by Lagrange in 1770.

In 1828, Jacobi wrote a monumentally important paper, in which he proved the four-squares theorem (and much more) as a consequence of an identity which became known as the triple product identity. This book starts with a proof of Jacobi's triple product identity, and within the first two chapters (30 pages), we will have proved Lagrange's four-squares theorem, a problem that challenged mankind for a century and a half! Thus we see that the theory of q-series is a powerful tool, and this explains the title, for which I am indebted to my wife Terri.

Before I go into more detail of what you will find in my book, let me tell you more about it.

The book truly is elementary, and should be accessible to a committed undergraduate or potential postgraduate, or to an interested layperson who is prepared to do a little hard (but not abstract) algebra. There are occasions also when we use complex numbers (roots of unity) and calculus (differentiation).

The book did not come about as a textbook for any course, but it ought to be possible to use at least parts of it as a text. There are plenty of exercises for the reader, some of them explicit, others to fill gaps in the exposition. The anonymous reviewer has prevailed upon me to include more exercises than I might have done, and accordingly, Chapters 14 and 26 are essentially all exercises.

The level of sophistication rises gradually through the book, with the occasional hard bit. The reader is invited to jump over anything too hard and proceed.

The subject matter of the book is essentially a selection of topics I have worked on over the years, as well as topics others have worked on that I wish I had. I hope that I have made my gratitude and appreciation sufficiently clear when I have used the work of others. In writing the book, I have in most cases reworked the material in my published and unpublished papers in order for the material to be more accessible, and, where I have found possible, arranged or explained more clearly.

One of my chief aims is to bring out the extreme beauty of some of the results. A mathematical formula exhibits beauty if it contains an element of surprise. This surprise might just be the simplicity of the formula. Perhaps the supreme example of this is the formula described by G.H. Hardy as most typical of Ramanujan, which we meet in Chapter 5.

As will become clear, the principal influence on my life's work has been Ramanujan. I have had contact with him now for more than 50 years, and yet I understand only a small part of his output. But there are many others who have had an influence on me, and it would be remiss of me not to acknowledge them. It is difficult to list the people who have influenced me in order of importance, so I am bound to offend someone. I wish particularly to acknowledge George E. Andrews, who was my Ph.D. supervisor, and gave me my career, Leonard Carlitz, who published my first paper, Dick Askey, who encouraged me to learn the subject, George Szekeres, Frank Garvan, Shaun Cooper, James Sellers, David Hunt, G.H. Hardy, Bruce Berndt, Heng Huat Chan, Krishna Alladi, Hershel Farkas and Elizabeth Loew. I would also like to acknowledge the many people with whom I have discussed matters over the years, and all my co-authors.

Now let me describe the material of the book in some detail. In introductory Chapter 1, you will find not only a proof of the triple product identity, but also the series expansion of Euler's product,

$$E(q) = (1-q)(1-q^2)(1-q^3)\cdots = \prod_{n\geq 1}(1-q^n),$$

as well as Jacobi's series expansion of the cube of Euler's product, and in Chapter 2 I prove the four-squares theorem and Gauss's two-squares theorem.

Next come 11 chapters on matters associated with (unrestricted) partitions and Ramanujan's partition congruences.

Euler was apparently the first to study $p(n)$, the number of partitions of n. If n is a positive integer, one can ask for the number of ways in which n can be partitioned, that is written as a sum of one or more positive integers. For example, the partitions of 4 are 4, $3+1$, $2+2$, $2+1+1$ and $1+1+1+1$. Thus, the number of partitions of 4 is 5 (note that 4 is itself a partition of 4 and that $1+3$ is considered the same partition as $3+1$), and we write

$$p(4) = 5.$$

Euler knew that

$$1 + \sum_{n\geq 1} p(n)q^n = \frac{1}{E(q)},$$

and that $p(n)$ can be calculated recursively (that is the value of $p(n)$ can be found from the values of $p(k)$ for $k<n$.)

Ramanujan was convinced that there was a closed formula for $p(n)$, and he and Hardy found such a formula. In checking their formula, Ramanujan discovered amazing, totally unexpected, arithmetic properties of $p(n)$. The simplest of these properties is that if N is 4 more than a multiple of 5, then $p(N)$ is a multiple of 5. This can be written

$$p(5n+4) \equiv 0 \pmod 5. \tag{P1}$$

Ramanujan also discovered that

$$p(7n+5) \equiv 0 \pmod 7 \tag{P2}$$

and that

$$p(11n+6) \equiv 0 \pmod{11}. \tag{P3}$$

In this book, you will find a handful of proofs of each of these. Indeed, one of the high points (for me) of the book is my recent simple proof of (P3) to be found in Chapter 3.

The congruence (P1) is an immediate consequence of the identity I alluded to earlier, which has become known as "Ramanujan's most beautiful identity",

$$\sum_{n\geq 0} p(5n+4)q^n = 5\frac{E(q^5)^5}{E(q)^6}. \tag{P4}$$

We will see proofs of this in Chapters 5 and 6.

In these chapters on partitions, you will meet some marvellous identities, including the quintuple product identity and Winquist's identity, the Atkin–Swinnerton-Dyer congruences, Ramanujan's 5-dissection of Euler's product and his factorisation of that identity, Ramanujan's 7-analogue of his most beautiful identity, as well as the Andrews–Garvan crank.

I make it one of my goals in these chapters to obtain as much as I can with as little as I can. Thus, for example I prove all of (P1)–(P4), as well as Ramanujan's partition congruences for powers of 5 and 7, making use only of the expansions of Euler's product and its cube.

Before I go on to detail the rest of the contents of this book, let me say something about a technique I use that runs as a unifying thread through the book.

Quite often we have under consideration a sum of the form

$$S = \sum_{k_1,\cdots,k_p=-\infty}^{\infty} (-1)^L q^Q,$$

where L is linear and Q is quadratic in the variables $k_1,\cdots,k_p$.

We separate S into a number of sums according to some property of Q, for example the residue of Q modulo m.

We then make use of linear transformations of the variables to write the separate sums in such a way that they can be summed by one of the several sum-to-product identities that we have established. This versatile, and as we shall see, powerful, technique of using linear transformations I learned from studying the papers of Leonard Carlitz. Apparently, the technique has its origin in the work of Jacobi. I illustrate this technique in §1.10.

In Chapter 14 we study partitions where even parts come in two colours, and see that, as discovered by Hei-Chi Chan, congruences hold modulo powers of 3, analogous to Ramanujan's partition congruences modulo powers of 5, 7 and 11.

The celebrated Rogers–Ramanujan identities and the Rogers–Ramanujan continued fraction feature in the next three chapters. We give Ramanujan's proof of the Rogers–Ramanujan identities, and observe the phenomenon, discovered by George Szekeres, that when the Rogers–Ramanujan continued fraction (or its reciprocal) is expanded as a series, the sign of the coefficients is periodic with period 5. We also complete the proof, begun in Chapter 9, of an identity sent by Ramanujan in his first letter to Hardy, which Hardy says "defeated me completely".

In Chapter 17, we briefly mention a famous set of forty identities of Ramanujan and make the observation that we find proofs of five of them in this book (two in this chapter).

In Chapter 18, we examine a continued fraction usually called the Ramanujan–Göllnitz-Gordon continued fraction which exhibits properties similar to the Rogers–Ramanujan continued fraction.

Jacobi discovered an identity which he described as *aequatio identica satis abstrusa"* ("a rather abstruse identity"), and which has gained importance in string and superstring theory. In Chapter 19 we give two simple proofs of this identity, and formulate it in two striking ways.

Chapter 20 is devoted to proving two modular identities, both of which have beautiful combinatorial interpretations as partition theorems. During the course of the book, we will give no fewer than five proofs of one of these, described by Venkatachaliengar as "a beautiful identity which looks extremely difficult to prove".

In 1917 or 1918, Ramanujan sent Hardy a letter from the nursing home Fitzroy House, in which he stated a striking identity. Chapter 21 is devoted to a proof of this identity, and this leads in the next chapter to a thorough investigation of the cubic theta function analogues introduced by Jon and Peter Borwein and Frank Garvan. These functions and the many relations between them prove to be very useful in our subsequent work.

The next seven chapters are devoted to results giving the number of representations of a number as a sum of certain combinations of figurate numbers, triangles, squares, pentagonal and octagonal numbers, extending the idea of two squares and so on. They culminate in Chapter 29, devoted to proving just three of a set of 298 such results discovered in a computer-assisted search by Ray Melham. In the course of this work we meet Lambert series, bilateral Lambert series, and the Jordan–Kronecker identity, useful in summing bilateral Lambert series.

In the next chapter, we return to the problem of four squares, and consider the problem of partitions, as opposed to representations. We find the generating function for the number of partitions of a number into four squares, $p_{4\square}(n)$, and spend considerable time and effort proving certain relations and Ramanujan-type identities satisfied by it. We also find a formula for $p_{4\square}(n)$ in terms of divisor functions, thus answering a quest of D.H. Lehmer.

In Chapter 31 we use the techniques introduced in the previous chapter to study partitions into four distinct squares of equal parity, a topic inspired by a conjecture made by Bill Gosper at the Ramanujan Centenary Conference in 1987, that if a number is the sum of four distinct odd squares then it is the sum of four distinct even squares. We prove the conjecture correct, and do much more.

In the next nine chapters, we consider various different partition-related functions, including partitions with odd parts distinct, partitions with even parts distinct, overpartitions (that is, partitions into two colours, the parts of one colour being distinct), bipartitions with odd parts distinct, overcubic partitions and generalised Frobenius partitions. In each case we prove certain identities, relations or congruences of the type we have seen in the context of unrestricted partitions.

On the way, we devote a chapter to gathering together many results involving Ramanujan's functions $\phi(q)$ and $\psi(q)$, which subsequently prove indispensable.

We also describe a parametrisation developed by Alaca, Alaca and Williams which has been put to good use in the proofs of certain partition-related results.

We finish with three chapters proving certain identities due to Ramanujan and others, and providing factorisations of these identities, and a final chapter on Ramanujan's tau function.

It is my hope that you will get inspiration from this book to investigate problems in q-series and their applications to partitions and other topics.

You might also read more widely. I suggest Bruce Berndt's "Number Theory in the Spirit of Ramanujan", Hei-Chi Chan's "Invitation to q-series", Shaun Cooper's "Ramanujan's Theta Functions", as well as Bruce Berndt's monumental five volume work "Ramanujan's Notebooks" and the four, soon to be five, volume work by Andrews and Berndt "Ramanujan's Lost Notebook". Of course, there are many other sources, including Ramanujan's Collected Papers, Hardy's book "Ramanujan Twelve lectures on subjects suggested by his life and work", the Notebooks of Ramanujan and "The Lost Notebook and Other Unpublished Papers". And then there are the many papers being produced on the subjects covered in this book and many other related areas.

There is one more thing I have to add, and that is Frank Garvan's computer package "qseries" has been invaluable to me, for which I am very grateful, and which I cannot recommend too highly.

In closing, I hope you enjoy dipping into my book and that you get a thrill from some of the remarkable and beautiful formulas.

Sydney, Australia Michael D. Hirschhorn

Contents

Chapter 1
Introduction

1.1 Introduction

In this chapter, we will do several important things.

We will begin by establishing the extremely powerful identity known as Jacobi's triple product identity,

$$\prod_{k\geq 1}(1+a^{-1}q^{2k-1})(1+aq^{2k-1})(1-q^{2k}) = \sum_{k=-\infty}^{\infty} a^k q^{k^2} \tag{1.1.1}$$

where $|q| < 1$ and $a \neq 0$, we will meet two functions, $\phi(q)$ and $\psi(q)$, introduced by Ramanujan, and give some simple relations between them, we will give the series expansion of Euler's product

$$E(q) = \prod_{n\geq 1}(1-q^n)$$

and of its cube,

$$J(q) = E(q)^3,$$

and we will demonstrate two techniques used often throughout the book, m-dissection of series and linear transformations of the indices in a multiple sum.

The original version of this chapter has been revised. The erratum to this chapter is available at https://doi.org/10.1007/978-3-319-57762-3_44.

M.D. Hirschhorn, *The Power of q*, Developments in Mathematics 49,
DOI 10.1007/978-3-319-57762-3_1

1.2 An Identity of Euler

We start by proving an important identity of Euler.

For $|q| < 1$,

$$(1+a)(1+aq)(1+aq^2)\ \cdots\ = 1 + \sum_{k\geq 1} a^k \frac{q^{(k^2-k)/2}}{(1-q)\cdots(1-q^k)}. \tag{1.2.1}$$

Suppose $|q| < 1$ and

$$(1+a)(1+aq)(1+aq^2)\ \cdots\ = \sum_{k\geq 0} a^k c_k \tag{1.2.2}$$

where $c_k = c_k(q)$.

(The condition $|q| < 1$ ensures the product converges.)

If in (1.2.2) we put aq for a, we find

$$(1+aq)(1+aq^2)(1+aq^3)\ \cdots\ = \sum_{k\geq 0} a^k q^k c_k. \tag{1.2.3}$$

It follows from (1.2.2) and (1.2.3) that

$$\begin{aligned}\sum_{k\geq 0} a^k c_k(q) &= (1+a)\sum_{k\geq 0} a^k q^k c_k(q)\\ &= c_0 + \sum_{k\geq 1} a^k (q^k c_k + q^{k-1} c_{k-1}).\end{aligned} \tag{1.2.4}$$

It follows from (1.2.2) that

$$c_0 = 1$$

and from (1.2.4) that for $k \geq 1$,

$$c_k = q^k c_k + q^{k-1} c_{k-1},$$

or,

$$c_k = \frac{q^{k-1}}{1-q^k} c_{k-1}. \tag{1.2.5}$$

By induction,

$$c_k = \frac{q^{(k^2-k)/2}}{(1-q)\ \cdots\ (1-q^k)}, \tag{1.2.6}$$

and (1.2.2) becomes

$$(1+a)(1+aq)(1+aq^2)\,\cdots = 1+\sum_{k\geq 1} a^k \frac{q^{(k^2-k)/2}}{(1-q)\,\cdots\,(1-q^k)},$$

which is (1.2.1).

We now write

$$(a;q)_k = \begin{cases} (1-a)(1-aq)\,\cdots\,(1-aq^{k-1}) & \text{for } k\geq 1,\\ 1 & \text{for } k=0, \end{cases} \tag{1.2.7}$$

and

$$(a;q)_\infty = (1-a)(1-aq)(1-aq^2)\,\cdots\,. \tag{1.2.8}$$

Then Euler's identity (1.2.1) can be written

$$(-a;q)_\infty = \sum_{k\geq 0} a^k \frac{q^{(k^2-k)/2}}{(q;q)_k}. \tag{1.2.9}$$

Euler's identity is a special case of an identity known as the q-binomial theorem, for which you are referred to the Appendix.

1.3 Our First Proof of Jacobi's Triple Product Identity

If in (1.2.9) we put q^2 for q and aq for a, we find

$$(-aq;q^2)_\infty = \sum_{k\geq 0} a^k \frac{q^{k^2}}{(q^2;q^2)_k}. \tag{1.3.1}$$

That is,

$$(1+aq)(1+aq^3)(1+aq^5)\,\cdots = \sum_{k\geq 0} a^k \frac{q^{k^2}}{(q^2;q^2)_k}. \tag{1.3.2}$$

We will now perform a three-step process that I will call "a pull-back of degree n".

If in (1.3.2) we suppose $q\neq 0$ and replace a by aq^{-2n}, we obtain

$$(1+aq^{1-2n})(1+aq^{3-2n})\cdots(1+aq^{-1})(1+aq)(1+aq^3)\;\cdots = \sum_{k\geq 0} a^k \frac{q^{k^2-2nk}}{(q^2;q^2)_k}. \tag{1.3.3}$$

Let us now multiply (1.3.3) by q^{n^2}, and recall that $n^2 = 1+3+\;\cdots\;+(2n-1)$. We obtain

$$(q^{2n-1}+a)(q^{2n-3}+a)\cdots(q+a)(1+aq)(1+aq^3)\;\cdots = \sum_{k\geq 0} a^k \frac{q^{k^2-2nk+n^2}}{(q^2;q^2)_k},$$

or,

$$(a+q)(a+q^3)\cdots(a+q^{2n-1})(1+aq)(1+aq^3)\;\cdots = \sum_{k\geq 0} a^k \frac{q^{(k-n)^2}}{(q^2;q^2)_k}. \tag{1.3.4}$$

Now suppose $a \neq 0$, and divide (1.3.4) by a^n, and we obtain

$$\begin{aligned}(1+a^{-1}q)(1+a^{-1}q^3)\cdots(1+a^{-1}q^{2n-1})&(1+aq)(1+aq^3)\;\cdots\\ &= \sum_{k\geq 0} a^{k-n} \frac{q^{(k-n)^2}}{(q^2;q^2)_k}\\ &= \sum_{k=-n}^{\infty} a^k \frac{q^{k^2}}{(q^2;q^2)_{n+k}}.\end{aligned} \tag{1.3.5}$$

We can write (1.3.5)

$$(-a^{-1}q;q^2)_n(-aq;q^2)_\infty = \sum_{k\geq -n} a^k \frac{q^{k^2}}{(q^2;q^2)_{n+k}}. \tag{1.3.6}$$

Incidentally, although we assumed $q \neq 0$ on the way to proving (1.3.6), we note that it is true for $q = 0$ also.

We remark that, alternatively, (1.3.6) can (more simply, perhaps) be proved by induction, starting with (1.2.9). The inductive step, namely

$$(1+a^{-1}q^{2n+1}) \sum_{k\geq -n} a^k \frac{q^{k^2}}{(q^2;q^2)_{n+k}} = \sum_{k\geq -(n+1)} a^k \frac{q^{k^2}}{(q^2;q^2)_{n+1+k}} \tag{1.3.7}$$

requires

$$q^{2n+1}\cdot q^{n^2} = q^{(n+1)^2} \tag{1.3.8}$$

and

$$\frac{q^{k^2}}{(q^2;q^2)_{n+k}}+\frac{q^{(k+1)^2+2n+1}}{(q^2;q^2)_{n+k+1}}=\frac{q^{k^2}}{(q^2;q^2)_{n+1+k}},$$

or,

$$q^{k^2}(1-q^{2n+2k+2})+q^{(k+1)^2+2n+1}=q^{k^2}, \tag{1.3.9}$$

which is trivial.

Exercise: Carry out the proof by induction of (1.3.6) starting from (1.2.9).

Returning to (1.3.6), we have, for $|q|<1$ and $a\neq 0$,

$$(-a^{-1}q;q^2)_n(-aq;q^2)_\infty=\sum_{k\geq -n}a^k\frac{q^{k^2}}{(q^2;q^2)_{n+k}}. \tag{1.3.10}$$

We now let $n\to\infty$ in (1.3.10). For each fixed k,

$$(q^2;q^2)_{n+k}\to(q^2;q^2)_\infty. \tag{1.3.11}$$

Although this convergence is not uniform ("the more highly negative the value of k, the longer it takes"), the factor q^{k^2} takes care of that, and ensures that in the limit,

$$(-a^{-1}q;q^2)_\infty(-aq;q^2)_\infty=\sum_{k=-\infty}^{\infty}a^k\frac{q^{k^2}}{(q^2;q^2)_\infty},$$

or,

$$(-a^{-1}q;q^2)_\infty(-aq;q^2)_\infty(q^2;q^2)_\infty=\sum_{k=-\infty}^{\infty}a^kq^{k^2},$$

and this is Jacobi's celebrated triple product identity (1.1.1), hereinafter referred to as JTP.

1.4 Our Second Proof of Jacobi's Triple Product Identity

We can give a proof of JTP along similar lines to our proof of Euler's identity (1.2.1).

Suppose

$$(-a^{-1}q;q^2)_\infty(-aq;q^2)_\infty=\sum_{k=-\infty}^{\infty}a^kc_k. \tag{1.4.1}$$

Note that by the symmetry $a \to a^{-1}$,

$$c_{-k} = c_k. \tag{1.4.2}$$

If in (1.4.1) we put aq^2 for a, we obtain

$$(-a^{-1}q^{-1};q^2)_\infty(-aq^3;q^2)_\infty = \sum_{k=-\infty}^{\infty} a^k q^{2k} c_k. \tag{1.4.3}$$

From (1.4.1) and (1.4.3) it follows that

$$(1+a^{-1}q^{-1}) \sum_{k=-\infty}^{\infty} a^k c_k = (1+aq) \sum_{k=-\infty}^{\infty} a^k q^{2k} c_k,$$

and, if we divide by $1+a^{-1}q^{-1}$,

$$\sum_{k=-\infty}^{\infty} a^k c_k = aq \sum_{k=-\infty}^{\infty} a^k q^{2k} c_k = \sum_{k=-\infty}^{\infty} a^k q^{2k-1} c_{k-1}. \tag{1.4.4}$$

It follows that, for all k,

$$c_k = q^{2k-1} c_{k-1}. \tag{1.4.5}$$

From (1.4.2) and (1.4.5) we deduce by induction that for all k,

$$c_k = q^{k^2} c_0$$

and

$$(-a^{-1}q;q^2)_\infty(-aq;q^2)_\infty = c_0 \sum_{k=-\infty}^{\infty} a^k q^{k^2}. \tag{1.4.6}$$

Exercise: Show that (1.4.5) yields $c_k = q^{k^2} c_0$ for all k, positive **and** negative.

We now need to determine c_0. We proceed as follows.
If in (1.4.6) we set $a = q$, we find

$$2(-q^2;q^2)_\infty(-q^2;q^2)_\infty = c_0(q) \sum_{k=-\infty}^{\infty} q^{k^2+k} = 2c_0(q) \sum_{k\geq 0} q^{k^2+k}, \tag{1.4.7}$$

while if in (1.4.6) we replace q by q^4 and set $a = q^2$, we find

$$(-q^2;q^8)_\infty(-q^6;q^8)_\infty = c_0(q^4)\sum_{k=-\infty}^{\infty} q^{4k^2+2k} = c_0(q^4)\sum_{k\geq 0} q^{k^2+k}. \tag{1.4.8}$$

where we have used the easily checked fact that

$$\sum_{k=-\infty}^{\infty} q^{2k^2+k} = \sum_{k\geq 0} q^{(k^2+k)/2} \tag{1.4.9}$$

Exercise: Verify (1.4.9).

We can write (1.4.7)

$$c_0(q)\sum_{k\geq 0} q^{k^2+k} = (-q^2;q^2)_\infty^2 = \frac{(q^4;q^4)_\infty^2}{(q^2;q^2)_\infty^2} \tag{1.4.10}$$

and we can write (1.4.8)

$$c_0(q^4)\sum_{k\geq 0} q^{k^2+k} = (-q^2;q^4)_\infty = \frac{(q^4;q^8)_\infty}{(q^2;q^4)_\infty} = \frac{(q^4;q^4)_\infty^2}{(q^2;q^2)_\infty(q^8;q^8)_\infty}. \tag{1.4.11}$$

If we divide (1.4.10) by (1.4.11) and rearrange, we find that

$$(q^2;q^2)_\infty c_0(q) = (q^8;q^8)_\infty c_o(q^4). \tag{1.4.12}$$

If we iterate (1.4.12), we find that

$$(q^2;q^2)_\infty c_0(q) = (q^{2^{2n+1}};q^{2^{2n+1}})_\infty c_0(q^{2^{2n}}). \tag{1.4.13}$$

If we now let $n \to \infty$ and invoke continuity, we find

$$(q^2;q^2)_\infty c_0(q) = 1,$$

or,

$$c_0 = \frac{1}{(q^2;q^2)_\infty}. \tag{1.4.14}$$

(1.4.14) together with (1.4.6) gives JTP.

1.5 Some Important Special Cases of Jacobi's Triple Product Identity

If in JTP we set $a = 1$, we find that

$$\begin{aligned}\sum_{-\infty}^{\infty} q^{k^2} &= (-q;q^2)_\infty^2 (q^2;q^2)_\infty \\ &= \frac{(q^2;q^4)_\infty^2}{(q;q^2)_\infty^2}(q^2;q^2)_\infty \\ &= \frac{(q^2;q^2)_\infty^4}{(q;q)_\infty^2 (q^4;q^4)_\infty^2}(q^2;q^2)_\infty \\ &= \frac{(q^2;q^2)_\infty^5}{(q;q)_\infty^2 (q^4;q^4)_\infty^2}. \end{aligned} \tag{1.5.1}$$

Likewise, if in JTP we replace q by q^2 and set $a = q$, we find

$$\begin{aligned}\sum_{-\infty}^{\infty} q^{2k^2+k} &= (-q;q^4)_\infty (-q^3;q^4)_\infty (q^4;q^4)_\infty \\ &= \frac{(-q;q)_\infty}{(-q^2;q^2)_\infty}(q^4;q^4)_\infty \\ &= \frac{(q^2;q^2)_\infty}{(q;q)_\infty}\frac{(q^2;q^2)_\infty}{(q^4;q^4)_\infty}(q^4;q^4)_\infty \\ &= \frac{(q^2;q^2)_\infty^2}{(q;q)_\infty}. \end{aligned} \tag{1.5.2}$$

By (1.4.9), (1.5.2) becomes

$$\sum_{k\geq 0} q^{(k^2+k)/2} = \frac{(q^2;q^2)_\infty^2}{(q;q)_\infty}. \tag{1.5.3}$$

I have a couple of comments to make. Although the steps I have taken above may seem obscure at first, with practice they should become less so. Indeed, some people describe such transformations as "trivial product manipulations". Note that I will often end with an expression which is a product/quotient of powers of terms of the form $(q^k;q^k)_\infty$.

Following Ramanujan, we will define

$$\phi(q) = \sum_{-\infty}^{\infty} q^{k^2} \tag{1.5.4}$$

and

$$\psi(q) = \sum_{k\geq 0} q^{(k^2+k)/2}. \tag{1.5.5}$$

Then (1.5.1) becomes

$$\phi(q) = \frac{(q^2;q^2)_\infty^5}{(q;q)_\infty^2 (q^4;q^4)_\infty^2} \tag{1.5.6}$$

and (1.5.3) becomes

$$\psi(q) = \frac{(q^2;q^2)_\infty^2}{(q;q)_\infty}. \tag{1.5.7}$$

Exercise: Show that

$$\phi(-q) = \frac{(q;q)_\infty^2}{(q^2;q^2)_\infty}, \tag{1.5.8}$$

$$\psi(-q) = \frac{(q;q)_\infty (q^4;q^4)_\infty}{(q^2;q^2)_\infty}, \tag{1.5.9}$$

$$\phi(-q)\psi(q) = (q;q)_\infty (q^2;q^2)_\infty, \tag{1.5.10}$$

$$\phi(-q)^2\psi(q) = (q;q)_\infty^3, \tag{1.5.11}$$

$$\phi(-q)\psi(q)^2 = (q^2;q^2)_\infty^3 = \phi(q)\psi(-q)^2, \tag{1.5.12}$$

$$\phi(q)\psi(q^2) = \psi(q)^2, \tag{1.5.13}$$

$$\phi(q)\phi(-q) = \phi(-q^2)^2, \tag{1.5.14}$$

$$\frac{\phi(-q)}{\phi(q)} = \left(\frac{\psi(-q)}{\psi(q)}\right)^2, \tag{1.5.15}$$

$$\frac{1}{\phi(-q)} = \phi(q)\phi(q^2)^2\phi(q^4)^4\cdots, \tag{1.5.16}$$

$$\frac{1}{(q;q)_\infty} = \psi(q)\psi(q^2)^2\psi(q^4)^4\cdots, \tag{1.5.17}$$

$$\phi(-q) = \frac{\phi(-q^2)\phi(-q^4)\phi(-q^8)\cdots}{\phi(q)\phi(q^2)\phi(q^4)\cdots}, \tag{1.5.18}$$

$$\psi(-q) = \frac{\phi(-q^4)\phi(-q^8)\phi(-q^{16})\cdots}{\psi(q)\psi(q^2)\psi(q^4)\cdots}. \tag{1.5.19}$$

1.6 Euler's Pentagonal Numbers Theorem

Our next application of JTP is to prove a celebrated result of Euler, namely

$$(q;q)_\infty = \sum_{-\infty}^{\infty}(-1)^k q^{(3k^2-k)/2}. \tag{1.6.1}$$

We have

$$(-a^{-1}q;q^2)_\infty(-aq;q^2)_\infty(q^2;q^2)_\infty = \sum_{-\infty}^{\infty} a^k q^{k^2}.$$

If we replace q by $q^{\frac{3}{2}}$ then set $a = -q^{\frac{1}{2}}$, we find

$$(q;q^3)_\infty(q^2;q^3)_\infty(q^3;q^3)_\infty = \sum_{-\infty}^{\infty}(-1)^k q^{(3k^2-k)/2},$$

which, since the left side can be seen to be $(q;q)_\infty$, is (1.6.1).

Explicitly, (1.6.1) is

$$(1-q)(1-q^2)(1-q^3)\cdots = 1-q-q^2+q^5+q^7-q^{12}-q^{15}++--\cdots. \tag{1.6.2}$$

The numbers 0, 1, 2, 5, 7, 12, 15, ..., given by $(3k^2-k)/2$ for $k = 0,\ 1,\ -1,\ 2,\ -2,\ 3,\ -3,\ \ldots$ are known as the pentagonal numbers, and (1.6.1) is known as Euler's pentagonal numbers theorem.

Incidentally, if the need arises, you can calculate the pentagonal numbers as follows. The triangular numbers $1,\ 1+2,\ 1+2+3,\ 1+2+3+4$ and so on are 1, 3, 6, 10, 15, 21, 28, 36, 45 and so on. Two of every three triangular numbers are divisible by 3, and if we divide these by 3 we obtain the pentagonal numbers.

Euler's pentagonal numbers theorem has a nice combinatorial interpretation, and has a very pretty combinatorial proof due to F. Franklin, for which see Hardy and Wright (1960) [60], pp. 285–287.

I point out that, by reversing the order of summation, that is, by replacing k by $-k$, we have

$$(q;q)_\infty = \sum_{-\infty}^{\infty}(-1)^k q^{(3k^2-k)/2} = \sum_{-\infty}^{\infty}(-1)^k q^{(3k^2+k)/2}.$$

It may be that in the course of the book I will reverse summations without comment.

Euler's pentagonal numbers theorem is simply the special case $a = 1$ of JTP when JTP is written in the equivalent form

$$(aq;q^3)_\infty(a^{-1}q^2;q^3)_\infty(q^3;q^3)_\infty=\sum_{-\infty}^{\infty}(-1)^k a^k q^{(3k^2-k)/2}. \qquad (1.6.3)$$

It seems to me that (1.6.3) might fall to a combinatorial proof, in which the power of a is equal to the residue modulo 3 of the attached part.

1.7 Jacobi's Formula for the Cube of Euler's Product

Another, extremely important, consequence of JTP is the identity, also due to Jacobi, which gives the expansion of the cube of Euler's product,

$$(q;q)_\infty^3=\sum_{k\geq 0}(-1)^k(2k+1)q^{(k^2+k)/2}. \qquad (1.7.1)$$

We have

$$(-a^{-1}q;q^2)_\infty(-aq;q^2)_\infty(q^2;q^2)_\infty=\sum_{-\infty}^{\infty}a^k q^{k^2}.$$

If we replace q by $q^{\frac{1}{2}}$ and replace a by $-aq^{\frac{1}{2}}$, we obtain

$$(a^{-1};q)_\infty(aq;q)_\infty(q;q)_\infty=\sum_{-\infty}^{\infty}(-1)^k a^k q^{(k^2+k)/2}. \qquad (1.7.2)$$

This can be written

$$(1-a^{-1})(a^{-1}q;q)_\infty(aq;q)_\infty(q;q)_\infty=\sum_{k\geq 0}(-1)^k(a^k-a^{-k-1})q^{(k^2+k)/2}. \qquad (1.7.3)$$

Next, multiply by $a^{\frac{1}{2}}$ to obtain the identity, invariant under $a\to -a^{-1}$,

$$(a^{\frac{1}{2}}-a^{-\frac{1}{2}})(a^{-1}q;q)_\infty(aq;q)_\infty(q^2;q^2)_\infty=\sum_{k\geq 0}(-1)^k\left(a^{k+\frac{1}{2}}-a^{-k-\frac{1}{2}}\right)q^{(k^2+k)/2}. \qquad (1.7.4)$$

If we suppose $a\neq 1$ and divide by $a^{\frac{1}{2}}-a^{-\frac{1}{2}}$, we obtain

$$\begin{aligned}(a^{-1}q;q)_\infty(aq;q)_\infty(q;q)_\infty&=1+\sum_{k\geq 1}(-1)^k\left(\frac{a^{k+\frac{1}{2}}-a^{-k-\frac{1}{2}}}{a^{\frac{1}{2}}-a^{-\frac{1}{2}}}\right)q^{(k^2+k)/2}\\&=1+\sum_{k\geq 1}(-1)^k\left(a^k+a^{k-1}+\cdots+a^{-k}\right)q^{(k^2+k)/2}.\end{aligned} \qquad (1.7.5)$$

If we now let $a \to 1$, we obtain (1.7.1).

Exercise: Put $a = -1$ in (1.7.5), and retrieve (1.5.7).

Exercise: Set $a = \omega$, where ω is a cube root of unity other than 1 in (1.7.5), and retrieve (1.6.1).

In the next chapter, we will see our first dramatic applications of JTP and of (1.7.1), Jacobi's formula for the cube of Euler's product.

1.8 Polynomial Versions of Earlier Results

We can give various polynomial identities that imply earlier results.

First, the so-called q-binomial coefficients $\begin{bmatrix} n \\ k \end{bmatrix}_q$ are defined somewhat analogously to the ordinary binomial coefficients $\binom{n}{k}$, as follows.

$$\begin{bmatrix} 0 \\ 0 \end{bmatrix}_q = 1, \quad \text{for } n \geq 1, \begin{bmatrix} n \\ 0 \end{bmatrix}_q = \begin{bmatrix} n \\ n \end{bmatrix}_q = 1,$$

and for $n > 1, 0 < k < n$,

$$\begin{bmatrix} n \\ k \end{bmatrix}_q = \begin{bmatrix} n-1 \\ k-1 \end{bmatrix}_q + q^k \begin{bmatrix} n-1 \\ k \end{bmatrix}_q.$$

The $\begin{bmatrix} n \\ k \end{bmatrix}_q$ are polynomials in q; they form a q-Pascal triangle. When $q = 1$, we have $\begin{bmatrix} n \\ k \end{bmatrix}_q = \binom{n}{k}$.

It is easy to prove by induction that, provided q is not a root of unity,

$$\begin{bmatrix} n \\ k \end{bmatrix}_q = \frac{(q;q)_n}{(q;q)_k (q;q)_{n-k}}$$

(If q is a root of unity, the right hand side may not be defined.)

Exercise: Show that the degree of $\begin{bmatrix} n \\ k \end{bmatrix}_q$ is $k(n-k)$.

Exercise: Show that

$$\begin{bmatrix} n \\ k \end{bmatrix}_q = \begin{bmatrix} n-1 \\ k \end{bmatrix}_q + q^{n-k} \begin{bmatrix} n-1 \\ k-1 \end{bmatrix}_q.$$

It is also easy to prove by induction that

$$(-a;q)_n = \sum_{k=0}^{n} a^k q^{(k^2-k)/2} \begin{bmatrix} n \\ k \end{bmatrix}_q, \tag{1.8.1}$$

which gives Euler's identity (1.2.1) on letting $n \to \infty$, provided that $|q| < 1$.

Polynomial versions of JTP are

$$(-a^{-1}q;q^2)_m(-aq;q^2)_n = \sum_{k=-m}^{n} a^k q^{k^2} \begin{bmatrix} m+n \\ m+k \end{bmatrix}_{q^2} \tag{1.8.2}$$

for $a \neq 0$, $m,\ n \geq 0$, which can be derived from (1.8.1) by replacing q by q^2, a by aq, n by $m+n$ then performing a pull–back of degree m (see §1.3), or by double induction, and

$$(-a^{-1}q;q^2)_n(-aq;q^2)_n = \sum_{k=-n}^{n} a^k q^{k^2} \begin{bmatrix} 2n \\ n+k \end{bmatrix}_{q^2}, \tag{1.8.3}$$

for $a \neq 0$, $n \geq 0$, which can be derived from (1.8.2) by setting $m = n$, or proved directly by induction.

Exercise: Prove (1.8.3) by induction.

JTP can be obtained from (1.8.2) by assuming $|q| < 1$ and letting $m,n \to \infty$, or from (1.8.3) by assuming $|q| < 1$ and letting $n \to \infty$.

I would like to add that there is a particularly nice polynomial version of (1.7.1), namely

$$(q;q)_n^2 = \sum_{k=0}^{n} (-1)^k (2k+1) q^{(k^2+k)/2} \begin{bmatrix} 2n+1 \\ n-k \end{bmatrix}_q. \tag{1.8.4}$$

Exercise: Prove (1.8.4) either from (1.8.2) or directly by induction.

Exercise: Prove the following polynomial version of Euler's pentagonal numbers theorem

$$\frac{(q;q)_{3n}}{(q^3;q^3)_n} = \sum_{k=-n}^{n} (-1)^k q^{(3k^2-k)/2} \begin{bmatrix} 2n \\ n+k \end{bmatrix}_{q^3}. \tag{1.8.5}$$

Exercise: Prove the following polynomial versions of (1.5.8) and (1.5.7).

$$\frac{(q;q)_{2n}}{(q^2;q^2)_n} = \sum_{k=-n}^{n} (-1)^k q^{k^2} \begin{bmatrix} 2n \\ n+k \end{bmatrix}_q \tag{1.8.6}$$

and

$$\frac{(q^2;q^2)_{2n}}{(q;q)_{2n}} = \sum_{k=-n}^{n} q^{2k^2+k} \begin{bmatrix} 2n \\ n+k \end{bmatrix}_{q^2}. \tag{1.8.7}$$

1.9 *m*-Dissection

A technique we use throughout the book is m-dissection. Given a series,

$$A(q) = \sum_n a(n)q^n, \tag{1.9.1}$$

and $m > 1$ an integer, the m-dissection of $A(q)$ is the expression of $A(q)$ as a sum of m sums,

$$A(q) = A_0(q) + A_1(q) + \cdots + A_{m-1}(q), \tag{1.9.2}$$

where for each r with $0 \leq r \leq m-1$,

$$A_r(q) = q^r \sum_n a(mn+r)q^{mn}. \tag{1.9.3}$$

For example, the 2-dissection of $\phi(q)$ is

$$\phi(q) = \phi(q^4) + 2q\psi(q^8). \tag{1.9.4}$$

We have

$$\begin{aligned}
\phi(q) &= \sum_{-\infty}^{\infty} q^{n^2} \\
&= 1 + 2q + 2q^4 + 2q^9 + 2q^{16} + 2q^{25} + 2q^{36} + 2q^{49} + \cdots \\
&= (1 + 2q^4 + 2q^{16} + 2q^{36} + \cdots) + 2q(1 + q^8 + q^{24} + q^{48} + \cdots) \\
&= \phi(q^4) + 2q\psi(q^8).
\end{aligned}$$

The even part of $\phi(q)$ is

$$\phi_0(q) = \sum_{-\infty}^{\infty} q^{(2n)^2} = \sum_{-\infty}^{\infty} q^{4n^2} = \phi(q^4),$$

while the odd part is

$$\phi_1(q) = \sum_{-\infty}^{\infty} q^{(2n+1)^2} = q\sum_{-\infty}^{\infty} q^{4n^2+4n} = 2q\sum_{n\geq 0} q^{4n^2+4n} = 2q\psi(q^8).$$

As a second example, the 3-dissection of $E(q)$ is

$$E(q) = (q^{12};q^{27})_\infty(q^{15};q^{27})_\infty(q^{27};q^{27})_\infty - q(q^6;q^{27})_\infty(q^{21};q^{27})_\infty(q^{27};q^{27})_\infty - q^2(q^3;q^{27})_\infty(q^{24};q^{27})_\infty(q^{27};q^{27})_\infty. \quad (1.9.5)$$

We have

$$\begin{aligned}
E(q) &= \sum_{-\infty}^{\infty}(-1)^n q^{(3n^2+n)/2} \\
&= 1-q-q^2+q^5+q^7-q^{12}-q^{15}+q^{22}+q^{26}-q^{35}-q^{40}+q^{51}+q^{57}-q^{70} \\
&\quad -q^{77}++--\cdots \\
&= (1-q^{12}-q^{15}+q^{51}+q^{57}--++\cdots)-q(1-q^6 \\
&\quad -q^{21}+q^{39}+q^{69}--++\cdots)-q^2(1-q^3-q^{24}+q^{33}+q^{75}--++\cdots) \\
&= \sum_{-\infty}^{\infty}(-1)^n q^{(27n^2-3n)/2} - q\sum_{-\infty}^{\infty}(-1)^n q^{(27n^2-15n)/2} - q^2\sum_{-\infty}^{\infty}(-1)^n q^{(27n^2-21n)/2} \\
&= (q^{12};q^{27})_\infty(q^{15};q^{27})_\infty(q^{27};q^{27})_\infty - q(q^6;q^{27})_\infty(q^{21};q^{27})_\infty(q^{27};q^{27})_\infty \\
&\quad - q^2(q^3;q^{27})_\infty(q^{24};q^{27})_\infty(q^{27};q^{27})_\infty,
\end{aligned}$$

by JTP, giving (1.9.5).

1.10 Linear Transformations

The technique of applying linear transformations to the indices in a multiple sum will often prove indispensable.

We illustrate the technique by providing the 2-dissection of $\phi(q)^2$,

$$\phi(q)^2 = \phi(q^2)^2 + 4q\psi(q^4)^2. \quad (1.10.1)$$

We have

$$\phi(q)^2 = \sum_{k,l=-\infty}^{\infty} q^{k^2+l^2}. \quad (1.10.2)$$

Now, $k^2+l^2 \equiv 0 \pmod 2$ if and only if $k \equiv l \pmod 2$, and $k^2+l^2 \equiv 1 \pmod 2$ if and only if $k \not\equiv l \pmod 2$.

In the first case, write $k+l=2r, k-l=2s$, and then $k=r+s, l=r-s$, and in the second case, $k+l=2r+1, k-l=2s+1$, and then $k=r+s+1$, $l=r-s$.

Then we have

$$\begin{aligned}\phi(q)^2 &= \sum_{r,s=-\infty}^{\infty} q^{(r+s)^2+(r-s)^2} + \sum_{r,s=-\infty}^{\infty} q^{(r+s+1)^2+(r-s)^2} \\ &= \sum_{r,s=-\infty}^{\infty} q^{2r^2+2s^2} + q \sum_{r,s=-\infty}^{\infty} q^{2r^2+2r+2s^2+2s} \\ &= \sum_{r,s=-\infty}^{\infty} q^{2r^2+2s^2} + 4q \sum_{r,s\geq 0} q^{2r^2+2r+2s^2+2s} \\ &= \phi(q^2)^2 + 4q\psi(q^4)^2,\end{aligned}$$

as claimed.

Note that in each case, the transformed quadratic has no cross-product (rs) term, and the new double sum splits into a product of sums.

Exercise: Use the same method to prove (1.5.13) and (1.5.14), thereby avoiding the use of JTP.

Endnotes.

Much of Chapter 1 appears in Hardy and Wright (1960) [60], Chapter 19, or is "mathematical folklore".

§1.4 The determination of c_0 is my recent take on the proof in Hardy and Wright (1960) [60] §19.8.

§1.8 Identity (1.8.2) appears in Hirschhorn (1976) [62], where it was proved by induction, but it was proved combinatorially by MacMahon (1916) [105], Art. 323. Versions of identities (1.8.4) and (1.8.5) appear in Hirschhorn (1987) [63].

§§1.9, 1.10 The methods of these sections have their roots in the work of Jacobi. They were used to great effect by such luminaries as Schröter, Rogers, Watson and Carlitz. I learnt them from reading the papers of Carlitz, among them [23, 24], so describe them as "Carlitzian".

References

23. L. Carlitz, M.V. Subbarao, A simple proof of the quintuple product identity. Proc. Am. Math. Soc. **32**, 42–44 (1972)
24. L. Carlitz, M.V. Subbarao, On a combinatorial identity of Winquist and its generalization. Duke Math. J. **39**, 165–172 (1972)
60. G.H. Hardy, E.M. Wright, *An Introduction to the Theory of Numbers*, 4th edn. (Oxford University Press, Oxford, 1960)

62. M.D. Hirschhorn, Simple proofs of identities of MacMahon and Jacobi. Discret. Math. **16**, 161–162 (1976)
63. M.D. Hirschhorn, Polynomial identities which imply identities of Euler and Jacobi. Acta Arith. **XXXII**, 73–78 (1977)
98. C.G.J. Jacobi, Fundamenta nova theoriae functionum ellipticarum, Königsberg (1829)
105. P.A. MacMahon, *Combinatory Analysis* (Cambridge University Press, Cambridge, 1916)

Chapter 2
Jacobi's Two-Squares and Four-Squares Theorems

2.1 Introduction

In 1640, Fermat stated that a prime p is the sum of two squares if and only if $p \equiv +1 \pmod 4$, and this was eventually proved by Euler in 1747. In 1801, Gauss showed that the number n is the sum of two squares if and only if the squarefree part of n has no divisor congruent to $-1 \pmod 4$. In 1829, Jacobi proved a result giving the number of representations of n as the sum of two squares, of which Gauss's result is, as we will see, a corollary.

Jacobi's two-squares theorem: The number of representations of $n \geq 1$ as the sum of two squares is given by

$$r\{\square+\square\}(n) = 4\left(d_{1,4}(n) - d_{3,4}(n)\right), \tag{2.1.1}$$

where $d_{r,m}(n)$ denotes the number of divisors d of n with $d \equiv r \pmod m$.

We will give two proofs of Jacobi's two-squares theorem, one directly from JTP, the other from (1.7.1). The second may be new.

In 1621 Bachet conjectured that every number is the sum of four squares, and the proof was completed by Lagrange in 1770. In 1829, Jacobi gave a formula for the number of representations of the number n as the sum of four squares, of which Lagrange's theorem is an immediate corollary.

Jacobi's four-squares theorem: The number of representations of $n \geq 1$ as the sum of four squares is given by

$$r\{\square+\square+\square+\square\}(n) = 8 \sum_{\substack{d \mid n \\ d \equiv 1,\, 2 \text{ or } 3 \pmod 4}} d. \tag{2.1.2}$$

We will give a proof of Jacobi's four-squares theorem from (1.7.1).

M.D. Hirschhorn, *The Power of q*, Developments in Mathematics 49,
DOI 10.1007/978-3-319-57762-3_2

2.2 Our First Proof of Jacobi's Two-Squares Theorem

We start by noting that (1.7.4) can be written

$$(a-a^{-1})(a^{-2}q;q)_\infty(a^2q;q)_\infty(q;q)_\infty=\sum_{-\infty}^{\infty}(-1)^k a^{2k+1}q^{(k^2+k)/2}. \tag{2.2.1}$$

Now consider k even, $k=2l$ and k odd, $k=2l+1$ in the right side of (2.2.1) and we find

$$\begin{aligned}(a-a^{-1})(a^{-2}q;q)_\infty(a^2q;q)_\infty(q;q)_\infty&=\sum_{-\infty}^{\infty}a^{4l+1}q^{2l^2+l}-\sum_{-\infty}^{\infty}a^{4l-1}q^{2l^2-l}\\ &=a(-a^{-4}q;q^4)_\infty(-a^4q^3;q^4)_\infty(q^4;q^4)_\infty-a^{-1}(-a^{-4}q^3;q^4)_\infty(-a^4q;q^4)_\infty(q^4;q^4)_\infty.\end{aligned} \tag{2.2.2}$$

If we differentiate (2.2.2) with respect to a using the product rule in the form

$$\left(\prod_k u_k\right)'=\left(\prod_k u_k\right)\sum_k\frac{u_k'}{u_k}$$

then multiply by a, we find (this is fairly tricky!)

$$\begin{aligned}&(a+a^{-1})(a^{-2}q;q)_\infty(a^2q;q)_\infty(q;q)_\infty\\ &\quad+2(a-a^{-1})(a^{-2}q;q)_\infty(a^2q;q)_\infty(q^2;q^2)_\infty\sum_{k\geq1}\left(\frac{a^{-2}q^k}{1-a^{-2}q^k}-\frac{a^2q^k}{1-a^2q^k}\right)\\ &\quad=a(a^{-4}q;q^4)_\infty(-a^4q^3;q^4)_\infty(q^4;q^4)_\infty\\ &\qquad\times\left(1-4\sum_{k\geq1}\left(\frac{a^{-4}q^{4k-3}}{1+a^{-4}q^{4k-3}}-\frac{a^4q^{4k-1}}{1+a^4q^{4k-1}}\right)\right)\\ &\quad+a^{-1}(a^{-4}q^3;q^4)_\infty(a^4q;q^4)_\infty(q^4;q^4)_\infty\\ &\qquad\times\left(1+4\sum_{k\geq1}\left(\frac{a^{-4}q^{4k-1}}{1+a^{-4}q^{4k-1}}-\frac{a^4q^{4k-3}}{1+a^4q^{4k-3}}\right)\right).\end{aligned} \tag{2.2.3}$$

If in (2.2.3) we set $a=1$ and divide by 2, we obtain

$$(q;q)_\infty^3=(-q;q^4)_\infty(-q^3;q^4)_\infty(q^4;q^4)_\infty\left(1-4\sum_{k\geq1}\left(\frac{q^{4k-3}}{1+q^{4k-3}}-\frac{q^{4k-1}}{1+q^{4k-1}}\right)\right). \tag{2.2.4}$$

If we invoke (1.4.9), (2.2.4) becomes

$$(q;q)_\infty^3 = \psi(q)\left(1-4\sum_{k\geq 1}\left(\frac{q^{4k-3}}{1+q^{4k-3}}-\frac{q^{4k-1}}{1+q^{4k-1}}\right)\right). \tag{2.2.5}$$

If we divide (2.2.5) by $\psi(q)$ and make use of (1.5.11), we obtain

$$\phi(-q)^2 = 1-4\sum_{k\geq 1}\left(\frac{q^{4k-3}}{1+q^{4k-3}}-\frac{q^{4k-1}}{1+q^{4k-1}}\right). \tag{2.2.6}$$

If in (2.2.6) we replace q by $-q$,

$$\phi(q)^2 = 1+4\sum_{k\geq 1}\left(\frac{q^{4k-3}}{1-q^{4k-3}}-\frac{q^{4k-1}}{1-q^{4k-1}}\right). \tag{2.2.7}$$

If we use the fact that $\dfrac{x}{1-x} = \sum_{l\geq 1} x^l$ for $|x|<1$, we can write

$$\begin{aligned}\left(\sum_{-\infty}^{\infty} q^{k^2}\right)^2 &= 1+4\sum_{k,l\geq 1}\left(q^{(4k-3)l}-q^{(4k-1)l}\right)\\ &= 1+4\left(\sum_{d\equiv +1 \pmod 4,\ l\geq 1} q^{dl} - \sum_{d\equiv -1 \pmod 4,\ l\geq 1} q^{dl}\right).\end{aligned} \tag{2.2.8}$$

If $n\geq 1$, and we compare coefficients of q^n on both sides of (2.2.8), we obtain Jacobi's two-squares theorem (2.1.1).

Now suppose the prime factorisation of $n\geq 1$ is

$$n = 2^\alpha \prod_i p_i^{\alpha_i}\prod_j q_j^{\beta_j}, \tag{2.2.9}$$

where the $p_i \equiv +1 \pmod 4$ and $q_j \equiv -1 \pmod 4$ are distinct primes, $\alpha\geq 0$ and $\alpha_i\geq 1,\ \beta_j\geq 1$. We observe that if $P\neq 0$,

$$\begin{aligned}&\sum_{\substack{d|n\\ d\equiv +1 \pmod 4}} d^P - \sum_{\substack{d|n\\ d\equiv -1 \pmod 4}} d^P\\ &= \prod_i (1+p_i^P+\cdots+p_i^{P\alpha_i})\prod_j (1-q_j^P+-\cdots+(-1)^{\beta_j}q_j^{P\beta_j})\\ &= \prod_i \frac{1-p_i^{(\alpha_i+1)P}}{1-p_i^P}\prod_j \frac{1+(-1)^{\beta_j}q_j^{(\beta_j+1)P}}{1+q_j^P}\end{aligned} \tag{2.2.10}$$

If we let $P \to 0$ in (2.2.10), we find that

$$d_{1,4}(n) - d_{3,4}(n) = \prod_i (\alpha_i + 1) \prod_j \frac{1 + (-1)^{\beta_j}}{2}. \tag{2.2.11}$$

It follows from (2.2.1) and (2.2.11) that for $n \geq 1$,

$$r\{\square + \square\}(n) = 4 \prod_i (\alpha_i + 1) \prod_j \frac{1 + (-1)^{\beta_j}}{2}. \tag{2.2.12}$$

Gauss's result is now clear. The number n is the sum of two squares if and only if all the β_j are even, that is, if and only if the squarefree part of n is divisible by no prime congruent to $-1 \pmod 4$, or, equivalently, has no divisor (prime or otherwise) congruent to $-1 \pmod 4$.

Before moving on to give our second proof of Jacobi's two-squares theorem, it is worth pointing out that the number of expressions, which we will denote by $p\{\square + \square\}(n)$, of $n \geq 1$ as the sum of two squares,

$$n = a^2 + b^2$$

with

$$a \geq b \geq 0$$

is given by

$$p\{\square + \square\}(n) = \begin{cases} 0 & \text{if any } \beta_j \text{ is odd} \\ \frac{1}{2} \prod_i (\alpha_i + 1) & \text{if all } \beta_j \text{ are even and any } \alpha_j \text{ is odd} \\ \frac{1}{2} \prod_i (\alpha_i + 1) + \frac{1}{2} & \text{if all } \beta_j \text{ and all } \alpha_i \text{ are even,} \\ & \text{that is, } n \text{ is a square or twice a square.} \end{cases} \tag{2.2.13}$$

2.3 Our Second Proof of Jacobi's Two-Squares Theorem

We start with (1.7.1),

$$(q;q)_\infty^3 = \sum_{k \geq 0} (-1)^k (2k+1) q^{(k^2+k)/2}.$$

Remarkably, (1.7.1) can be written

$$(q;q)_\infty^3 = \sum_{-\infty}^{\infty}(4k+1)q^{2k^2+k}. \tag{2.3.1}$$

Exercise: Verify (2.3.1).

This leads automatically to what follows.

$$\begin{aligned}
(q;q)_\infty^3 &= \left[\frac{d}{da}\left(\sum_{-\infty}^{\infty} a^{4k+1}q^{2k^2+k}\right)\right]_{a=1} \\
&= \left[\frac{d}{da}\left(a(-a^{-4}q;q^4)_\infty(-a^4q^3;q^4)_\infty(q^4;q^4)_\infty\right)\right]_{a=1} \\
&= \Bigg[(-a^{-4}q;q^4)_\infty(-a^4q^3;q^4)_\infty(q^4;q^4)_\infty \\
&\quad \times \left(1-4\sum_{k\geq 1}\left(\frac{a^{-4}q^{4k-3}}{1+a^{-4}q^{4k-3}} - \frac{a^4q^{4k-1}}{1+a^4q^{4k-1}}\right)\right)\Bigg]_{a=1} \\
&= (-q;q^4)_\infty(-q^3;q^4)_\infty(q^4;q^4)_\infty\left(1-4\sum_{k\geq 1}\left(\frac{q^{4k-3}}{1+q^{4k-3}} - \frac{q^{4k-1}}{1+q^{4k-1}}\right)\right).
\end{aligned} \tag{2.3.2}$$

We note that (2.3.2) is identical with (2.2.4), so we can complete the proof of Jacobi's two-squares theorem as in §2.2.

2.4 A Proof of Jacobi's Four-Squares Theorem

We start with (1.7.1),

$$(q;q)_\infty^3 = \sum_{k\geq 0}(-1)^k(2k+1)q^{(k^2+k)/2}.$$

We can write this

$$(q;q)_\infty^3 = \frac{1}{2}\sum_{-\infty}^{\infty}(-1)^k(2k+1)q^{(k^2+k)/2}. \tag{2.4.1}$$

Exercise: Verify (2.4.1).

If we square (2.4.1), we find

$$(q;q)_\infty^6 = \frac{1}{4}\sum_{k,l=-\infty}^{\infty}(-1)^{k+l}(2k+1)(2l+1)q^{(k^2+k+l^2+l)/2}. \tag{2.4.2}$$

We now split the sum in two, according as $k+l$ is even or odd, and obtain

$$(q;q)_\infty^6 = \frac{1}{4}\left(\sum_{k\equiv l \pmod 2}(2k+1)(2l+1)q^{(k^2+k+l^2+l)/2}\right.$$
$$\left. - \sum_{k\not\equiv l \pmod 2}(2k+1)(2l+1)q^{(k^2+k+l^2+l)/1}\right). \tag{2.4.3}$$

In the first sum, let $r=\frac{1}{2}(k+l)$, $s=\frac{1}{2}(k-l)$, $k=r+s$, $l=r-s$, and in the second let $r=\frac{1}{2}(k-l-1)$, $s=\frac{1}{2}(k+l+1)$, $k=r+s$, $l=s-r-1$, and we find

$$\begin{aligned}
(q;q)_\infty^6 &= \frac{1}{4}\left(\sum_{r,s=-\infty}^{\infty}(2r+2s+1)(2r-2s+1)q^{((r+s)^2+(r+s)+(r-s)^2+(r-s))/2}\right.\\
&\qquad\left. - \sum_{r,s=-\infty}^{\infty}(2r+2s+1)(2s-2r-1)q^{((r+s)^2+(r+s)+(s-r-1)^2+(s-r-1))/2}\right)\\
&= \frac{1}{4}\left(\sum_{r,s=-\infty}^{\infty}((2r+1)^2-(2s)^2)q^{r^2+r+s^2} - \sum_{r,s=-\infty}^{\infty}((2s)^2-(2r+1)^2)q^{r^2+r+s^2}\right)\\
&= \frac{1}{2}\sum_{r,s=-\infty}^{\infty}((2r+1)^2-(2s)^2)q^{r^2+r+s^2}\\
&= \frac{1}{2}\left(\sum_{-\infty}^{\infty}q^{s^2}\sum_{r=-\infty}^{\infty}(2r+1)^2q^{r^2+r} - \sum_{r=-\infty}^{\infty}q^{r^2+r}\sum_{-\infty}^{\infty}(2s)^2q^{s^2}\right)\\
&= (-q;q^2)_\infty^2(q^2;q^2)_\infty\left(1+4q\frac{d}{dq}\right)((-q^2;q^2)_\infty^2(q^2;q^2)_\infty)\\
&\qquad - (-q^2;q^2)_\infty^2(q^2;q^2)_\infty\cdot 4\frac{d}{dq}((-q;q^2)_\infty^2(q^2;q^2)_\infty)
\end{aligned}$$

$$= (-q;q^2)_\infty^2(-q^2;q^2)_\infty^2(q^2;q^2)_\infty^2 \left(1+4\left(2\sum_{k\ge1}\frac{2kq^{2k}}{1+q^{2k}}-\sum_{k\ge1}\frac{2kq^{2k}}{1-q^{2k}}\right)\right.$$
$$\left.-4\left(2\sum_{k\ge1}\frac{(2k-1)q^{2k-1}}{1+q^{2k-1}}-\sum_{k\ge1}\frac{2kq^{2k}}{1-q^{2k}}\right)\right)$$
$$= (-q;q)_\infty^4(q;q)_\infty^2\left(1-8\sum_{k\ge1}\left(\frac{(2k-1)q^{2k-1}}{1+q^{2k-1}}-\frac{2kq^{2k}}{1+q^{2k}}\right)\right). \tag{2.4.4}$$

If we divide (2.4.4) by $(-q;q)_\infty^4(q;q)_\infty^2$ and use (1.5.8), we obtain

$$\phi(-q)^4 = 1-8\sum_{k\ge1}\left(\frac{(2k-1)q^{2k-1}}{1+q^{2k-1}}-\frac{2kq^{2k}}{1+q^{2k}}\right). \tag{2.4.5}$$

If in (2.4.5) we put $-q$ for q, we find

$$\begin{aligned}
\phi(q)^4 &= 1+8\sum_{k\ge1}\left(\frac{(2k-1)q^{2k-1}}{1-q^{2k-1}}+\frac{2kq^{2k}}{1+q^{2k}}\right)\\
&= 1+8\sum_{k\ge1}\left(\frac{(2k-1)q^{2k-1}}{1-q^{2k-1}}+\frac{2kq^{2k}}{1-q^{2k}}\right)-8\sum_{k\ge1}\left(\frac{2kq^{2k}}{1-q^{2k}}-\frac{2kq^{2k}}{1+q^{2k}}\right)\\
&= 1+8\sum_{k\ge1}\frac{kq^k}{1-q^k}-8\sum_{k\ge1}\frac{4kq^{4k}}{1-q^{4k}}\\
&= 1+8\sum_{\substack{k\ge1\\ k\equiv1,\,2\text{ or }3 \pmod 4}}\frac{kq^k}{1-q^k}\\
&= 1+8\sum_{\substack{k,l\ge1\\ k\equiv1,\,2\text{ or }3 \pmod 4}}kq^{kl}.
\end{aligned} \tag{2.4.6}$$

If $n \ge 1$ and we compare the coefficients of q^n on both sides of (2.4.6), we obtain Jacobi's four-squares theorem (2.1.2).

Endnotes.

§2.2 This proof of Jacobi's two-squares theorem appears in Hirschhorn (1985) [65].

§2.3 This proof of Jacobi's two-squares theorem has not been published.

§2.4 An earlier version of this proof of Jacobi's four-squares theorem appeared in Hirschhorn (1987) [66]. A somewhat clumsy proof, starting from JTP, appeared earlier, in Hirschhorn (1982) [64].

§2.4 It is not possible to give a formula for $p\{\Box+\Box+\Box+\Box\}(n)$, the number of expressions of n, $n = a^2+b^2+c^2+d^2$ with $a \geq b \geq c \geq d \geq 0$, comparable in simplicity to that for $p\{\Box+\Box\}(n)$ in §2.2, but see Chapter 30.

References

64. M.D. Hirschhorn, A simple proof of Jacobi's four-square theorem. J. Aust. Math. Soc. Ser. A. **32**, 61–67 (1982)
65. M.D. Hirschhorn, A simple proof of Jacobi's two-square theorem. Am. Math. Mon. **92**, 579–580 (1985)
66. M.D. Hirschhorn, A simple proof of Jacobi's four-square theorem. Proc. Am. Math. Soc. **101**, 436–438 (1987)
98. C.G.J. Jacobi, Fundamenta nova theoriae functionum ellipticarum, Königsberg (1829)

Chapter 3
Ramanujan's Partition Congruences

3.1 Introduction

If n is a non-negative integer, a partition of n is an expression of the form

$$n = n_1 + n_2 + \cdots + n_k,$$

with $n_1 \geq n_2 \geq \cdots \geq n_k$ (and $k \geq 1$), and the number of partitions of n is denoted by $p(n)$.

Thus, for example, $p(4) = 5$ since 4 can be written 4, $3+1$, $2+2$, $2+1+1$, $1+1+1+1$.

We define $p(0) = 1$.

If $n = n_1 + n_2 + \cdots + n_k$ is a partition of n, we refer to $n_1, n_2, \ldots n_k$ as the "parts".

Euler knew that

$$\sum_{n \geq 0} p(n) q^n = \frac{1}{(q;q)_\infty}. \tag{3.1.1}$$

By way of explanation,

$$\begin{aligned}
&\frac{1}{(q;q)_\infty} = \frac{1}{1-q} \cdot \frac{1}{1-q^2} \cdot \frac{1}{1-q^3} \cdot \cdots \\
&= \left(1 + q^1 + q^{1+1} + q^{1+1+1} + \cdots\right)\left(1 + q^2 + q^{2+2} + \cdots\right)\left(1 + q^3 + q^{3+3} + \cdots\right)\cdots \\
&= 1 + q^1 + q^{1+1} + q^2 + q^{1+1+1} + q^{1+2} + q^3 + \cdots && (3.1.2)\\
&= 1 + q^1 + \left(q^{1+1} + q^2\right) + \left(q^{1+1+1} + q^{1+2} + q^3\right) + \cdots && (3.1.3)\\
&= 1 + p(1)q + p(2)q^2 + p(3)q^3 + \cdots . && (3.1.4)
\end{aligned}$$

The original version of this chapter has been revised. The erratum to this chapter is available at https://doi.org/10.1007/978-3-319-57762-3_44.

M.D. Hirschhorn, *The Power of q*, Developments in Mathematics 49,
DOI 10.1007/978-3-319-57762-3_3

In (3.1.2), each partition appears (though written in reverse order) precisely once as an index, and the coefficient is 1. In (3.1.3), the terms are grouped according to the integer partitioned (the index). In (3.1.4), for each n, the number of such terms is denoted by $p(n)$.

We remark that $(q;q)_\infty$ is often referred to as "Euler's product", and we may in future write

$$(q;q)_\infty = E(q). \tag{3.1.5}$$

Euler knew also that

$$\begin{aligned}(q;q)_\infty &= 1+\sum_{n\geq 1}(-1)^n\left(q^{(3n^2-n)/2}+q^{(3n^2+n)/2}\right)\\ &= 1-q-q^2+q^5+q^7-q^{12}-q^{15}++--\cdots.\end{aligned} \tag{3.1.6}$$

As we saw in §1.6, (3.1.6) is an easy consequence of JTP.

As a consequence of (3.1.1) and (3.1.6), we have that

$$\sum_{n\geq 0}p(n)q^n = 1\Big/\left(1-q-q^2+q^5+q^7-q^{12}-q^{15}++--\cdots\right)$$

or,

$$\left(1-q-q^2+q^5+q^7-q^{12}-q^{15}++--\cdots\right)\sum_{n\geq 0}p(n)q^n = 1. \tag{3.1.7}$$

In other words, the partition generating function and Euler's product (regarded as a series) are reciprocals of one another.

From (3.1.7) we obtain the following recurrence for $p(n)$ for $n\geq 1$:

$$\begin{aligned}p(n)-p(n-1)-p(n-2)+p(n-5)+p(n-7)-p(n-12)&\\ -p(n-15)++--\cdots &= 0,\end{aligned}$$

or,

$$\begin{aligned}p(n) = p(n-1)+p(n-2)-p(n-5)-p(n-7)+p(n-12)&\\ +p(n-15)--++\cdots&.\end{aligned} \tag{3.1.8}$$

where the sum on the right terminates when the arguments become negative. (3.1.8) enables us to calculate $p(n)$ as far as we like.

It seems that P. A. MacMahon, in around 1916, was the first to calculate the sequence $p(n)$ to any great distance. He calculated $p(n)$ for n up to 200, and rather serendipitously for all those whose careers depend on such things, displayed the results in a table with columns of five numbers, thus:

1	7	42	176	627	...
1	11	56	231	792	...
2	15	77	297	1002	...
3	22	101	385	1255	...
5	30	135	490	1575	...

Ramanujan noticed that the bottom entry in each column is a multiple of 5, that is,

$$p(5n+4) \equiv 0 \pmod 5, \tag{3.1.9}$$

a somewhat surprising fact, and went on to further conjecture that

$$p(7n+5) \equiv 0 \pmod 7, \tag{3.1.10}$$

$$p(11n+6) \equiv 0 \pmod{11}, \tag{3.1.11}$$

and indeed that if δ is the reciprocal modulo 24 of $A = 5^\alpha 7^\beta 11^\gamma$, then

$$p(An+\delta) \equiv 0 \pmod A. \tag{3.1.12}$$

In Ramanujan's Collected Papers, [59] p. xxxv, Hardy says "[Ramanujan] worked, far more than the majority of modern mathematicians, by induction from numerical examples; all of his congruence properties of partitions, for example, were discovered in this way. But with his memory, his patience, and his power of calculation, he combined a power of generalisation, a feeling for form, and a capacity for rapid modification of his hypotheses, that were often really startling, ..."

Chowla discovered in 1934 [32] that Ramanujan's conjecture was not quite correct. The modification that one needs to make to (3.1.12) is that if $\beta \geq 1$, the term 7^β in the modulus should be replaced by $7^{1+\lfloor \beta/2 \rfloor}$.

Thus,

$$p(5^\alpha 7^\beta 11^\gamma n+\delta) \equiv 0 \pmod{5^\alpha 7^{\beta'} 11^\gamma}. \tag{3.1.13}$$

where, if $\beta \geq 1$,

$$\beta' = 1 + \lfloor \beta/2 \rfloor. \tag{3.1.14}$$

Over the intervening years, many people have given varied proofs of some or all of these "Ramanujan's partition congruences". It was discovered not long ago that Ramanujan, in his Lost Notebook [112], essentially had a proof of the congruences for powers of 5, but the first published proofs for powers of 5 and 7 were by Watson (1938) [123], and the proof for powers of 11 was given by Atkin (1967) [11], thus completing the proof of (3.1.13).

In this book, I will give several proofs of some of these congruences, but not the congruence for general powers of 11. On the way, we will come across many surprising identities, including the identity that Hardy [59], pp. xxxiv–xxxv, regarded as Ramanujan's most beautiful.

Before we embark on our proofs, we need to make a definition and some observations.

If $A = \sum_k a_k q^k$ and $B = \sum_k b_k q^k$ with the a_k and b_k integers, we will say $A \equiv B \pmod{p}$ if and only if for all k, $a_k \equiv b_k \pmod{p}$.

If $A = 1 + \sum_{k \geq 1} a_k q^k$ with the a_k integers, then there is a series $A^{-1} = 1 + \sum_{k \geq 1} c_k q^k$ with the c_k integers, such that $AA^{-1} = 1$.

If $A = 1 + \sum_{k \geq 1} a_k q^k$ and $B = 1 + \sum_{k \geq 1} b_k q^k$ with the a_k and b_k integers and if $A \equiv B \pmod{p}$ then $A^{-1} \equiv B^{-1} \pmod{p}$.

These observations follow from the equations

$$c_1 + a_1 = 0,$$

and for $k > 1$,

$$c_k + a_1 c_{k-1} + \cdots + a_{k-1} c_1 + a_k = 0.$$

3.2 $p(5n+4) \equiv 0 \pmod{5}$

Our first proof of (3.1.9), $p(5n+4) \equiv 0 \pmod{5}$, is a streamlined version of a proof due to Ramanujan [59], p. 211, [60], p. 288.

First we need the fact that

$$(1-q)^5 \equiv 1 - q^5 \pmod{5}. \tag{3.2.1}$$

Indeed, for prime p,

$$(1-q)^p \equiv 1 - q^p \pmod{p} \tag{3.2.2}$$

since $p \left| \binom{p}{k} \right.$ for $1 \leq k \leq p-1$.

If in (3.2.1) we put q^k for q, and form the product over $k \geq 1$, we find that, modulo 5,

$$(q;q)_\infty^5 = \prod_{k \geq 1} (1-q^k)^5 \equiv \prod_{k \geq 1} (1-q^{5k}) = (q^5;q^5)_\infty. \tag{3.2.3}$$

From our remarks at the end of §3.1, we have, modulo 5,

$$\frac{1}{(q;q)_\infty^5} \equiv \frac{1}{(q^5;q^5)_\infty}. \tag{3.2.4}$$

We start with Euler's result (1.6.1). Note that $(3k^2-k)/2 \equiv 0,\ 1 \text{ or } 2 \pmod 5$, so we can write

$$\begin{aligned}(q;q)_\infty &= \sum_{-\infty}^{\infty}(-1)^k q^{(3k^2-k)/2}\\ &= 1-q-q^2+q^5+q^7-q^{12}-q^{15}+q^{22}+q^{26}-q^{35}-q^{40}++--\ \cdots\\ &= (1+q^5-q^{15}-q^{35}\ \cdots)-q(1-q^{25}+\ \cdots)-q^2(1-q^5+q^{10}-q^{20}\ \cdots)\\ &= E_0+E_1+E_2 \end{aligned}\tag{3.2.5}$$

where E_i consists of those terms of E in which the power of q is congruent to i modulo 5.

This is another example of an "m-dissection", where we have expressed a series as a sum of series, in each of which the powers fall in just one residue class modulo m. Here $m=5$.

Also, from Jacobi's result (1.7.1), noting that $(k^2+k)/2 \equiv 0,\ 1 \text{ or } 3 \pmod 5$, and **importantly** that $(k^2+k)/2 \equiv 3 \pmod 5$ if and only if $k \equiv 2 \pmod 5$ and then the coefficient $2k+1 \equiv 0 \pmod 5$, we find that, modulo 5,

$$\begin{aligned}(q;q)_\infty^3 &= \sum_{k\ge 0}(-1)^k(2k+1)q^{(k^2+k)/2}\\ &= 1-3q+5q^3-7q^6+9q^{10}-11q^{15}+13q^{21}-15q^{28}+17q^{36}\\ &\quad -19q^{45}+21q^{55}-+\ \cdots\\ &\equiv 1-3q+3q^6-q^{10}-q^{15}+3q^{21}-3q^{36}+q^{45}+q^{55}-3q^{66}+\ \cdots\\ &= (1-q^{10}-q^{15}+q^{45}+---+\ \cdots)-3q(1-q^5-q^{20}+q^{35}+q^{65}--++\ \cdots)\\ &= J_0+J_1 \end{aligned}\tag{3.2.6}$$

where J_i consists of terms in which the power of q is congruent to i modulo 5.

So now we have, modulo 5, by (3.2.4)–(3.2.6),

$$\begin{aligned}\sum_{n\ge 0} p(n)q^n &= \frac{1}{(q;q)_\infty} = \frac{(q;q)_\infty^4}{(q;q)_\infty^5} = \frac{(q;q)_\infty(q;q)_\infty^3}{(q;q)_\infty^5}\\ &\equiv \frac{1}{(q^5;q^5)_\infty}\Big(E_0+E_1+E_2\Big)\Big(J_0+J_1\Big)\\ &= \frac{1}{(q^5;q^5)_\infty}\Big(E_0J_0+(E_0J_1+E_1J_0)+(E_1J_1+E_2J_0)+E_2J_1\Big). \end{aligned}\tag{3.2.7}$$

There are no terms on the right side in which the exponent on q is congruent to 4 (mod 5), so

$$\sum_{n\geq 0} p(5n+4)q^{5n+4} \equiv 0 \pmod 5, \tag{3.2.8}$$

and hence

$$p(5n+4) \equiv 0 \pmod 5,$$

which is (3.1.9).

3.3 $p(5n+4) \equiv 0 \pmod 5$ Again

In this proof we use only (3.2.4) and (3.2.6). We avoid the use of (3.2.5).

We have, modulo 5,

$$\begin{aligned}\sum_{n\geq 0} p(n)q^n = \frac{1}{(q;q)_\infty} = \frac{(q;q)_\infty^9}{(q;q)_\infty^{10}} = \frac{\left((q;q)_\infty^3\right)^3}{\left((q;q)_\infty^5\right)^2}\\ \equiv \frac{(J_0+J_1)^3}{(q^5;q^5)_\infty^2}\\ = \frac{J_0^3+3J_0^2J_1+3J_0J_1^2+J_1^3}{(q^5;q^5)_\infty^2}.\end{aligned} \tag{3.3.1}$$

There are no terms on the right side in which the exponent on q is congruent to 4 modulo 5, and (3.1.9) follows.

3.4 $p(7n+5) \equiv 0 \pmod 7$

This proof is along the same lines as the proof that $p(5n+4) \equiv 0 \pmod 5$ in §3.3.

We start by noting that, referring to Jacobi's result (1.7.1), $(k^2+k)/2 \equiv 0,\ 1,\ 3$ or $6 \pmod 7$, and $(k^2+k)/2 \equiv 6 \pmod 7$ if and only if $k \equiv 3 \pmod 7$, and then the coefficient $2k+1 \equiv 0 \pmod 7$.

It follows that

$$\begin{aligned}(q;q)_\infty^3 &= \sum_{k\geq 0}(-1)^k(2k+1)q^{(k^2+k)/2}\\ &= 1-3q+5q^3-7q^6+9q^{10}-11q^{15}+13q^{21}-15q^{28}+17q^{36}\\ &\quad -19q^{45}+21q^{55}-+\cdots\\ &\equiv 1-3q+5q^3-5q^{10}+3q^{15}-q^{21}-q^{28}+3q^{36}-5q^{45}+5q^{66}-+\cdots\\ &= (1-q^{21}-q^{28}+\cdots)-3q(1-q^{14}-q^{35}+\cdots)+5q^3(1-q^7-q^{42}+\cdots)\\ &= J_0+J_1+J_3, \end{aligned} \tag{3.4.1}$$

where J_i consists of terms in which the exponent on q is congruent to i modulo 7.

It follows that, modulo 7,

$$\begin{aligned}\sum_{n\geq 0}p(n)q^n &= \frac{1}{(q;q)_\infty} = \frac{(q;q)_\infty^6}{(q;q)_\infty^7} = \frac{\left((q;q)_\infty^3\right)^2}{(q;q)_\infty^7}\\ &\equiv \frac{(J_0+J_1+J_3)^2}{(q^7;q^7)_\infty}\\ &= \frac{J_0^1+2J_0J_1+J_1^2+2J_0J_3+2J_1J_3+J_3^2}{(q^7;q^7)_\infty}. \end{aligned} \tag{3.4.2}$$

There are no terms on the right side in which the exponent on q is congruent to 5 modulo 7, and (3.1.10) follows.

3.5 $p(11n+6) \equiv 0 \pmod{11}$

We start by noting that $(k^2+k)/2 \equiv 0,\ 1,\ 3,\ 6,\ 10$ or $4 \pmod{11}$, and $(k^2+k)/2 \equiv 4 \pmod{11}$ if and only if $k \equiv 5 \pmod{11}$, and then $2k+1 \equiv 0 \pmod{11}$, so, modulo 11,

$$(q;q)_\infty^3 \equiv J_0+J_1+J_3+J_6+J_{10}.$$

where J_i consists of terms in which the exponent on q is i modulo 11.

Thus we have, modulo 11,

$$\begin{aligned}\sum_{n\geq 0}p(n)q^n &= \frac{1}{(q;q)_\infty} = \frac{(q;q)_\infty^{21}}{(q;q)_\infty^{22}} = \frac{\left((q;q)_\infty^3\right)^7}{\left((q;q)_\infty^{11}\right)^2}\\ &\equiv \frac{(J_0+J_1+J_3+J_6+J_{10})^7}{(q^{11};q^{11})_\infty^2}. \end{aligned} \tag{3.5.1}$$

If we extract those terms in which the exponent on q is 6 modulo 11, we find

$$\sum_{n\geq 0} p(11n+6)q^{11n+6} \equiv \frac{P}{(q^{11};q^{11})_\infty^2} \tag{3.5.2}$$

where

$$\begin{aligned} P \equiv\ & 7J_0^6J_6+10J_0^5J_3^2+J_0^4J_1J_6J_{10}+8J_0^3J_1^3J_3+2J_0^3J_1J_3^2J_{10}+8J_0^3J_6^3J_{10} \\ &+3J_0^2J_1^2J_3J_6^2+3J_0^2J_1^2J_6J_{10}^2+3J_0^2J_3^2J_6^2J_{10}+2J_0^2J_3J_6J_{10}^3+10J_0^2J_{10}^5+7J_0J_1^6 \\ &+J_0J_1^4J_3J_{10}+2J_0J_1^2J_3^3J_6+3J_0J_1^2J_3^2J_{10}^2+J_0J_1J_3J_6^4+2J_0J_1J_6^3J_{10}^2+J_0J_3^4J_6J_{10} \\ &+8J_0J_3^3J_{10}^3+10J_1^5J_6^2+2J_1^3J_3J_6^2J_{10}+8J_1^3J_6J_{10}^3+10J_1^2J_3^5+8J_1J_3^3J_6^3 \\ &+3J_1J_3^2J_6^2J_{10}^2+J_1J_3J_6J_{10}^4+7J_1J_{10}^6+7J_3^6J_{10}+7J_3J_6^6+10J_6^5J_{10}^2. \end{aligned} \tag{3.5.3}$$

We want to show that $P \equiv 0 \pmod{11}$. Here is the trick.

We have

$$\begin{aligned} (J_0+J_1+J_3+J_6+J_{10})^4 &\equiv \left(E(q)^3\right)^4 = E(q)^{12} = E(q)^{11}E(q) \equiv E(q^{11})E(q) \\ &\equiv E(q^{11})\left(E_0+E_1+E_2+E_4+E_5+E_7\right), \end{aligned} \tag{3.5.4}$$

where E_i consists of terms in which the exponent on q is i modulo 11.

If we expand the left side of (3.5.4) and observe that there are no terms on the right in which the exponent on q is 3, 6, 8, 9 or 10 modulo 11, we obtain five polynomials in the J_i which are congruent to 0 modulo 11. For example,

$$4J_0^3J_3+4J_0J_1^3+24J_0J_1J_3J_{10}+6J_1^2J_6^2+12J_1J_6^2J_{10}+4J_6J_{10}^3 \equiv 0. \tag{3.5.5}$$

If we multiply (3.5.5) by 3, many of the coefficients become 1 modulo 11.

$$J_0^3J_3+J_0J_1^3+6J_0J_1J_3J_{10}+7J_1^2J_6^2+3J_1J_6^2J_{10}+J_6J_{10}^3 \equiv 0. \tag{3.5.6}$$

Similarly,

$$J_0^3J_6+7J_0^2J_3^2+6J_0J_1J_6J_{10}+J_1^3J_3+3J_1J_3^2J_{10}+J_6^3J_{10} \equiv 0, \tag{3.5.7}$$

$$3J_0J_1^2J_6+6J_0J_3J_6J_{10}+J_0J_{10}^3+7J_1^2J_3^2+J_1J_6^3+J_3^3J_{10} \equiv 0, \tag{3.5.8}$$

$$3J_0^2J_3J_6+7J_0^2J_{10}^2+J_0J_3^3+6J_1J_3J_6J_{10}+J_1^3J_6+J_1J_{10}^3 \equiv 0, \tag{3.5.9}$$

$$J_0^3J_{10}+6J_0J_1J_3J_6+3J_0J_1J_{10}^2+J_1J_3^3+J_3J_6^3+7J_6^2J_{10}^2 \equiv 0. \tag{3.5.10}$$

We now seek $a,\ b,\ c$ so that

$$\begin{aligned}P \equiv& \left(aJ_1J_3J_{10}+bJ_0^2J_3+cJ_1^3\right)\left(J_0^3J_3+J_0J_1^3+6J_0J_1J_3J_{10}+7J_1^2J_6^2+3J_1J_6^2J_{10}+J_6J_{10}^3\right)\\&+\left(cJ_0^3+aJ_0J_1J_{10}+bJ_6^2J_{10}\right)\left(J_0^3J_6+7J_0^2J_3^2+6J_0J_1J_6J_{10}+J_1^3J_3+3J_1J_3^2J_{10}+J_6^3J_{10}\right)\\&+\left(aJ_0J_3J_6+bJ_0J_{10}^2+cJ_3^3\right)\left(3J_0J_1^2J_6+6J_0J_3J_6J_{10}+J_0J_{10}^3+7J_1^2J_3^2+J_1J_6^3+J_3^3J_{10}\right)\\&+\left(bJ_1^2J_6+aJ_3J_6J_{10}+cJ_{10}^3\right)\left(3J_0^2J_3J_6+7J_0^2J_{10}^2+J_0J_3^3+6J_1J_3J_6J_{10}+J_1^3J_6+J_1J_{10}^3\right)\\&+\left(aJ_0J_1J_6+bJ_1J_3^2+cJ_6^3\right)\left(J_0^3J_{10}+6J_0J_1J_3J_6+3J_0J_1J_{10}^2+J_1J_3^3+J_3J_6^3+7J_6^2J_{10}^2\right),\end{aligned} \tag{3.5.11}$$

because if such $a,\ b,\ c$ exist we will have, by virtue of (3.5.6)–(3.5.10),

$$P \equiv 0 \pmod{11}, \tag{3.5.12}$$

which is what we want.

It turns out that, indeed, $a,\ b$ and c do exist!

$$a=7,\ \ b=5,\ \ c=7. \tag{3.5.13}$$

(The calculation or verification is left as an exercise.)

So we are done!

3.6 The Atkin–Swinnerton-Dyer Congruences for the Modulus 5

We start by introducing some new notation, which generalises (1.2.8). We will write

$$(a_1,a_2,\ldots,a_n;q)_\infty = (a_1;q)_\infty(a_2;q)_\infty\cdots(a_n;q)_\infty. \tag{3.6.1}$$

Thus, for example, JTP can be written

$$(-a^{-1}q,-aq,q^2;q^2)_\infty = \sum_{-\infty}^{\infty} a^k q^{k^2} \tag{3.6.2}$$

and, as a second example, relevant to what we are about to see,

$$(q,q^4,q^5;q^5)_\infty = \sum_{-\infty}^{\infty}(-1)^k q^{(5k^2-3k)/2},\quad (q^2,q^3,q^5;q^5)_\infty = \sum_{-\infty}^{\infty}(-1)^k q^{(5k^2-k)/2}. \tag{3.6.3}$$

We saw in (3.2.6) that, modulo 5,

$$(q;q)_\infty^3 \equiv F(q^5) - 3qG(q^5) \tag{3.6.4}$$

where

$$F(q^5) = 1 - q^{10} - q^{15} + q^{45} + q^{55} - - + + \cdots ,$$
$$G(q^5) = 1 - q^5 - q^{20} + q^{35} + q^{65} - - + + \cdots .$$

It can be checked (by seeing how we arrived at (3.2.6)) that

$$F(q^5) = \sum_{-\infty}^{\infty} (-1)^k q^{(25k^2-5k)/2} = (q^{10}, q^{15}, q^{25}; q^{25})_\infty,$$
$$G(q^5) = \sum_{-\infty}^{\infty} (-1)^k q^{(25k^2-15k)/2} = (q^5, q^{20}, q^{25}; q^{25})_\infty. \tag{3.6.5}$$

Exercise: Verify (3.6.5).

Now, making use of (3.2.4) and (3.6.4), we have, modulo 5,

$$\begin{aligned}
\sum_{n\geq 0} p(n)q^n &= \frac{1}{(q;q)_\infty} = \frac{(q;q)_\infty^9}{(q;q)_\infty^{10}} = \frac{((q;q)_\infty^3)^3}{((q;q)_\infty^5)^2} \\
&\equiv \frac{(F(q^5) - 3qG(q^5))^3}{(q^5;q^5)_\infty^2} \\
&\equiv \frac{F(q^5)^3 - 9qF(q^5)^2G(q^5) + 27q^2F(q^5)G(q^5)^2 - 27q^3G(q^5)^3}{(q^5;q^5)_\infty^2} \\
&\equiv \frac{F(q^5)^3 + qF(q^5)^2G(q^5) + 2q^2F(q^5)G(q^5)^2 + 3q^3G(q^5)^3}{(q^5;q^5)_\infty^2}. \qquad (3.6.6)
\end{aligned}$$

We find, using (3.6.5), that the following marvellous congruences hold modulo 5.

$$\sum_{n\geq 0} p(5n)q^n \equiv \frac{F(q)^3}{(q;q)_\infty^2} = \frac{(q^2,q^3,q^5;q^5)_\infty^3}{(q;q)_\infty^2},$$
$$\sum_{n\geq 0} p(5n+1)q^n \equiv \frac{(q^2,q^3,q^5;q^5)_\infty^2 (q,q^4,q^5;q^5)_\infty}{(q;q)_\infty^2},$$
$$\sum_{n\geq 0} p(5n+2)q^n \equiv 2\frac{(q^2,q^3,q^5;q^5)_\infty (q,q^4,q^5;q^5)_\infty^2}{(q;q)_\infty^2},$$
$$\sum_{n\geq 0} p(5n+3)q^n \equiv 3\frac{(q,q^4,q^5;q^5)_\infty^3}{(q;q)_\infty^2}$$

and

$$\sum_{n\geq 0} p(5n+4)q^n \equiv 0. \tag{3.6.7}$$

The congruences (3.6.7) are known as the Atkin–Swinnerton-Dyer congruences for the modulus 5.

We will see in §3.7 and §3.8 that similar congruences hold with respect to the moduli 7 and 11.

Amusingly, it follows from (3.6.7) that, modulo 5,

$$\sum_{n\geq 0} p(5n)q^n \sum_{n\geq 0} p(5n+3)q^n \equiv 3(q;q)_\infty^{14}, \tag{3.6.8}$$

and

$$\sum_{n\geq 0} p(5n+1)q^n \sum_{n\geq 0} p(5n+2)q^n \equiv 2(q;q)_\infty^{14}. \tag{3.6.9}$$

3.7 The Atkin–Swinnerton-Dyer Congruences for the Modulus 7

If we re-examine (3.4.1), we see that, modulo 7,

$$(q;q)_\infty^3 \equiv H(q^7) - 3qJ(q^7) + 5q^3K(q^7), \tag{3.7.1}$$

where

$$\begin{aligned}
H(q^7) &= \sum_{-\infty}^{\infty}(-1)^k q^{(49k^2-7k)/2} = (q^{21},q^{28},q^{49};q^{49})_\infty,\\
J(q^7) &= \sum_{-\infty}^{\infty}(-1)^k q^{(49k^2-21k)/2} = (q^{14},q^{35},q^{49};q^{49})_\infty,\\
K(q^7) &= \sum_{-\infty}^{\infty}(-1)^k q^{(49k^2-35k)/2} = (q^{7},q^{42},q^{49};q^{49})_\infty.
\end{aligned} \tag{3.7.2}$$

Exercise: Verify (3.7.2).

Then we have

$$\begin{aligned}
\sum_{n\geq 0} p(n)q^n &= \frac{1}{(q;q)_\infty} = \frac{(q;q)_\infty^6}{(q;q)_\infty^7} = \frac{((q;q)_\infty^3)^2}{(q;q)_\infty^7} \\
&\equiv \frac{(H(q^7)-3qJ(q^7)+5q^3K(q^7))^2}{(q^7;q^7)_\infty} \\
&= \frac{H(q^7)^2-6qH(q^7)J(q^7)+9q^2J(q^7)^2+10q^3H(q^7)K(q^7)-30q^4J(q^7)K(q^7)+25q^6K(q^7)^2}{(q^7;q^7)_\infty} \\
&\equiv \frac{H(q^7)^2+qH(q^7)J(q^7)+2q^2J(q^7)^2+3q^3H(q^7)K(q^7)+5q^4J(q^7)K(q^7)+11q^6K(q^7)^2}{(q^7;q^7)_\infty}.
\end{aligned} \tag{3.7.3}$$

It follows that, modulo 7,

$$\begin{aligned}
\sum_{n\geq 0} p(7n)q^n &\equiv \frac{(q^3,q^4,q^7;q^7)_\infty^2}{(q;q)_\infty}, \\
\sum_{n\geq 0} p(7n+1)q^n &\equiv \frac{(q^2,q^3,q^4,q^5,q^7,q^7;q^7)_\infty}{(q;q)_\infty}, \\
\sum_{n\geq 0} p(7n+2)q^n &\equiv 2\frac{(q^2,q^5,q^7;q^7)_\infty^2}{(q;q)_\infty}, \\
\sum_{n\geq 0} p(7n+3)q^n &\equiv 3\frac{(q,q^3,q^4,q^6,q^7,q^7;q^7)_\infty}{(q;q)_\infty}, \\
\sum_{n\geq 0} p(7n+4)q^n &\equiv 5\frac{(q,q^2,q^5,q^6,q^7,q^7;q^7)_\infty}{(q;q)_\infty}, \\
\sum_{n\geq 0} p(7n+5)q^n &\equiv 0
\end{aligned}$$

and

$$\sum_{n\geq 0} p(7n+6)q^n \equiv 11\frac{(q,q^6,q^7;q^7)_\infty^2}{(q;q)_\infty}. \tag{3.7.4}$$

and these are the Atkin–Swinnerton-Dyer congruences for the modulus 7.

It follows from (3.7.4) that, modulo 7,

$$\sum_{n\geq 0} p(n)q^n \sum_{n\geq 0} p(7n+2)q^n \sum_{n\geq 0} p(7n+6)q^n \equiv (q;q)_\infty^{27} \tag{3.7.5}$$

and

$$\sum_{n\geq 0} p(7n+1)q^n \sum_{n\geq 0} p(7n+3)q^n \sum_{n\geq 0} p(7n+4)q^n \equiv (q;q)_\infty^{27}. \tag{3.7.6}$$

Exercise: Verify (3.7.5) and (3.7.6).

3.8 The Atkin–Swinnerton-Dyer Congruences for the Modulus 11

The Atkin–Swinnerton-Dyer congruences for modulus 11 are as follows. Our proofs will have to wait till §13.4, when we have Winquist's identity.

$$\sum_{n\geq 0} p(11n)q^n \equiv \frac{(q^2,q^3,q^4,q^5,q^6,q^7,q^8,q^9,q^{11},q^{11};q^{11})_\infty}{(q;q)_\infty},$$

$$\sum_{n\geq 0} p(11n+1)q^n \equiv \frac{(q,q^4,q^5,q^5,q^6,q^6,q^7,q^{10},q^{11},q^{11};q^{11})_\infty}{(q;q)_\infty},$$

$$\sum_{n\geq 0} p(11n+2)q^n \equiv 2\frac{(q^2,q^3,q^3,q^5,q^6,q^8,q^8,q^9,q^{11},q^{11};q^{11})_\infty}{(q;q)_\infty},$$

$$\sum_{n\geq 0} p(11n+3)q^n \equiv 3\frac{(q^2,q^2,q^4,q^5,q^6,q^7,q^9,q^9,q^{11},q^{11};q^{11})_\infty}{(q;q)_\infty},$$

$$\sum_{n\geq 0} p(11n+4)q^n \equiv 5\frac{(q,q^3,q^4,q^5,q^6,q^7,q^8,q^{10},q^{11},q^{11};q^{11})_\infty}{(q;q)_\infty},$$

$$\sum_{n\geq 0} p(11n+5)q^n \equiv 7\frac{(q,q^3,q^4,q^4,q^7,q^7,q^8,q^{10},q^{11},q^{11};q^{11})_\infty}{(q;q)_\infty},$$

$$\sum_{n\geq 0} p(11n+6)q^n \equiv 0,$$

$$\sum_{n\geq 0} p(11n+7)q^n \equiv 4\frac{(q,q^2,q^4,q^5,q^6,q^7,q^9,q^{10},q^{11},q^{11};q^{11})_\infty}{(q;q)_\infty},$$

$$\sum_{n\geq 0} p(11n+8)q^n \equiv 6q\frac{(q,q,q^2,q^3,q^8,q^9,q^{10},q^{10},q^{11},q^{11};q^{11})_\infty}{(q;q)_\infty},$$

$$\sum_{n\geq 0} p(11n+9)q^n \equiv 8\frac{(q,q^2,q^3,q^5,q^6,q^8,q^9,q^{10},q^{11},q^{11};q^{11})_\infty}{(q;q)_\infty},$$

and

$$\sum_{n\geq 0} p(11n+10)q^n \equiv 9\frac{(q,q^2,q^3,q^4,q^7,q^8,q^9,q^{10},q^{11},q^{11};q^{11})_\infty}{(q;q)_\infty}. \tag{3.8.1}$$

We note that these yield

$$\sum_{n\geq 0} p(11n)q^n \sum_{n\geq 0} p(11n+4)q^n \sum_{n\geq 0} p(11n+7)q^n \sum_{n\geq 0} p(11n+9)q^n \sum_{n\geq 0} p(11n+10)q^n \equiv -(q;q)_\infty^{65} \tag{3.8.2}$$

and

$$\sum_{n\geq 0} p(11n+1)q^n \sum_{n\geq 0} p(11n+2)q^n \sum_{n\geq 0} p(11n+3)q^n \sum_{n\geq 0} p(11n+5)q^n \sum_{n\geq 0} p(11n+8)q^n \equiv -q(q;q)_\infty^{65}. \tag{3.8.3}$$

Endnotes.

§3.1 There are several other recurrences for $p(n)$, none of which is as efficient as (3.1.8).

The first I will mention appears on the very last page of Ramanujan's third notebook [111].

We have

$$\sum_{n\geq 0} p(n)q^n = \prod_{n\geq 1}(1-q^n)^{-1}.$$

If we differentiate this with respect to q and multiply by q, we obtain

$$\begin{aligned}\sum_{n\geq 0} np(n)q^n &= \prod_{n\geq 1}(1-q^n)^{-1}\cdot\sum_{n\geq 1}\frac{nq^n}{1-q^n}\\ &= \sum_{n\geq 1}\sigma(n)q^n\cdot\sum_{n\geq 0}p(n)q^n,\end{aligned}$$

where $\sigma(n)$ is the sum of the divisors of n.

It follows that for $n \geq 1$,

$$np(n) = \sum_{k=1}^{n} \sigma(k)p(n-k).$$

With $p(0) = 1$, we can calculate $p(n)$ recursively.

In January 2016, I discovered the following recurrence for $p(n)$.

We have

$$\sum_{n\geq 0} p(n)q^n = \frac{1}{(q;q)_\infty} = \frac{\psi(q)}{(q^2;q^2)_\infty^2} = \psi(q)\left(\sum_{n\geq 0} p(n)q^{2n}\right)^2.$$

If we extract the even part and replace q^2 by q, we obtain

$$\sum_{n\geq 0} p(2n)q^n = \left(1 + \sum_{n\geq 1}\left(q^{4n^2-n} + q^{4n^2+n}\right)\right)\sum_{n\geq 0}\left(\sum_{k+l=n} p(k)p(l)\right)q^n$$

while if we extract the odd part, divide by q and replace q^2 by q, we obtain

$$\sum_{n\geq 0} p(2n+1)q^n = \left(1 + \sum_{n\geq 1}\left(q^{4n^2-3n} + q^{4n^2+3n}\right)\right)\sum_{n\geq 0}\left(\sum_{k+l=n} p(k)p(l)\right)q^n.$$

It follows that

$$\begin{aligned} p(2n) = &\sum_{k+l=n} p(k)p(l) + \sum_{k+l=n-3} p(k)p(l) + \sum_{k+l=n-5} p(k)p(l) \\ &+ \sum_{k+l=n-14} p(k)p(l) + \cdots \end{aligned}$$

where the numbers $3, 5, 14, \ldots$ are numbers of the form $4n^2 \pm n$, and

$$\begin{aligned} p(2n+1) = &\sum_{k+l=n} p(k)p(l) + \sum_{k+l=n-1} p(k)p(l) + \sum_{k+l=n-7} p(k)p(l) \\ &+ \sum_{k+l=n-10} p(k)p(l) + \cdots \end{aligned}$$

where the numbers $1, 7, 10, \ldots$ are numbers of the form $4n^2 \pm 3n$.

For example, if $n = 5$,

$$p(10) = 7\times 1 + 5\times 1 + 3\times 2 + 2\times 3 + 1\times 5 + 1\times 7 + 2\times 1 + 1\times 1 + 1\times 2 + 1\times 1 = 42$$

and

$$\begin{aligned} p(11) = &7\times 1 + 5\times 1 + 3\times 2 + 2\times 3 + 1\times 5 + 1\times 7 + 5\times 1 + 3\times 1 + 2\times 2 \\ &+ 1\times 3 + 1\times 5 = 56. \end{aligned}$$

Exercise: Find the recurrence for $p(n)$ which comes from

$$(q^2;q^2)_\infty \sum_{n\geq 0} p(n)q^n = \psi(q) \sum_{n\geq 0} p(2n)q^{2n}.$$

We find yet another recurrence for $p(n)$ in Chapter 22.

§3.1 The first counterexample to Ramanujan's conjecture (3.1.12) occurs at 243, beyond MacMahon's table, and was discovered in 1934 by S. Chowla [32] from H. Gupta's table of $p(n)$ for $n \leq 300$. (The author is the proud possessor of a table of $p(n)$ for $0 \leq n \leq 10699$, calculated by Frank Garvan in 1980.)

§3.2 Ramanujan's proof appears in Ramanujan (1919) [109], and is reproduced in Hardy and Wright (1960) [60], §19.12. The difference between my version and Ramanujan's is that I examine congruences before multiplying the series, whereas Ramanujan multiplies the series first.

§3.5 This proof appears in Hirschhorn (2014) [89].

§§3.6–3.8 The Atkin–Swinnerton-Dyer congruences appear in Atkin–Swinnerton-Dyer (1954) [12].

References

11. A.O.L. Atkin, Proof of a conjecture of Ramanujan. Glasg. Math. J. **8**, 14–32 (1967)
12. A.O.L. Atkin, H.P.F. Swinnerton-Dyer, Some properties of partitions. Proc. Lond. Math. Soc. **4**, 84–106 (1954)
32. S. Chowla, Congruence properties of partitions. J. Lond. Math. Soc. **9**, 247 (1934)
60. G.H. Hardy, E.M. Wright, *An Introduction to the Theory of Numbers*, 4th edn. (Oxford University Press, Oxford, 1960)
89. M.D. Hirschhorn, A short and simple proof of Ramanujan's mod 11 partition congruence. J. Number Theory **139**, 205–209 (2014)
109. S. Ramanujan, Some properties of $p(n)$, the number of partitions of n. Proc. Camb. Philos. Soc. **XIX**, 207–210 (1919)
111. S. Ramanujan, Notebooks, II, Tata Institute of Fundamental Research, Mumbai (2012)
112. S. Ramanujan, *The Lost Notebook and Other Unpublished Papers* (Springer, Berlin, 1988)
123. G.N. Watson, Ramanujans Vermutung über Zerfällungsanzahlen. J. Reine Angew. Math. **179**, 97–128 (1938)

Chapter 4
Ramanujan's Partition Congruences—A Uniform Proof

4.1 Introduction

In this chapter, we give uniform proofs of the three partition congruences

$$p(5n+4) \equiv 0 \pmod{5},$$
$$p(7n+5) \equiv 0 \pmod{7}$$

and

$$p(11n+6) \equiv 0 \pmod{11}.$$

That we can do so is perhaps rather surprising, since for a long time it was felt that the congruence modulo 11 was "more difficult", Hardy and Wright (1960), §19.12.

The proofs depend on a great idea of Garvan and Stanton (1990) and Garvan, Kim and Stanton (1990), with a small modification by me. My proofs appear in Hirschhorn (1994).

4.2 $p(5n+4) \equiv 0 \pmod{5}$

We begin with JTP by writing

$$\frac{\sum_{-\infty}^{\infty} a^k q^{k^2}}{(q^2;q^2)_\infty} = (-a^{-1}q, -aq; q^2)_\infty$$
$$= (-a^{-1}q, -a^{-1}q^3, -a^{-1}q^5, -a^{-1}q^7, -a^{-1}q^9, -aq, -aq^3, -aq^5, -aq^7, -aq^9; q^{10})_\infty$$

M.D. Hirschhorn, *The Power of q*, Developments in Mathematics 49,
DOI 10.1007/978-3-319-57762-3_4

$$= \frac{1}{(q^{10};q^{10})_\infty^5}(-a^{-1}q^9,-aq,q^{10};q^{10})_\infty(-a^{-1}q^7,-aq^3,q^{10};q^{10})_\infty$$
$$\times (-a^{-1}q^5,-aq^5,q^{10};q^{10})_\infty(-a^{-1}q^3,-aq^7,q^{10};q^{10})_\infty(-a^{-1}q,-aq^9,q^{10};q^{10})_\infty$$
$$= \frac{1}{(q^{10};q^{10})_\infty^5}\sum_{-\infty}^{\infty} a^k q^{5k^2-4k}\sum_{-\infty}^{\infty} a^k q^{5k^2-2k}\sum_{-\infty}^{\infty} a^k q^{5k^2}\sum_{-\infty}^{\infty} a^k q^{5k^2+2k}\sum_{-\infty}^{\infty} a^k q^{5k^2+4k}. \tag{4.2.1}$$

If from (4.2.1) we extract the constant term (the term independent of a) and replace q by $q^{\frac{1}{2}}$, we obtain

$$\sum_{n\geq 0} p(n)q^n = \frac{1}{(q^5;q^5)_\infty^5}\sum q^{Q(\mathbf{k})}, \tag{4.2.2}$$

where

$$Q(\mathbf{k}) = (5k_1^2-4k_1+5k_2^2-2k_2+5k_3^2+5k_4^2+2k_4+5k_5^2+4k_5)/2$$
$$= \frac{1}{10}\Big((5k_1-2)^2+(5k_2-1)^2+(5k_3)^2+(5k_4+1)^2+(5k_5+2)^2\Big)-1, \tag{4.2.3}$$

and where the sum is taken over the set of all $\mathbf{k}=(k_1,k_2,k_3,k_4,k_5)\in\mathbb{Z}^5$ for which $k_1+k_2+k_3+k_4+k_5=0$.

Let $u_1=5k_4+1,\ u_2=5k_5+2,\ u_3=5k_1-2,\ u_4=5k_2-1,\ u_5=5k_3$. Then, if we multiply (4.2.3) by q, we find

$$\sum_{n\geq 0} p(n)q^{n+1} = \frac{1}{(q^5;q^5)_\infty^5}\sum q^{(u_1^2+u_2^2+u_3^2+u_4^2+u_5^2)/10}$$
$$= \frac{1}{(q^5;q^5)_\infty^5}\sum q^{||\mathbf{u}||^2/10}, \tag{4.2.4}$$

where the sum is taken over the set of all $\mathbf{u}=(u_1,u_2,u_3,u_4,u_5)\in\mathbb{Z}^5$ with $u_i\equiv i \pmod 5$, $i=1,\ \dots\ ,5$ and $u_1+u_2+u_3+u_4+u_5=0$.

If from (4.2.4) we extract those terms in which the power of q is divisible by 5, then replace q by $q^{\frac{1}{5}}$, we obtain

$$\sum_{n\geq 0} p(5n+4)q^{n+1} = \frac{1}{(q;q)_\infty^5}\sum_{U_5} q^{||\mathbf{u}||^2/50}, \tag{4.2.5}$$

where

$$U_5 = \{\mathbf{u}\in\mathbb{Z}^5 : u_i\equiv i \pmod 5,\ \ \mathbf{u}\perp\mathbf{e}=(1,1,1,1,1),\ 50\,\Big|\,||\mathbf{u}||^2\}.$$

The five vectors $\mathbf{u}\in U_5$ that give rise to the first term, $p(4)=5$, (they are found by trial and error, but it helps to know they exist), are

$$\mathbf{u_1} = (1,2,-2,4,-5) = \mathbf{u_6},\ \ \mathbf{u_2} = (1,2,3,-6,0),\ \ \mathbf{u_3} = (-4,-3,3,4,0),$$
$$\mathbf{u_4} = (6,-3,-2,-1,0),\ \ \mathbf{u_5} = (-4,2,-2,-1,5). \tag{4.2.6}$$

I have written the five vectors in a particular order. It is easy to check that the angle between each pair of neighbours, $\mathbf{u_i}$, $\mathbf{u_{i+1}}$, $i = 1, \ldots, 5$, is given by

$$\cos\theta_{i,i+1} = \frac{\mathbf{u}_i \cdot \mathbf{u}_{i+1}}{||\mathbf{u}_i||||\mathbf{u}_{i+1}||} = -\frac{1}{2} \tag{4.2.7}$$

while otherwise

$$\cos\theta_{i,j} = \frac{\mathbf{u}_i \cdot \mathbf{u}_j}{||\mathbf{u}_i||||\mathbf{u}_j||} = 0. \tag{4.2.8}$$

Let us now consider the linear transformation M defined by

$$M\mathbf{u}_i = \mathbf{u}_{i+1},\ \ i = 1, \ldots 4 \ \text{ and } \ M\mathbf{e} = \mathbf{e}, \tag{4.2.9}$$

that is, where now, suddenly, I'm writing the vectors as columns, because that's what I'm familiar with,

$$M\begin{pmatrix} 1 & 1 & -4 & 6 & 1 \\ 2 & 2 & -3 & -3 & 1 \\ -2 & 3 & 3 & -2 & 1 \\ 4 & -6 & 4 & -1 & 1 \\ -5 & 0 & 0 & 0 & 1 \end{pmatrix} = \begin{pmatrix} 1 & -4 & 6 & -4 & 1 \\ 2 & -3 & -3 & 2 & 1 \\ 3 & 3 & -2 & -2 & 1 \\ -6 & 4 & -1 & -1 & 1 \\ 0 & 0 & 0 & 5 & 1 \end{pmatrix} \tag{4.2.10}$$

We have

$$\begin{aligned} M &= \begin{pmatrix} 1 & -4 & 6 & -4 & 1 \\ 2 & -3 & -3 & 2 & 1 \\ 3 & 3 & -2 & -2 & 1 \\ -6 & 4 & -1 & -1 & 1 \\ 0 & 0 & 0 & 5 & 1 \end{pmatrix} \begin{pmatrix} 1 & 1 & -4 & 6 & 1 \\ 2 & 2 & -3 & -3 & 1 \\ -2 & 3 & 3 & -2 & 1 \\ 4 & -6 & 4 & -1 & 1 \\ -5 & 0 & 0 & 0 & 1 \end{pmatrix}^{-1} \\ &= \begin{pmatrix} 1 & -4 & 6 & -4 & 1 \\ 2 & -3 & -3 & 2 & 1 \\ 3 & 3 & -2 & -2 & 1 \\ -6 & 4 & -1 & -1 & 1 \\ 0 & 0 & 0 & 5 & 1 \end{pmatrix} \cdot \frac{1}{25} \begin{pmatrix} 1 & 1 & 1 & 1 & -4 \\ 1 & 0 & 4 & -2 & -3 \\ 0 & -3 & 4 & 1 & -2 \\ 3 & -3 & 1 & 0 & -1 \\ 5 & 5 & 5 & 5 & 5 \end{pmatrix} \\ &= \frac{1}{5} \begin{pmatrix} -2 & 0 & 2 & 4 & 1 \\ 2 & 2 & -3 & 2 & 2 \\ 1 & 4 & 2 & 0 & -2 \\ 0 & 1 & 2 & -2 & 4 \\ 4 & -2 & 2 & 1 & 0 \end{pmatrix}. \end{aligned} \tag{4.2.11}$$

It is also true that

$$M\mathbf{u_5} = \mathbf{u_1}. \tag{4.2.12}$$

It is easy to check that $M^T M = I$, $M^5 = I$, and that the fixed point set of M is span$\{\mathbf{e}\}$. In other words, M is a rotation of $\mathbb{R}^5$ of order 5 about the axis $\mathbf{e}$.

We now prove the remarkable fact that M acts as a permutation on U_5.

Suppose $\mathbf{u} \in U_5$ and $\mathbf{v} = M\mathbf{u}$. First, $\mathbf{u} \perp \mathbf{e}$, so $\mathbf{v} \perp \mathbf{e}$

$$\mathbf{v} \cdot \mathbf{e} = M\mathbf{u} \cdot M\mathbf{e} = \mathbf{u}^T M^T M\mathbf{e} = \mathbf{u}^T \mathbf{e} = \mathbf{u} \cdot \mathbf{e} = 0.$$

We need only show that $v_i \in \mathbb{Z}$ and $v_i \equiv i \pmod 5$.

We have $||\mathbf{u}||^2 \equiv 0 \pmod{50}$, so $||\mathbf{u}||^2 \equiv 0 \pmod{25}$. Suppose $u_i = 5m_i + i$, $i = 1, \ldots 5$.

We have

$$(5m_1+1)^2 + (5m_2+2)^2 + (5m_3+3)^2 + (5m_4+4)^2 + (5m_5+5)^2 \equiv 0 \pmod{25},$$

from which it follows that

$$10(m_1 + 2m_2 + 3m_3 + 4m_4) + 30 \equiv 0 \pmod{25},$$

$$10(m_1 + 2m_2 + 3m_3 + 4m_4) \equiv -30 \pmod{25},$$

$$m_1 + 2m_2 + 3m_3 + 4m_4 \equiv -3 \equiv 2 \pmod 5,$$

$$\frac{1}{5}(u_1 - 1) + 2 \times \frac{1}{5}(u_2 - 2) + 3 \times \frac{1}{5}(u_3 - 3) + 4 \times \frac{1}{5}(u_4 - 4) \equiv 2 \pmod 5$$

and

$$\frac{1}{5}(u_1 + 2u_2 + 3u_3 + 4u_4) \equiv 3 \pmod 5. \tag{4.2.13}$$

We have, modulo 5, using the fact that $u_1 + u_2 + u_3 + u_4 + u_5 = 0$,

$$\mathbf{v} = M\mathbf{u} = \begin{pmatrix} \frac{1}{5}(-2u_1+2u_3+4u_4+u_5) \\ \frac{1}{5}(2u_1+2u_2-3u_3+2u_4+2u_5) \\ \frac{1}{5}(u_1+4u_2+2u_3-2u_5) \\ \frac{1}{5}(u_2+2u_3-2u_4+4u_5) \\ \frac{1}{5}(4u_1-2u_2+2u_3+u_4) \end{pmatrix}$$

$$= \begin{pmatrix} \frac{1}{5}(-3u_1-u_2+u_3+3u_4) \\ \frac{1}{5}(-5u_3) \\ \frac{1}{5}(3u_1+6u_2+4u_3+2u_4) \\ \frac{1}{5}(-4u_1-3u_2-2u_3-6u_4) \\ \frac{1}{5}(4u_1-2u_2+2u_3+u_4) \end{pmatrix}$$

$$= \begin{pmatrix} -3\times\frac{1}{5}(u_1+2u_2+3u_3+4u_4)+u_2+2u_3+3u_4 \\ -u_3 \\ 3\times\frac{1}{5}(u_1+2u_2+3u_3+4u_4)-u_3-2u_4 \\ -4\times\frac{1}{5}(u_1+2u_2+3u_3+4u_4)+u_2+2u_3+2u_4 \\ 4\times\frac{1}{5}(u_1+2u_2+3u_3+4u_4)-2u_2-2u_3-3u_4 \end{pmatrix}$$

$$\equiv \begin{pmatrix} -3\times 3+2+2\times 3+3\times 4 \\ -3 \\ 3\times 3-3-2\times 4 \\ -4\times 3+2+2\times 3+2\times 4 \\ 4\times 3-2\times 2-2\times 3-3\times 4 \end{pmatrix}$$

$$\equiv \begin{pmatrix} 1 \\ 2 \\ 3 \\ 4 \\ 5 \end{pmatrix} \tag{4.2.14}$$

as required.

The orbits of M in U_5 are sets of five points, on which $||\mathbf{u}||^2/50$ is constant. It follows that

$$5 \Big| \sum_{\mathbf{u}\in U_5} q^{||\mathbf{u}||^2/50},$$

$$5 \Big| \sum_{n\geq 0} p(5n+4)q^{n+1}$$

and

$$p(5n+4) \equiv 0 \pmod 5.$$

4.3 $p(7n+5) \equiv 0 \pmod 7$

As we did in §4.2, we can show that

$$\sum_{n\geq 0} p(n)q^{n+2} = \frac{1}{(q^7;q^7)^7_\infty} \sum q^{(u_1^2+\cdots+u_7^2)/14}, \tag{4.3.1}$$

where the sum is taken over all those $\mathbf{u} = (u_1, \ldots u_7) \in \mathbb{Z}^7$ with $u_i \equiv i \pmod 7$, $i = 1, \ldots 7$ and $u_1 + \cdots + u_7 = 0$ and

$$\sum_{n\geq 0} p(7n+5)q^{n+1} = \frac{1}{(q;q)^7_\infty} \sum_{\mathbf{u}\in U_7} q^{||\mathbf{u}||^2/98}, \tag{4.3.2}$$

where

$$U_7 = \{\mathbf{u} \in \mathbb{Z}^7,\ u_i \equiv i \pmod 7,\ \mathbf{u} \perp \mathbf{e} = (1,1,1,1,1,1,1),\ 98 \,\Big|\, ||\mathbf{u}||^2\}.$$

The seven vectors that contribute to the first term, $p(5) = 7$ are

$$\begin{aligned} &\mathbf{u_1} = (1,2,3,-3,5,-1,-7) = \mathbf{u_8},\ \mathbf{u_2} = (1,-5,-4,4,-2,6,0), \\ &\mathbf{u_3} = (1,2,3,4,-2,-8,0), \qquad \mathbf{u_4} = (-6,2,3,-3,-2,6,0), \\ &\mathbf{u_5} = (8,2,-4,-3,-2,-1,0),\ \mathbf{u_6} = (-6,2,-4,4,5,-1,0), \\ &\mathbf{u_7} = (1,-5,3,-3,-2,-1,7). \end{aligned} \tag{4.3.3}$$

Then we have

$$\cos\theta_{i,i+1} = \frac{\mathbf{u}_i \cdot \mathbf{u}_{i+1}}{||\mathbf{u}_i||||\mathbf{u}_{i+1}||} = -\frac{1}{2}, \tag{4.3.4}$$

while otherwise

$$\cos\theta_{i,j} = \frac{\mathbf{u}_i \cdot \mathbf{u}_j}{||\mathbf{u}_i||||\mathbf{u}_j||} = 0. \tag{4.3.5}$$

As in §4.2, we consider the linear transformation M defined by

$$M\mathbf{u}_i = \mathbf{u}_{i+1},\ \ i = 1, \ldots 6 \text{ and } M\mathbf{e} = \mathbf{e}. \tag{4.3.6}$$

It turns out that

$$M = \frac{1}{7}\begin{pmatrix} -3 & 0 & 3 & -1 & 2 & 5 & 1 \\ 2 & 2 & 2 & 2 & -5 & 2 & 2 \\ 0 & -3 & 1 & 5 & 2 & -1 & 3 \\ 5 & -1 & 0 & 1 & 2 & 3 & -3 \\ 3 & 1 & -1 & -3 & 2 & 0 & 5 \\ 1 & 3 & 5 & 0 & 2 & -3 & -1 \\ -1 & 5 & -3 & 3 & 2 & 1 & 0 \end{pmatrix}. \tag{4.3.7}$$

Then $M\mathbf{u_7} = \mathbf{u_1}$ also, $M^T M = I$, $M^7 = I$, and the fixed point set of M is span$\{\mathbf{e}\}$. M is a rotation of $\mathbb{R}^7$ of order 7 about $\mathbf{e}$.

Once again, we prove the remarkable fact that M acts as a permutation on U_7.

Suppose $\mathbf{u} \in U_7$ and $\mathbf{v} = M\mathbf{u}$. Again, $\mathbf{u} \perp \mathbf{e}$, so $\mathbf{v} \perp \mathbf{e}$.

We have $||\mathbf{u}||^2 \equiv 0 \pmod{98}$, so $||\mathbf{u}||^2 \equiv 0 \pmod{49}$. Suppose $u_i = 7m_i + i$, $i = 1, \ldots 7$. Then

$$\begin{aligned} (7m_1+1)^2 + (7m_2+2)^2 + \cdots + (7m_7+7)^2 &\equiv 0 \pmod{49}, \\ 14(m_1 + 2m_2 + \cdots + 6m_6) &\equiv -140 \equiv 7 \pmod{49}, \\ m_1 + 2m_2 + \cdots + 6m_6 &\equiv -10 \equiv 4 \pmod 7, \end{aligned}$$

$$\frac{1}{7}(u_1 - 1) + 2 \times \frac{1}{7}(u_2 - 2) + \cdots + 6 \times \frac{1}{7}(u_6 - 6) \equiv 4 \pmod 7$$

and

$$\frac{1}{7}(u_1 + 2u_2 + \cdots + 6u_6) \equiv 3 \pmod 7. \tag{4.3.8}$$

Then we have, modulo 7,

$$\mathbf{v} = M\mathbf{u} = \begin{pmatrix} \frac{1}{7}(-3u_1 + 3u_3 - u_4 + 2u_5 + 5u_6 + u_7) \\ \frac{1}{7}(2u_1 + 2u_2 + 2u_3 + 2u_4 - 5u_5 + 2u_6 + 2u_7) \\ \frac{1}{7}(-3u_2 + u_3 + 5u_4 + 2u_5 - u_6 + 3u_7) \\ \frac{1}{7}(5u_1 - u_2 + u_4 + 2u_5 + 3u_6 - 3u_7) \\ \frac{1}{7}(3u_1 + u_2 - u_3 - 3u_4 + 2u_5 + 5u_7) \\ \frac{1}{7}(u_1 + 3u_2 + 5u_3 + 2u_5 - 3u_6 - u_7) \\ \frac{1}{7}(-u_1 + 5u_2 - 3u_3 + 3u_4 + 2u_5 + u_6) \end{pmatrix}$$

$$= \begin{pmatrix} \frac{1}{7}(-4u_1 - u_2 + 2u_3 - 2u_4 + u_5 + 4u_6) \\ \frac{1}{7}(-7u_5) \\ \frac{1}{7}(-3u_1 - 6u_2 - 2u_3 + 2u_4 - u_5 - 4u_6) \\ \frac{1}{7}(8u_1 + 2u_2 + 3u_3 + 4u_4 + 5u_5 + 6u_6) \\ \frac{1}{7}(-2u_1 - 4u_2 - 6u_3 - 8u_4 - 3u_5 - 5u_6) \\ \frac{1}{7}(2u_1 + 4u_2 + 6u_3 + u_4 + 3u_5 - 2u_6) \\ \frac{1}{7}(-u_1 + 5u_2 - 3u_3 + 3u_4 + 2u_5 + u_6) \end{pmatrix}$$

$$= \begin{pmatrix} -4 \times \frac{1}{7}(u_1 + \cdots + 6u_6) + u_2 + 2u_3 + 2u_4 + 3u_5 + 4u_6 \\ -u_5 \\ -3 \times \frac{1}{7}(u_1 + \cdots + 6u_6) + u_3 + 2u_4 + 2u_5 + 2u_6 \\ 8 \times \frac{1}{7}(u_1 + \cdots + 6u_6) - 2u_2 - 3u_3 - 4u_4 - 5u_5 - 6u_6 \\ -2 \times \frac{1}{7}(u_1 + \cdots + 6u_6) + u_5 + u_6 \\ 2 \times \frac{1}{7}(u_1 + \cdots + 6u_6) - u_4 - u_5 - 2u_6 \\ -1 \times \frac{1}{7}(u_1 + \cdots + 6u_6) + u_2 + u_4 + u_5 + u_6 \end{pmatrix}$$

$$\equiv \begin{pmatrix} -4 \times 3 + 2 + 2 \times 3 + 2 \times 4 + 3 \times 5 + 4 \times 6 \\ -5 \\ -3 \times 3 + 3 + 2 \times 4 + 2 \times 5 + 2 \times 6 \\ 8 \times 3 - 2 \times 2 - 3 \times 3 - 4 \times 4 - 5 \times 5 - 6 \times 6 \\ -2 \times 3 + 5 + 6 \\ 2 \times 3 - 4 - 5 - 2 \times 6 \\ -1 \times 3 + 2 + 4 + 5 + 6 \end{pmatrix}$$

$$\equiv \begin{pmatrix} 1 \\ 2 \\ 3 \\ 4 \\ 5 \\ 6 \\ 7 \end{pmatrix}, \tag{4.3.9}$$

as required.

The orbits of M in U_7 are sets of seven points on which $||\mathbf{u}||^2/98$ is constant, so

$$7 \Big| \sum_{\mathbf{u} \in U_7} q^{||\mathbf{u}||^2/98},$$

$$7 \Big| \sum_{n \geq 0} p(7n+5) q^{n+1}$$

and

$$p(7n+5) \equiv 0 \pmod 7.$$

4.4 $p(11n+6) \equiv 0 \pmod{11}$

Once again, as we did in §§4.2, 4.3, we can show that

$$\sum_{n\geq 0} p(n)q^{n+5} = \frac{1}{(q^{11};q^{11})_\infty^{11}} \sum q^{(u_1^2+\cdots+u_{11}^2)/22}, \tag{4.4.1}$$

where the sum is taken over all those $\mathbf{u} = (u_1, \ldots u_{11}) \in \mathbb{Z}^{11}$ with $u_i \equiv i \pmod{11}$, $i = 1, \ldots 11$ and $u_1 + \cdots + u_{11} = 0$, and

$$\sum_{n\geq 0} p(11n+6)q^{n+1} = \frac{1}{(q;q)_\infty^{11}} \sum_{\mathbf{u}\in U_{11}} q^{||\mathbf{u}||^2/242}, \tag{4.4.2}$$

where

$$U_{11} = \{\mathbf{u} \in \mathbb{Z}^{11} : u_i \equiv i \pmod{11}, \mathbf{u} \perp \mathbf{e} = (1, \ldots 1), 242 \,\Big|\, ||\mathbf{u}||^2\}.$$

The eleven vectors that contribute to the first term, $p(6) = 11$, are

$$\begin{aligned}
\mathbf{u}_1 &= (1,2,3,-7,-6,6,-4,-3,9,-1,0) = \mathbf{u}_{12},\\
\mathbf{u}_2 &= (1,-9,3,4,5,-5,-4,8,-2,-1,0),\\
\mathbf{u}_3 &= (1,2,-8,-7,5,6,7,-3,-2,-1,0),\\
\mathbf{u}_4 &= (1,2,3,4,-6,-5,-4,-3,-2,-1,11),\\
\mathbf{u}_5 &= (1,2,3,4,5,6,-4,-3,-2,-1,11),\\
\mathbf{u}_6 &= (1,2,3,-7,-6,-5,7,8,-2,-1,0),\\
\mathbf{u}_7 &= (1,2,-8,4,5,-5,-4,-3,9,-1,0),\\
\mathbf{u}_8 &= (1,-9,3,4,-6,6,7,-3,-2,-1,0),\\
\mathbf{u}_9 &= (1,2,3,-7,5,-5,-4,-3,-2,10,0),\\
\mathbf{u}_{10} &= (1,2,-8,4,-6,6,-4,8,-2,-1,0),\\
\mathbf{u}_{11} &= (-10,2,3,4,5,-5,7,-3,-2,-1,0).
\end{aligned} \tag{4.4.3}$$

Again we have

$$\cos\theta_{i,i+1} = \frac{\mathbf{u}_i \cdot \mathbf{u}_{i+1}}{||\mathbf{u}_i|| ||\mathbf{u}_{i+1}||} = -\frac{1}{2}, \tag{4.4.4}$$

while otherwise

$$\cos\theta_{i,j} = \frac{\mathbf{u}_i \cdot \mathbf{u}_j}{||\mathbf{u}_i|| ||\mathbf{u}_j||} = 0. \tag{4.4.5}$$

If as before we define the linear transformation M by

$$M\mathbf{u}_i = \mathbf{u}_{i+1},\ i = 1,\ \ldots\ 10 \text{ and } M\mathbf{e} = \mathbf{e}, \tag{4.4.6}$$

then

$$M = \frac{1}{11}\begin{pmatrix} 6 & 0 & 5 & -1 & 4 & -2 & 3 & -3 & 2 & -4 & 1 \\ 2 & 2 & 2 & 2 & 2 & 2 & 2 & 2 & -9 & 2 & 2 \\ -2 & 4 & -1 & 5 & 0 & 6 & 1 & -4 & 2 & -3 & 3 \\ 5 & 6 & -4 & -3 & -2 & -1 & 0 & 1 & 2 & 3 & 4 \\ 1 & -3 & 4 & 0 & -4 & 3 & -1 & 6 & 2 & -2 & 5 \\ -3 & -1 & 1 & 3 & 5 & -4 & -2 & 0 & 2 & 4 & 6 \\ 4 & 1 & -2 & 6 & 3 & 0 & -3 & 5 & 2 & -1 & -4 \\ 0 & 3 & 6 & -2 & 1 & 4 & -4 & -1 & 2 & 5 & -3 \\ -4 & 5 & 3 & 1 & -1 & -3 & 6 & 4 & 2 & 0 & -2 \\ 3 & -4 & 0 & 4 & -3 & 1 & 5 & -2 & 2 & 6 & -1 \\ -1 & -2 & -3 & -4 & 6 & 5 & 4 & 3 & 2 & 1 & 0 \end{pmatrix}, \tag{4.4.7}$$

$M\mathbf{u_{11}} = \mathbf{u_1}$, $M^T M = I$, $M^{11} = I$, the fixed point set of M is span$\{\mathbf{e}\}$, M is a rotation of $\mathbb{R}^{11}$ of order 11 about $\mathbf{e}$, and M acts as a permutation on U_{11}. For, suppose $\mathbf{u} \in U_{11}$ and $u_i = 11m_i + i$, $i = 1,\ \ldots\ 11$. As before, if $\mathbf{v} = M\mathbf{u}$, then $\mathbf{v} \perp \mathbf{e}$. Also, $||\mathbf{u}||^2 \equiv 0 \pmod{242}$, so $||\mathbf{u}||^2 \equiv 0 \pmod{121}$, that is,

$$\begin{aligned} (11m_1+1)^2 + \cdots + (11m_{11}+11)^2 &\equiv 0 \pmod{121}, \\ 22(m_1 + \cdots + 10m_{10}) &\equiv -22 \pmod{121}, \\ m_1 + \cdots + 10m_{10} &\equiv -1 \pmod{11}, \end{aligned}$$

$$\frac{1}{11}(u_1 - 1) + 2 \times \frac{1}{11}(u_2 - 2) + \cdots + 10 \times \frac{1}{11}(u_{10} - 10) \equiv -1 \pmod{11},$$

and

$$\frac{1}{11}(u_1 + 2u_2 + \cdots + 10u_{10}) \equiv 1 \pmod{11}. \tag{4.4.8}$$

If we abbreviate matters by writing $\frac{1}{11}(u_1+2u_2+\cdots+10u_{10}) = s$ then we have, modulo 11,

$$
\begin{aligned}
\mathbf{v} &= M\mathbf{u} \\
&= \begin{pmatrix}
\frac{1}{11}(5u_1-u_2+4u_3-2u_4+3u_5-3u_6+2u_7-4u_8+u_9-5u_{10} \\
\frac{1}{11}(-11u_9) \\
\frac{1}{11}(-5u_1+u_2-4u_3+2u_4-3u_5+3u_6-2u_7-7u_8-u_9-6u_{10}) \\
\frac{1}{11}(u_1+2u_2-8u_3-7u_4-6u_5-5u_6-4u_7-3u_8-2u_9-u_{10}) \\
\frac{1}{11}(-4u_1-8u_2-u_3-5u_4-9u_5-2u_6-6u_7+u_8-3u_9-7u_{10}) \\
\frac{1}{11}(-9u_1-7u_2-5u_3-3u_4-u_5-10u_6-8u_7-6u_8-4u_9-2u_{10}) \\
\frac{1}{11}(8u_1+5u_2+2u_3+10u_4+7u_5+4u_6+u_7+9u_8+6u_9+3u_{10}) \\
\frac{1}{11}(3u_1+6u_2+9u_3+u_4+4u_5+7u_6-u_7+2u_8+5u_9+8u_{10}) \\
\frac{1}{11}(-2u_1+7u_2+5u_3+3u_4+u_5-u_6+8u_7+6u_8+4u_9+2u_{10}) \\
\frac{1}{11}(4u_1-3u_2+u_3+5u_4-2u_5+2u_6+6u_7-u_8+3u_9+7u_{10}) \\
\frac{1}{11}(-u_1-2u_2-3u_3-4u_4+6u_5+5u_6+4u_7+3u_8+2u_9+u_{10})
\end{pmatrix} \\
&= \begin{pmatrix}
5s-u_2-u_3-2u_4-2u_5-3u_6-3u_7-4u_8-4u_9-5u_{10} \\
-u_9 \\
-5s+u_2+u_3+2u_4+2u_5+3u_6+3u_7+3u_8+4u_9+4u_{10} \\
s-u_3-u_4-u_5-u_6-u_7-u_8-u_9-u_{10} \\
-4s+u_3+u_4+u_5+2u_6+2u_7+3u_8+3u_9+3u_{10} \\
-9s+u_2+2u_3+3u_4+4u_5+4u_6+5u_7+6u_8+7u_9+8u_{10} \\
8s-u_2-2u_3-2u_4-3u_5-4u_6-5u_7-5u_8-6u_9-7u_{10} \\
3s-u_4-u_5-u_6-2u_7-2u_8-2u_9-2u_{10} \\
-2s+u_2+u_3+u_4+u_5+u_6+2u_7+2u_8+2u_9+2u_{10} \\
4s-u_2-u_3-u_4-2u_5-2u_6-2u_7-3u_8-3u_9-3u_{10} \\
-s+u_5+u_6+u_7+u_8+u_9+u_{10}
\end{pmatrix} \\
&\equiv \begin{pmatrix} 1 \\ 2 \\ 3 \\ 4 \\ 5 \\ 6 \\ 7 \\ 8 \\ 9 \\ 10 \\ 11 \end{pmatrix},
\end{aligned}
\qquad (4.4.9)
$$

as required.

The orbits of M in U_{11} are sets of eleven points on which $||\mathbf{u}||^2/242$ is constant, so

$$11 \Big| \sum_{\mathbf{u}\in U_{11}} q^{||\mathbf{u}||^2/242},$$

$$11 \Big| \sum_{n\geq 0} p(11n+6)q^{n+1}$$

and

$$p(11n+6) \equiv 0 \pmod{11}.$$

Endnotes.

§§4.2, 4.3, 4.4 My contribution was to "complete the square" in $Q(\mathbf{k})$ in (4.2.3), (4.3.1) and (4.4.1). This means that in each case the map is linear, whereas in Garvan–Stanton (1990) and Garvan–Kim–Stanton (1990), the map is affine.

§4.2 In Hirschhorn (1994) [70] the detail of the proof is a little different.

References

49. F. Garvan, D. Stanton, Sieved partition functions and q-binomial coefficients. Math. Comput. **55**, 299–311 (1990)
50. F. Garvan, D. Kim, D. Stanton, Cranks and t-cores. Inventiones Mathematicae **101**, 1–17 (1990)
60. G.H. Hardy, E.M. Wright, *An Introduction to the Theory of Numbers*, 4th edn. (Oxford University Press, Oxford, 1960)
70. M.D. Hirschhorn, Ramanujan's partition congruences. Discret. Math. **131**, 351–355 (1994)

Chapter 5
Ramanujan's Most Beautiful Identity

5.1 Introduction

In this chapter, we will give our first proof of Ramanujan's result

$$\sum_{n\geq 0} p(5n+4)q^n = 5\frac{E(q^5)^5}{E(q)^6}. \tag{5.1.1}$$

Hardy (1927) [59], pp. xxxiv–xxxv, says of this result "It would be difficult to find more beautiful formulae than the 'Rogers–Ramanujan' identities, but here Ramanujan must take second place to Prof. Rogers; and if I had to select one formula from all Ramanujan's work, I would agree with Major MacMahon in selecting [this formula].

Because of this, and because of its supreme elegance and simplicity, I and other writers often refer to (5.1.1) as "Ramanujan's most beautiful identity".

Of course, (5.1.1) gives us yet another proof (the fourth in this book) of (3.1.9),

$$p(5n+4) \equiv 0 \pmod{5}.$$

5.2 The 5-Dissection of the Partition Generating Function

We have

$$\sum_{n\geq 0} p(n)q^n = \frac{1}{E(q)} \tag{5.2.1}$$

with, according to (3.1.1), (3.1.6) and (3.2.5),

M.D. Hirschhorn, *The Power of q*, Developments in Mathematics 49,
DOI 10.1007/978-3-319-57762-3_5

$$E(q) = (q;q)_\infty = \sum_{-\infty}^{\infty}(-1)^k q^{(3k^2-k)/2} = E_0 + E_1 + E_2. \tag{5.2.2}$$

It follows that, with ζ a fifth root of unity (other than 1),

$$\sum_{n\geq 0} p(n)q^n = \frac{1}{E(q)} = \frac{E(\zeta q)E(\zeta^2 q)E(\zeta^3 q)E(\zeta^4 q)}{E(q)E(\zeta q)E(\zeta^2 q)E(\zeta^3 q)E(\zeta^4 q)}. \tag{5.2.3}$$

The denominator in (5.2.3) is

$$\begin{aligned} D(q) &= E(q)E(\zeta q)E(\zeta^2 q)E(\zeta^3 q)E(\zeta^4 q) \\ &= \prod_{n\geq 1}(1-q^n)(1-\zeta^n q^n)(1-\zeta^{2n}q^n)(1-\zeta^{3n}q^n)(1-\zeta^{4n}q^n) \\ &= \prod_{n\equiv 0 \pmod 5}(1-q^n)^5 \cdot \prod_{n\not\equiv 0 \pmod 5}(1-q^{5n}) \\ &= (q^5;q^5)_\infty^5 (q^5,q^{10},q^{15},q^{20};q^{25})_\infty \\ &= \frac{(q^5;q^5)_\infty^6}{(q^{25};q^{25})_\infty} \\ &= \frac{E(q^5)^6}{E(q^{25})}. \end{aligned} \tag{5.2.4}$$

The numerator in (5.2.3) is

$$\begin{aligned} N(q) &= E(\zeta q)E(\zeta^2 q)E(\zeta^3 q)E(\zeta^4 q) \\ &= \left(E_0+\zeta E_1+\zeta^2 E_2\right)\left(E_0+\zeta^2 E_1+\zeta^4 E_2\right) \\ &\quad \times \left(E_0+\zeta^3 E_1)+\zeta^6 E_2\right)\left(E_0+\zeta^4 E_1+\zeta^8 E_2)\right) \\ &= \left(E_0^4+2E_0E_1E_2^2-E_1^3E_2\right)+\left(-E_0^3E_1-E_0E_2^3+E_1^2E_2^2\right) \\ &\quad +\left(-E_0^3E_2+E_0^2E_1^2-E_1E_2^3\right)+\left(2E_0^2E_1E_2-E_0E_1^3+E_2^4\right) \\ &\quad +\left(E_0^2E_2^2-3E_0E_1^2E_2+E_1^4\right). \end{aligned} \tag{5.2.5}$$

It follows from (5.2.4) and (5.2.5) that

$$\begin{aligned} \sum_{n\geq 0} p(5n)q^{5n} &= \frac{E(q^{25})}{E(q^5)^6}\left(E_0^4+2E_0E_1E_2^2-E_1^3E_2\right), \\ \sum_{n\geq 0} p(5n+1)q^{5n+1} &= \frac{E(q^{25})}{E(q^5)^6}\left(-E_0^3E_1-E_0E_2^3+E_1^2E_2^2\right), \\ \sum_{n\geq 0} p(5n+2)q^{5n+2} &= \frac{E(q^{25})}{E(q^5)^6}\left(-E_0^3E_2+E_0^2E_1^2-E_1E_2^3\right), \end{aligned}$$

$$\sum_{n\geq 0} p(5n+3)q^{5n+3} = \frac{E(q^{25})}{E(q^5)^6}\left(2E_0^2E_1E_2 - E_0E_1^3 + E_2^4\right)$$

and, most particularly,

$$\sum_{n\geq 0} p(5n+4)q^{5n+4} = \frac{E(q^{25})}{E(q^5)^6}\left(E_0^2E_2^2 - 3E_0E_1^2E_2 + E_1^4\right). \tag{5.2.6}$$

In the next section, we will obtain two surprising results that will lead to a dramatic simplification of (5.2.6).

5.3 Two Surprising Results

We will prove two results which lead to a dramatic simplification of (5.2.6),

$$E_0E_2 = -E_1^2 \tag{5.3.1}$$

and

$$E_1 = -qE(q^{25}). \tag{5.3.2}$$

We have (3.2.5),

$$E = E_0 + E_1 + E_2$$

and (1.7.1),

$$E^3 = \sum_{n\geq 0}(-1)^n(2n+1)q^{(n^2+n)/2}.$$

It follows that

$$\left(E_0 + E_1 + E_2\right)^3 = \sum_{n\geq 0}(-1)^n(2n+1)q^{(n^2+n)/2}. \tag{5.3.3}$$

That is,

$$\begin{aligned}&\left(E_0^3 + 3E_1E_2^2\right) + \left(3E_0^2E_1 + E_2^3\right) + 3E_0\left(E_0E_2 + E_1^2\right)\\ &\quad + \left(6E_0E_1E_2 + E_1^3\right) + 3E_2\left(E_0E_2 + E_1^2\right) = \sum_{n\geq 0}(-1)^n(2n+1)q^{(n^2+n)/2}.\end{aligned} \tag{5.3.4}$$

Now, as we observed in §3.2, $(n^2+n)/2 \not\equiv 2$ or $4 \pmod 5$, and consequently, if we extract those terms in which the exponent on q is 2 or 4 modulo 5, we obtain

$$E_0E_2 + E_1^2 = 0,$$

which is (5.3.1).

We now turn to (5.3.2). We saw in (1.6.1) that

$$E = \sum_{-\infty}^{\infty} (-1)^k q^{(3k^2-k)/2}. \tag{5.3.5}$$

Now $(3k^2-k)/2 \equiv 1 \pmod 5$ if and only if $k \equiv 1 \pmod 5$, $k = 5l+1$, say, and so

$$E_1 = \sum_{-\infty}^{\infty} (-1)^{5l+1} q^{\left(3(5l+1)^2-(5l+1)\right)/2} = -\sum_{-\infty}^{\infty} (-1)^l q^{(75l^2+25l+2)/2} = -qE(q^{25}),$$

where we have used (1.6.1) with q replaced by q^{25}, and (5.3.2) is proved.

Exercise: Obtain (5.3.2) by considering those terms in (5.3.4) in which the exponent on q is 3 modulo 5 and making use of (5.3.1).

5.4 Ramanujan's Most Beautiful Identity

We saw in (5.2.6) that

$$\sum_{n\geq 0} p(5n+4)q^{5n+4} = \frac{E(q^{25})}{E(q^5)^6}\left(E_0^2E_2^2 - 3E_0E_1^2E_2 + E_1^4\right).$$

If in turn we apply (5.3.1) and (5.3.2), we find

$$\sum_{n\geq 0} p(5n+4)q^{5n+4} = \frac{E(q^{25})}{E(q^5)^6}\left(5E_1^4\right) = 5q^4\frac{E(q^{25})^5}{E(q^5)^6}.$$

If we divide by q^4 and replace q^5 by q, we obtain (5.1.1), Ramanujan's most beautiful identity!

Reference

59. G.H. Hardy, P.V. Seshu Aiyar, B.M. Wilson, *Collected Papers of Srinivasa Ramanujan* (AMS Chelsea, Providence, 2000)

Chapter 6
Ramanujan's Partition Congruences for Powers of 5

6.1 Introduction

The object of this chapter is to give a proof, in considerable detail, of Ramanujan's partition congruences for powers of 5,

$$p(5^\alpha n+\delta_\alpha)\equiv 0 \pmod{5^\alpha}, \tag{6.1.1}$$

where $\alpha\geq 1$ and δ_α is the reciprocal of 24 modulo 5^α.

In §6.2, we will establish the modular equation of degree 5.

In §6.3, we will demonstrate how the modular equation can be used to obtain Ramanujan's most beautiful identity.

In §6.4, we will obtain the generating function

$$\sum_{n\geq 0} p(5^\alpha n+\delta_\alpha)q^n \tag{6.1.2}$$

for $\alpha\geq 1$, in terms of $E(q)=(q;q)_\infty$.

In §6.5, we will give a proof of (6.1.1), and in §6.6, we will confirm a conjecture made by Ramanujan in his Lost Notebook [112].

My presentation is based on that in Hirschhorn–Hunt [93], but is highly influenced by Garvan [47].

The original version of this chapter has been revised. The erratum to this chapter is available at https://doi.org/10.1007/978-3-319-57762-3_44.

M.D. Hirschhorn, *The Power of q*, Developments in Mathematics 49,
DOI 10.1007/978-3-319-57762-3_6

6.2 The Modular Equation of Degree 5

We have that (3.2.5)

$$E(q) = E_0 + E_1 + E_2, \tag{6.2.1}$$

where (5.3.1)

$$E_0E_2 + E_1^2 = 0. \tag{6.2.2}$$

and (5.3.2)

$$E_1 = -qE(q^{25}). \tag{6.2.3}$$

We can write (6.2.1)

$$E(q) = qE(q^{25})(\alpha - 1 + \beta), \tag{6.2.4}$$

where

$$\alpha = -\frac{E_0}{E_1}, \quad \beta = -\frac{E_2}{E_1}. \tag{6.2.5}$$

Then (6.2.2) becomes

$$\alpha\beta = -1. \tag{6.2.6}$$

If η is a fifth root of unity (other than 1), and in (6.2.4) we set q, ηq, $\eta^2 q$, $\eta^3 q$ and $\eta^4 q$ for q and multiply the results, we obtain

$$\begin{aligned}\frac{E(q^5)^6}{E(q^{25})} &= q^5E(q^{25})^5(\alpha-1+\beta)(\alpha-\eta+\eta^2\beta)(\alpha-\eta^2+\eta^4\beta)(\alpha-\eta^3+\eta^6\beta)\\ &\qquad\times(\alpha-\eta^4+\eta^8\beta)\\ &= q^5E(q^{25})^5(\alpha^5-11+\beta^5).\end{aligned} \tag{6.2.7}$$

Now write

$$\zeta = \frac{E(q)}{qE(q^{25})} = \alpha - 1 + \beta \tag{6.2.8}$$

and

$$T = \frac{E(q^5)^6}{q^5E(q^{25})^6} = \alpha^5 - 11 + \beta^5. \tag{6.2.9}$$

It is easy to calculate, given $\alpha\beta = -1$, that

$$\zeta^2 = \alpha^2 - 2\alpha - 1 - 2\beta + \beta^2, \tag{6.2.10}$$

$$\zeta^3 = \alpha^3 - 3\alpha^2 + 5 - 3\beta^2 + \beta^3, \tag{6.2.11}$$

$$\zeta^4 = \alpha^4 - 4\alpha^3 + 2\alpha^2 + 8\alpha - 5 + 8\beta + 2\beta^2 - 4\beta^3 + \beta^4 \tag{6.2.12}$$

and

$$\zeta^5 = \alpha^5 - 5\alpha^4 + 5\alpha^3 + 10\alpha^2 - 15\alpha - 11 - 15\beta + 10\beta^2 + 5\beta^3 - 5\beta^4 + \beta^5. \tag{6.2.13}$$

It follows from (6.2.9)–(6.2.13) that

$$T = \zeta^5 + 5\zeta^4 + 15\zeta^3 + 25\zeta^2 + 25\zeta, \tag{6.2.14}$$

and this is the modular equation of degree 5.

6.3 Ramanujan's Most Beautiful Identity Again

We start by writing (6.2.14)

$$\frac{1}{\zeta} = \frac{1}{T}(25 + 25\zeta + 15\zeta^2 + 5\zeta^3 + \zeta^4). \tag{6.3.1}$$

Now we introduce the "huffing" operator H, which operates on a series by extracting those terms in which the power of q is a multiple of 5, and huffing (that is, blowing) the rest away.

That is,

$$H\left(\sum a_n q^n\right) = \sum a_{5n} q^{5n}.$$

We find that

$$H\left(\frac{1}{\zeta}\right) = \frac{1}{T}\left(25 + 25H(\zeta) + 15H(\zeta^2) + 5H(\zeta^3) + H(\zeta^4)\right). \tag{6.3.2}$$

Now, from (6.2.8) and (6.2.10)–(6.2.12)

$$H(\zeta) = -1,\ H(\zeta^2) = -1,\ H(\zeta^3) = 5,\ H(\zeta^4) = -5 \tag{6.3.3}$$

and (6.3.2) becomes

$$H\left(\frac{1}{\zeta}\right) = \frac{1}{T}(25 + 25\times -1 + 15\times -1 + 5\times 5 - 5) = \frac{5}{T}, \tag{6.3.4}$$

which yields, successively,

$$H\left(\frac{qE(q^{25})}{E(q)}\right) = 5\frac{q^5E(q^{25})^6}{E(q^5)^6}, \tag{6.3.5}$$

$$H\left(\frac{q}{E(q)}\right) = 5q^5\frac{E(q^{25})^5}{E(q^5)^6}, \tag{6.3.6}$$

$$H\left(\sum_{n\geq 0} p(n)q^{n+1}\right) = 5q^5\frac{E(q^{25})^5}{E(q^5)^6} \tag{6.3.7}$$

and

$$\sum_{n\geq 0} p(5n+4)q^{5n+5} = 5q^5\frac{E(q^{25})^5}{E(q^5)^6}, \tag{6.3.8}$$

which becomes Ramanujan's most beautiful identity (5.1.1) if we divide by q^5 and replace q by $q^{\frac{1}{5}}$.

6.4 The Generating Function $\sum_{n\geq 0} p(5^\alpha n + \delta_\alpha)q^n$

In this section, we will show that

$$\sum_{n\geq 0} p(5^\alpha n+\delta_\alpha)q^n = \begin{cases} \sum_{i\geq 1} x_{\alpha,i}q^{i-1}\dfrac{E(q^5)^{6i-1}}{E(q)^{6i}}, & \alpha \text{ odd}, \\ \sum_{i\geq 1} x_{\alpha,i}q^{i-1}\dfrac{E(q^5)^{6i}}{E(q)^{6i+1}}, & \alpha \text{ even}, \end{cases} \tag{6.4.1}$$

where the coefficient vectors $\mathbf{x}_\alpha = (x_{\alpha,i})_{i\geq 1}$ are given recursively by

$$\mathbf{x}_1 = (5,0,0,0,\,\ldots) \tag{6.4.2}$$

and for $\alpha \geq 1$,

$$\mathbf{x}_{\alpha+1} = \begin{cases} \mathbf{x}_\alpha A, & \alpha \text{ odd}, \\ \mathbf{x}_\alpha B, & \alpha \text{ even}, \end{cases} \tag{6.4.3}$$

where $A = (a_{i,j})_{i,j\geq 1}$ and $B = (b_{i,j})_{i,j\geq 1}$ are defined by

$$a_{i,j} = m_{6i,i+j}, \quad b_{i,j} = m_{6i+1,i+j}, \tag{6.4.4}$$

where $M=(m_{i,j})_{i,j\geq 1}$ is defined as follows. The first five rows of M are

$$\begin{pmatrix} 5 & 0 & 0 & 0 & 0 & 0 & \dots \\ 2\times 5 & 5^3 & 0 & 0 & 0 & 0 & \dots \\ 9 & 3\times 5^3 & 5^5 & 0 & 0 & 0 & \dots \\ 4 & 22\times 5^2 & 4\times 5^5 & 5^7 & 0 & 0 & \dots \\ 1 & 4\times 5^3 & 8\times 5^5 & 5^8 & 5^9 & 0 & \dots \end{pmatrix} \tag{6.4.5}$$

and for $i\geq 6$, $m_{i,1}=0$, and for $j\geq 2$,

$$m_{i,j}=25m_{i-1,j-1}+25m_{i-2,j-1}+15m_{i-3,j-1}+5m_{i-4,j-1}+m_{i-5,j-1}. \tag{6.4.6}$$

From (6.3.1) it follows that for $i\geq 1$,

$$\frac{1}{\zeta^i}=\frac{1}{T}\left(\frac{25}{\zeta^{i-1}}+\frac{25}{\zeta^{i-2}}+\frac{15}{\zeta^{i-3}}+\frac{5}{\zeta^{i-4}}+\frac{1}{\zeta^{i-5}}\right), \tag{6.4.7}$$

and so

$$H\left(\frac{1}{\zeta^i}\right)=\frac{1}{T}\left(25H\left(\frac{1}{\zeta^{i-1}}\right)+25H\left(\frac{1}{\zeta^{i-2}}\right)+15H\left(\frac{1}{\zeta^{i-3}}\right)+5H\left(\frac{1}{\zeta^{i-4}}\right)+H\left(\frac{1}{\zeta^{i-5}}\right)\right). \tag{6.4.8}$$

It is easy to prove using (6.3.4) and (6.4.8) that for $i\geq 1$,

$$H\left(\frac{1}{\zeta^i}\right)=\sum_{j\geq 1}\frac{m_{i,j}}{T^j}, \tag{6.4.9}$$

where the $m_{i,j}$ satisfy (6.4.5) and (6.4.6).

Indeed, it is easy to check that $m_{6i,j}=0$ and $m_{6i+1,j}=0$ for $1\leq j\leq i$, so we can write

$$H\left(\frac{1}{\zeta^{6i}}\right)=\sum_{j\geq 1}\frac{a_{i,j}}{T^{i+j}} \tag{6.4.10}$$

and

$$H\left(\frac{1}{\zeta^{6i+1}}\right)=\sum_{j\geq 1}\frac{b_{i,j}}{T^{i+j}}, \tag{6.4.11}$$

where the $a_{i,j}$ and $b_{i,j}$ are defined by (6.4.4).

We can write (6.4.10)

$$H\left(\left(\frac{qE(q^{25})}{E(q)}\right)^{6i}\right) = \sum_{j\geq 1} a_{i,j}\left(\frac{q^5E(q^{25})^6}{E(q^5)^6}\right)^{i+j}. \tag{6.4.12}$$

It follows that

$$H\left(q^{i-5}\frac{E(q^5)^{6i-1}}{E(q)^{6i}}\right) = \sum_{j\geq 1} a_{i,j}q^{5j-5}\frac{E(q^{25})^{6j}}{E(q^5)^{6j+1}}. \tag{6.4.13}$$

Similarly, we can write (6.4.11)

$$H\left(\left(\frac{qE(q^{25})}{E(q)}\right)^{6i+1}\right) = \sum_{j\geq 1} b_{i,j}\left(\frac{q^5E(q^{25})^6}{E(q^5)^6}\right)^{i+j}, \tag{6.4.14}$$

from which it follows that

$$H\left(q^{i-4}\frac{E(q^5)^{6i}}{E(q)^{6i+1}}\right) = \sum_{j\geq 1} b_{i,j}q^{5j-5}\frac{E(q^{25})^{6j-1}}{E(q^5)^{6j}}. \tag{6.4.15}$$

We note also that if $\delta_1 = 4$ and

$$\delta_{\alpha+1} = \begin{cases} 4\times 5^\alpha + \delta_\alpha \text{ if } \alpha \text{ is odd,} \\ 3\times 5^\alpha + \delta_\alpha \text{ if } \alpha \text{ is even,} \end{cases} \tag{6.4.16}$$

then

$$\delta_\alpha = \begin{cases} \dfrac{19\times 5^\alpha + 1}{24} \text{ if } \alpha \text{ is odd,} \\ \dfrac{23\times 5^\alpha + 1}{24} \text{ if } \alpha \text{ is even} \end{cases} \tag{6.4.17}$$

and

$$1 \leq \delta_\alpha < 5^\alpha \text{ and } 24\delta_\alpha \equiv 1 \pmod{5^\alpha}. \tag{6.4.18}$$

That is to say, δ_α is the reciprocal of 24 modulo 5^α.

We know that

$$\sum_{n\geq 0} p(5n+4)q^n = 5\frac{E(q^5)^5}{E(q)^6}. \tag{6.4.19}$$

Suppose α is odd, and that

$$\sum_{n\geq 0} p(5^\alpha n+\delta_\alpha)q^n = \sum_{i\geq 1} x_{\alpha,i}q^{i-1}\frac{E(q^5)^{6i-1}}{E(q)^{6i}}. \tag{6.4.20}$$

Then

$$\sum_{n\geq 0} p(5^\alpha n+\delta_\alpha)q^{n-4} = \sum_{i\geq 1} x_{\alpha,i}q^{i-5}\frac{E(q^5)^{6i-1}}{E(q)^{6i}}. \tag{6.4.21}$$

If we apply H to this, we find

$$\begin{aligned}\sum_{n\geq 0} p\left(5^\alpha(5n+4)+\delta_\alpha\right)q^{5n} &= \sum_{i\geq 1} x_{\alpha,i}H\left(q^{i-5}\frac{E(q^5)^{6i-1}}{E(q)^{6i}}\right)\\ &= \sum_{i\geq 1} x_{\alpha,i}\sum_{j\geq 1} a_{i,j}q^{5j-5}\frac{E(q^{25})^{6j}}{E(q^5)^{6j+1}}\\ &= \sum_{i,j\geq 1} x_{\alpha,i}a_{i,j}q^{5j-5}\frac{E(q^{25})^{6j}}{E(q^5)^{6j+1}}\\ &= \sum_{j\geq 1} x_{\alpha+1,j}q^{5j-5}\frac{E(q^{25})^{6j}}{E(q^5)^{6j+1}} \end{aligned}\tag{6.4.22}$$

If we now replace q by $q^{\frac{1}{5}}$, we obtain

$$\sum_{n\geq 0} p(5^{\alpha+1}n+\delta_{\alpha+1})q^n = \sum_{j\geq 1} x_{\alpha+1,j}q^{j-1}\frac{E(q^5)^{6j}}{E(q)^{6j+1}}. \tag{6.4.23}$$

Next suppose α is even, and that

$$\sum_{n\geq 0} p(5^\alpha n+\delta_\alpha)q^n = \sum_{i\geq 1} x_{\alpha,i}q^{i-1}\frac{E(q^5)^{6i}}{E(q)^{6i+1}}. \tag{6.4.24}$$

Then

$$\sum_{n\geq 0} p(5^\alpha n+\delta_\alpha)q^{n-3} = \sum_{i\geq 1} x_{\alpha,i}q^{i-4}\frac{E(q^5)^{6i}}{E(q)^{6i+1}}. \tag{6.4.25}$$

If we apply H to this, we find

$$\begin{aligned}\sum_{n\geq 0} p(5^\alpha(5n+3)+\delta_\alpha)q^{5n} &= \sum_{i\geq 1} x_{\alpha,i} H\left(q^{i-4}\frac{E(q^5)^{6i}}{E(q)^{6i+1}}\right)\\ &= \sum_{i\geq 1} x_{\alpha,i}\sum_{j\geq 1} b_{i,j}q^{5j-5}\frac{E(q^{25})^{6j-1}}{E(q^5)^{6j}}\\ &= \sum_{i,j\geq 1} x_{\alpha,i}b_{i,j}q^{5j-5}\frac{E(q^{25})^{6j-1}}{E(q^5)^{6j}}\\ &= \sum_{j\geq 1} x_{\alpha+1,j}q^{5j-5}\frac{E(q^{25})^{6j-1}}{E(q^5)^{6j}}. \end{aligned} \tag{6.4.26}$$

If we replace q by $q^{\frac{1}{5}}$, we obtain

$$\sum_{n\geq 0} p(5^{\alpha+1}n+\delta_{\alpha+1})q^n = \sum_{j\geq 1} x_{\alpha+1,j}q^{j-1}\frac{E(q^5)^{6j-1}}{E(q)^{6j}}, \tag{6.4.27}$$

and thc proof of (6.4.1) by induction is complete.

6.5 Ramanujan's Partition Congruences for Powers of 5

It is not hard to prove that if $\nu(n)$ denotes the highest power of 5 that divides n then

$$\nu(m_{i,j}) \geq \lfloor\frac{1}{2}(5j-i-1)\rfloor,\quad \nu(a_{i,j}) \geq \lfloor\frac{1}{2}(5j-i-1)\rfloor,\quad \nu(b_{i,j}) \geq \lfloor\frac{1}{2}(5j-i-2)\rfloor, \tag{6.5.1}$$

and that

$$\nu(x_{\alpha,i}) \geq \begin{cases} \alpha + \lfloor\frac{1}{2}(5i-5)\rfloor & \text{if } \alpha \text{ is odd,}\\ \alpha + \lfloor\frac{1}{2}(5i-4)\rfloor & \text{if } \alpha \text{ is even,}\end{cases} \tag{6.5.2}$$

while

$$\nu(x_{\alpha,1}) = \alpha. \tag{6.5.3}$$

Ramanujan's partition congruence (6.1.1) is immediate.

6.6 A Conjecture of Ramanujan

By virtue of (6.5.2) and (6.5.3), we have that for each $\alpha \geq 1$, there is a number c_α such that, modulo $5^{\alpha+1}$,

$$\sum_{n\geq 0} p(5^\alpha n+\delta_\alpha)q^n \equiv \begin{cases} c_\alpha 5^\alpha \dfrac{E(q^5)^5}{E(q)^6} & \text{if } \alpha \text{ is odd,} \\ c_\alpha 5^\alpha \dfrac{E(q^5)^6}{E(q)^7} & \text{if } \alpha \text{ is even.} \end{cases} \tag{6.6.1}$$

It follows that, modulo $5^{\alpha+1}$,

$$\sum_{n\geq 0} p(5^\alpha n+\delta_\alpha)q^n \equiv \begin{cases} c_\alpha 5^\alpha E(q)^{19} & \text{if } \alpha \text{ is odd,} \\ c_\alpha 5^\alpha E(q)^{23} & \text{if } \alpha \text{ is even.} \end{cases} \tag{6.6.2}$$

Moreover,

$$c_1 = 1 \tag{6.6.3}$$

and, since modulo $5^{\alpha+1}$,

$$x_{\alpha+1,1} \equiv \begin{cases} a_{1,1}x_{\alpha,1} = m_{6,2}x_{\alpha,1} = 5\times 63x_{\alpha,1} & \text{if } \alpha \text{ is odd,} \\ b_{1,1}x_{\alpha,1} = m_{7,2}x_{\alpha,1} = 5\times 28x_{\alpha,1} & \text{if } \alpha \text{ is even,} \end{cases} \tag{6.6.4}$$

it follows that, modulo 5,

$$c_{\alpha+1} = \begin{cases} 63c_\alpha \equiv -2c_\alpha & \text{if } \alpha \text{ is odd,} \\ 28c_\alpha \equiv -2c_\alpha & \text{if } \alpha \text{ is even,} \end{cases} \tag{6.6.5}$$

and so

$$c_\alpha \equiv (-2)^{\alpha-1} \pmod 5, \tag{6.6.6}$$

which confirms a conjecture of Ramanujan [112], p. 241.

Thus we have, modulo $5^{\alpha+1}$,

$$\sum_{n\geq 0} p(5^\alpha n+\delta_\alpha)q^n \equiv \begin{cases} (-2)^{\alpha-1}5^\alpha E(q)^{19} & \text{if } \alpha \text{ is odd,} \\ (-2)^{\alpha-1}5^\alpha E(q)^{23} & \text{if } \alpha \text{ is even.} \end{cases} \tag{6.6.7}$$

Incidentally, the number of terms in the sum on the right of (6.4.1) is

$$\frac{5^{\alpha+1}-1}{24} \text{ if } \alpha \text{ is odd,} \quad \frac{5^{\alpha+1}-5}{24} \text{ if } \alpha \text{ is even.} \tag{6.6.8}$$

Exercise

We know that if α is even then

$$\sum_{n\geq 0} p(5^{\alpha}n+\delta_{\alpha})q^{n} \equiv (-2)^{\alpha-1}5^{\alpha}E(q)^{23} \pmod{5^{\alpha+1}}.$$

Use this, (3.2.3) and (3.2.6) to show that if α is odd and greater than 1, then

$$p(5^{\alpha}n+\delta_{\alpha}\pm 5^{\alpha-1}) \equiv 0 \pmod{5^{\alpha}}. \tag{6.6.9}$$

It follows that $p(n)$ is divisible by 125 at least $\dfrac{3}{125}$ of the time!

6.7 Some Comments

With the insight provided by (6.2.4) and (6.2.5), it is possible to redo Chapter 5.

Thus, (5.2.3) is

$$\sum_{n\geq 0} p(n)q^{n} = \frac{N(q)}{D(q)}, \tag{6.7.1}$$

where

$$D(q) = \frac{E(q^{5})^{6}}{E(q^{25})} \tag{6.7.2}$$

and

$$\begin{aligned} N(q) &= E(\eta q)E(\eta^{2}q)E(\eta^{3}q)E(\eta^{4}q) \\ &= q^{4}E(q^{25})^{4}(\alpha-1+\beta)(\alpha-\eta+\eta^{2}\beta)(\alpha-\eta^{3}+\eta^{6}\beta)(\alpha-\eta^{4}+\eta^{8}\beta) \\ &= q^{4}E(q^{25})^{4}\left(\alpha^{4}+\alpha^{3}+2\alpha^{2}+3\alpha+5+3\beta+2\beta^{2}+\beta^{3}+\beta^{4}\right). \end{aligned} \tag{6.7.3}$$

Thus

$$\sum_{n\geq 0} p(n)q^{n} = q^{4}\frac{E(q^{25})^{5}}{E(q^{5})^{6}}\left(\alpha^{4}+\alpha^{3}+2\alpha^{2}+3\alpha+5+3\beta+2\beta^{2}+\beta^{3}+\beta^{4}\right). \tag{6.7.4}$$

It follows that

$$\sum_{n\geq 0} p(5n)q^{5n} = q^4 \frac{E(q^{25})^5}{E(q^5)^6}\left(\alpha^4 + 3\beta\right), \tag{6.7.5}$$

$$\sum_{n\geq 0} p(5n+1)q^{5n+1} = q^4 \frac{E(q^{25})^5}{E(q^5)^6}\left(\alpha^3 + 2\beta^2\right), \tag{6.7.6}$$

$$\sum_{n\geq 0} p(5n+2)q^{5n+2} = q^4 \frac{E(q^{25})^5}{E(q^5)^6}\left(2\alpha^2 + \beta^3\right), \tag{6.7.7}$$

$$\sum_{n\geq 0} p(5n+3)q^{5n+3} = q^4 \frac{E(q^{25})^5}{E(q^5)^6}\left(3\alpha + \beta^4\right) \tag{6.7.8}$$

and

$$\sum_{n\geq 0} p(5n+4)q^{5n+4} = 5q^4 \frac{E(q^{25})^5}{E(q^5)^6}. \tag{6.7.9}$$

Exercise: Show that

$$\sum_{n\geq 0} p(5n)q^n \sum_{n\geq 0} p(5n+3)q^n = 3\frac{E(q^5)^4}{E(q)^6} + 25q\frac{E(q^5)^{10}}{E(q)^{12}}, \tag{6.7.10}$$

and

$$\sum_{n\geq 0} p(5n+1)q^n \sum_{n\geq 0} p(5n+2)q^n = 2\frac{E(q^5)^4}{E(q)^6} + 25q\frac{E(q^5)^{10}}{E(q)^{12}}. \tag{6.7.11}$$

These imply (3.6.8) and (3.6.9).

Exercise: For $0 \leq i \leq 4$ let $P_i = \sum_{n\geq 0} p(5n+i)q^n$.

Show, following Kolberg [101], that

$$P_0P_4 + P_1P_3 = 2P_2^2, \tag{6.7.12}$$

$$P_0P_2 + P_3P_4 = 2P_1^2 \tag{6.7.13}$$

and

$$3P_1P_2 - 2P_0P_3 = P_4^2. \tag{6.7.14}$$

Endnotes.

§6.2 From (6.2.4) we obtain

$$\sum_{n\geq 0} p(n)q^n = \frac{1}{E(q)} = \frac{1}{qE(q^{25})} \cdot \frac{\alpha^{-1}}{1-\alpha^{-1}-\alpha^{-2}} = \frac{1}{qE(q^{25})} \cdot \sum_{n\geq 0} F_n \alpha^{-n},$$

where the F_n are the Fibonacci numbers.

From the fact that $F_{5n} \equiv 0 \pmod 5$, we deduce (3.1.9).

§§6.4 The derivation of the modular equation of degree 5 and the proof of (6.4.1) presented here are far simpler than those of Watson [123], Hirschhorn–Hunt [93] and Hirschhorn [72].

§6.4 In Hirschhorn–Hunt [93], we stated the values of $\mathbf{x}_\alpha$ for $\alpha \leq 4$. Ramanujan [112], p. 239 calculated $\mathbf{x}_2$ (as did I, by hand).

§6.6 The conjecture (6.6.6) appears in [112], p. 241. It was not widely known until recently that Ramanujan had outlined a proof of the partition congruences for powers of 5 [112], pp. 240–241.

The point of (6.6.8) is that Ramanujan and Watson have the number of terms equal to $5^{\alpha-1}$, and this is wrong for $\alpha \geq 3$.

§6.6 The congruence (6.6.9) appears in [112], p. 241. It was thought to be original with Watson [123].

References

47. F.G. Garvan, A simple proof of Watson's partition congruences for powers of 7. J. Aust. Math. Soc. Ser. A **36**, 316–334 (1984)
72. M.D. Hirschhorn, An Identity of Ramanujan, and Applications, in *q-Series from a Contemporary Perspective*, vol. 254, Contemporary Mathematics (AMS, Providence, 2000), pp. 229–234
93. M.D. Hirschhorn, D.C. Hunt, A simple proof of the Ramanujan conjecture for powers of 5. J. Reine Angew. Math. **326**, 1–17 (1981)
101. O. Kolberg, Some identities involving the partition function. Math. Scand. **5**, 77–92 (1957)
112. S. Ramanujan, *The Lost Notebook and Other Unpublished Papers* (Springer, Berlin, 1988)
123. G.N. Watson, Ramanujans Vermutung über Zerfällungsanzahlen. J. Reine Angew. Math. **179**, 97–128 (1938)

Chapter 7
Ramanujan's Partition Congruences for Powers of 7

7.1 Introduction

In this chapter, we follow Frank Garvan's proof, Garvan (1984) [47], of Ramanujan's partition congruences for powers of 7,
for $\beta \geq 1$,

$$p(7^{\beta}n+\delta_{\beta}) \equiv 0 \pmod{7^{1+\lfloor \beta/2 \rfloor}}, \tag{7.1.1}$$

where δ_{β} is the reciprocal of 24 modulo 7^{β}.

In §7.2, we obtain the 7-dissection of Euler's product.

In §7.3, we obtain the 7-analogue of Ramanujan's most beautiful identity,

In §7.4, we establish the modular equation of degree 7.

In §7.5, we give the general form of the generating function

$$\sum_{n\geq 0} p(7^{\beta}n+\delta_{\beta})q^{n} \tag{7.1.2}$$

analogous to (6.4.1), and give the proof of (7.1.1).

In §7.6, we state two identities I discovered while writing the book.

7.2 The 7-Dissection of Euler's Product

We can write

$$E = E_0 + E_1 + E_2 + E_5, \tag{7.2.1}$$

where E_i consists of those terms in which the exponent on q is congruent to i modulo 7.

M.D. Hirschhorn, *The Power of q*, Developments in Mathematics 49,
DOI 10.1007/978-3-319-57762-3_7

Now, $(3n^2+n)/2$ is congruent to 2 modulo 7 if and only if $n \equiv 1 \pmod 7$, and

$$E_2 = \sum_{-\infty}^{\infty}(-1)^{7k+1} q^{\left(3(7k+1)^2+(7k+1)\right)/2} = -q^2 E(q^{49}). \tag{7.2.2}$$

We write (7.2.1)

$$E(q) = q^2 E(q^{49})\,(\alpha+\beta-1+\gamma), \tag{7.2.3}$$

where

$$\alpha = -\frac{E_0}{E_2}, \quad \beta = -\frac{E_1}{E_2} \quad \text{and} \quad \gamma = -\frac{E_5}{E_2}. \tag{7.2.4}$$

Jacobi's identity (1.7.1) gives

$$q^6 E(q^{49})^3\,(\alpha+\beta-1+\gamma)^3 = \sum_{n\ge 0}(-1)^n(2n+1)q^{(n^2+n)/2}. \tag{7.2.5}$$

That is,

$$\begin{aligned} q^6 E(q^{49})^3\,&((\alpha^3-6\alpha\gamma+3\beta^2\gamma)+(3\alpha^2\beta-6\beta\gamma+\gamma^3)+3(-\alpha^2+\alpha\beta^2+\gamma)\\ &+(-6\alpha\beta+3\alpha\gamma^2+\beta^3)+3(\alpha-\beta^2+\beta\gamma^2)+3(\alpha^2\gamma+\beta-\gamma^2)\\ &+(6\alpha\beta\gamma-1)) = \sum_{n\ge 0}(-1)^n(2n+1)q^{(n^2+n)/2}. \end{aligned} \tag{7.2.6}$$

Now, $(n^2+n)/2 \equiv 6 \pmod 7$ if and only if $n \equiv 3 \pmod 7$, so if we consider those terms in (7.2.6) in which the exponent on q is 6 modulo 7, we find

$$\begin{aligned} q^6 E(q^{49})^3\,(6\alpha\beta\gamma-1) &= \sum_{k\ge 0}(-1)^{7k+3}(2(7k+3)+1)q^{\left((7k+3)^2+(7k+3)\right)/2}\\ &= -7q^6E(q^{49})^3. \end{aligned} \tag{7.2.7}$$

It follows that

$$\alpha\beta\gamma = -1. \tag{7.2.8}$$

Also, since $(n^2+n)/2 \not\equiv 2,\ 4,\ 5 \pmod 7$, we find

$$-\alpha^2+\alpha\beta^2+\gamma = 0, \tag{7.2.9}$$

$$\alpha-\beta^2+\beta\gamma^2 = 0, \tag{7.2.10}$$

$$\alpha^2\gamma+\beta-\gamma^2 = 0. \tag{7.2.11}$$

If we multiply (7.2.9), (7.2.10) and (7.2.11), we obtain

$$3(\alpha^2\beta^3+\beta^2\gamma^3+\gamma^2\alpha^3)-3(\alpha^3\beta+\beta^3\gamma+\gamma^3\alpha)-(\alpha\beta^5+\beta\gamma^5+\gamma\alpha^5)-6=0. \tag{7.2.12}$$

Let

$$\sigma=\alpha^3\beta+\beta^3\gamma+\gamma^3\alpha. \tag{7.2.13}$$

Then

$$\begin{aligned}\alpha^2\beta^3+\beta^2\gamma^3+\gamma^2\alpha^3&=\alpha\beta(\alpha^2-\beta)+\beta\gamma(\beta^2-\alpha)+\gamma\alpha(\gamma^2-\beta)\\&=(\alpha^3\beta+\beta^3\gamma+\gamma^3\alpha)-3\alpha\beta\gamma\\&=\sigma+3.\end{aligned} \tag{7.2.14}$$

Then (7.2.12) becomes

$$3(\sigma+3)-3\sigma-(\alpha\beta^5+\beta\gamma^5+\gamma\alpha^5)-6=0,$$

or,

$$\alpha\beta^5+\beta\gamma^5+\gamma\alpha^5=3. \tag{7.2.15}$$

We note also that

$$\begin{aligned}\alpha^7+\beta^7+\gamma^7&=\alpha^5(\alpha\beta^2+\gamma)+\beta^5(\beta\gamma^2+\alpha)+\gamma^5(\gamma\alpha^2+\beta)\\&=(\alpha^6\beta^2+\beta^6\gamma^2+\gamma^6\alpha^2)+(\alpha\beta^5+\beta\gamma^5+\gamma\alpha^5)\\&=(\alpha^3\beta+\beta^3\gamma+\gamma^3\alpha)^2-2\alpha\beta\gamma(\alpha^2\beta^3+\beta^2\gamma^3+\gamma^2\alpha^3)+3\\&=\sigma^2+2(\sigma+3)+3\\&=\sigma^2+2\sigma+9.\end{aligned} \tag{7.2.16}$$

7.3 The 7-Analogue of Ramanujan's Most Beautiful Identity

In this section, we prove Ramanujan's beautiful identity,

$$\sum_{n\geq 0} p(7n+5)q^n=7\frac{(q^7;q^7)_\infty^3}{(q;q)_\infty^4}+49q\frac{(q^7;q^7)_\infty^7}{(q;q)_\infty^8}. \tag{7.3.1}$$

We have, with η a seventh root of unity other than 1,

$$\sum_{n\geq 0} p(n)q^n = \frac{1}{E(q)}$$
$$= \frac{E(\eta q)E(\eta^2 q)E(\eta^3 q)E(\eta^4 q)E(\eta^5 q)E(\eta^6 q)}{E(q)E(\eta q)E(\eta^2 q)E(\eta^3 q)E(\eta^4 q)E(\eta^5 q)E(\eta^6 q)}. \tag{7.3.2}$$

The denominator is both (see §5.2)

$$D(q) = \frac{E(q^7)^8}{E(q^{49})}, \tag{7.3.3}$$

and, by (7.2.3),

$$\begin{aligned} D(q) &= q^{14}E(q^{49})^7(\alpha+\beta-1+\gamma)(\alpha+\eta\beta-\eta^2+\eta^5\gamma)(\alpha+\eta^2\beta-\eta^4+\eta^{10}\gamma) \\ &\qquad (\alpha+\eta^3\beta-\eta^6+\eta^{15}\gamma)(\alpha+\eta^4\beta-\eta^8+\eta^{20}\gamma) \\ &\qquad (\alpha+\eta^5\beta-\eta^{10}+\eta^{25}\gamma)(\alpha+\eta^6\beta-\eta^{12}+\eta^{30}\gamma) \\ &= q^{14}E(q^{49})^7((\alpha^7+\beta^7+\gamma^7)+7(\alpha^5\beta+\beta^5\gamma+\gamma^5\alpha) \\ &\qquad +14(\alpha^2\beta^3+\beta^2\gamma^3+\gamma^2\alpha^3)-8) \\ &= q^{14}E(q^{49})^7\left((\sigma^2+2\sigma+9)+7\times 3+14(\sigma+3)-8\right) \\ &= q^{14}E(q^{49})^7\left(\sigma^2+16\sigma+64\right) \\ &= q^{14}E(q^{49})^7(\sigma+8)^2. \end{aligned} \tag{7.3.4}$$

If we now equate the two expressions we have for $D(q)$ in (7.3.3) and (7.3.4), we discover the remarkable fact that

$$(\sigma+8)^2 = \frac{E(q^7)^8}{q^{14}E(q^{49})^8} = \left(\frac{E(q^7)^4}{q^7E(q^{49})^4}\right)^2. \tag{7.3.5}$$

It follows that

$$\sigma+8 = -\frac{E(q^7)^4}{q^7E(q^{49})^4},$$

since both sides begin $-q^{-7}+\cdots$, or,

$$\sigma = -\frac{E(q^7)^4}{q^7E(q^{49})^4} - 8. \tag{7.3.6}$$

The numerator in (7.3.2) is

$$\begin{aligned}
N(q) = q^{12}E(q^{49})^6(\alpha+\eta\beta-\eta^2+\eta^5\gamma)(\alpha+\eta^2\beta-\eta^4+\eta^{10}\gamma)\\
\times(\alpha+\eta^3\beta-\eta^6+\eta^{15}\gamma)(\alpha+\eta^4\beta-\eta^8+\eta^{20}\gamma)\\
\times(\alpha+\eta^5\beta-\eta^{10}+\eta^{25}\gamma)(\alpha+\eta^6\beta-\eta^{12}+\eta^{30}\gamma)
\end{aligned}$$

$$\begin{aligned}
&= q^{12}E(q^{49})^6(\alpha^6-\alpha^5\beta+(\alpha^5+\alpha^4\beta^2)+(-2\alpha^4\beta-\alpha^3\beta^3)\\
&\quad+(\alpha^4+3\alpha^3\beta^2+\alpha^2\beta^4)+(-\alpha^5\gamma-3\alpha^3\beta-4\alpha^2\beta^3-\alpha\beta^5)\\
&\quad+(-\alpha^3+6\alpha^2\beta^2+5\alpha\beta^4+\beta^6)+(5\alpha^4\gamma-\alpha^2\beta+4\alpha\beta^3+\beta^5)\\
&\quad+(2\alpha^2+6\alpha\beta^2+\beta^4)+(4\alpha^3\gamma-\beta^3)+(\alpha^4\gamma^2+3\alpha-\beta^5\gamma+2\beta^2)\\
&\quad+(6\alpha^2\gamma-2\beta^4\gamma+3\beta)+(-4\alpha^3\gamma^2-3\beta^3\gamma+8)-\beta^2\gamma\\
&\quad+(6\alpha^2\gamma^2+\beta^4\gamma^2)+(-\alpha^3\gamma^3+3\beta^3\gamma^2+3\gamma)+(-\alpha\gamma^2+6\beta^2\gamma^2)\\
&\quad+(3\alpha^2\gamma^3+6\beta\gamma^2)+(-\beta^3\gamma^3+2\gamma^2)+(-3\alpha\gamma^3-4\beta^2\gamma^3)\\
&\quad+(\alpha^2\gamma^4+4\beta\gamma^3)-\gamma^3+(-2\alpha\gamma^4+\beta^2\gamma^4)+5\beta\gamma^4+\gamma^4-\alpha\gamma^5-\beta\gamma^5\\
&\quad+\gamma^5+\gamma^6). \qquad (7.3.7)
\end{aligned}$$

If we use (7.3.3) and (7.3.7) and extract from (7.3.2) those terms in which the exponent is 5 modulo 7, we obtain

$$\begin{aligned}
\sum_{n\geq 0} p(7n+5)q^{7n+5} &= q^{12}\frac{E(q^{49})^7}{E(q^7)^8}(-4(\alpha^2\beta^3+\beta^2\gamma^3+\gamma^2\alpha^3)-(\alpha\beta^5+\beta\gamma^5+\gamma\alpha^5)\\
&\qquad -3\,(\alpha^3\beta+\beta^2\gamma+\gamma^3\alpha)+8)\\
&= q^{12}\frac{E(q^{49})^7}{E(q^7)^8}(-4(\sigma+3)-3-3\sigma+8)\\
&= q^{12}\frac{E(q^{49})^7}{E(q^7)^8}(-7\sigma-7)\\
&= q^{12}\frac{E(q^{49})^7}{E(q^7)^8}\left(7\frac{E(q^7)^4}{q^7E(q^{49})^4}+49\right)\\
&= 7q^5\frac{E(q^{49})^3}{E(q^7)^4}+49q^{12}\frac{E(q^{49})^7}{E(q^7)^8},
\end{aligned}$$

which leads immediately to (7.3.1).

7.4 The Modular Equation of Degree 7

Let

$$\zeta = \frac{E(q)}{q^2E(q^{49})}, \quad T = \frac{E(q^7)^4}{q^7E(q^{49})^4}. \tag{7.4.1}$$

The purpose of this section is to establish the modular equation of degree 7,

$$T^2 = (7\zeta^3+35\zeta^2+49\zeta)T+\zeta^7+7\zeta^6+21\zeta^5+49\zeta^4+147\zeta^3+343\zeta^2+343\zeta. \tag{7.4.2}$$

We start by noting that in view of (7.3.6), (7.2.13)–(7.2.16) become

$$\alpha^3\beta+\beta^3\gamma+\gamma^3\alpha = -T-8 \tag{7.4.3}$$

$$\alpha^2\beta^3+\beta^2\gamma^3+\gamma^2\alpha^3 = -T-5, \tag{7.4.4}$$

$$\alpha\beta^5+\beta\gamma^5+\gamma\alpha^5 = 3. \tag{7.4.5}$$

and

$$\alpha^7+\beta^7+\gamma^7 = T^2+14T+57. \tag{7.4.6}$$

We have

$$\zeta = \alpha+\beta-1+\gamma, \tag{7.4.7}$$

$$\zeta^2 = \alpha^2+2\alpha\beta+(\beta^2-2\alpha)-2\beta+1+2\alpha\gamma+2\beta\gamma-2\gamma+\gamma^2, \tag{7.4.8}$$

$$\zeta^3 = \alpha^3+3\alpha^2\beta+\cdots-7+\cdots+\gamma^3, \tag{7.4.9}$$

$$\zeta^4 = \alpha^4+4\alpha^3\beta+\cdots+(4\beta^3\gamma+25)+\cdots+4\gamma^3\alpha+\cdots+\gamma^4, \tag{7.4.10}$$

$$\begin{aligned}\zeta^5 &= \alpha^5+5\alpha^4\beta+\cdots+(10\alpha^2\beta^3-20\alpha^3\beta)+\cdots+(10\gamma^2\alpha^3-20\beta^3\gamma-61)\\ &\quad+\cdots+(10\beta^2\gamma^3-20\gamma^3\alpha^2)+\cdots+\gamma^5,\end{aligned} \tag{7.4.11}$$

$$\begin{aligned}\zeta^6 &= \alpha^6+6\alpha^5\beta+\cdots+(6\alpha\beta^5+6\gamma\alpha^5-60\alpha^2\beta^3+60\alpha^3\beta)\\ &\quad+\cdots+(-60\gamma^2\alpha^3+60\beta^3\gamma+211)+\cdots+(-60\beta^2\gamma^3+60\gamma^3\alpha)\\ &\quad+\cdots+6\beta\gamma^5\cdots+\gamma^6.\end{aligned} \tag{7.4.12}$$

So, if H is the huffing operator modulo 7 (see §6.3 for an explanation), we have

$$H(\zeta) = -1, \tag{7.4.13}$$

$$H(\zeta^2) = 1, \tag{7.4.14}$$

$$H(\zeta^3) = -7, \tag{7.4.15}$$

$$H(\zeta^4) = 4(\alpha^3\beta+\beta^3\gamma+\gamma^3\alpha)+25 = -4T-7, \tag{7.4.16}$$

$$\begin{aligned} H(\zeta^5) &= 10(\alpha^2\beta^3 + \beta^2\gamma^3 + \gamma^2\alpha^3) - 20(\alpha^3\beta + \beta^3\gamma + \gamma^3\alpha) - 61 \\ &= 10(-T-5) - 20(-T-8) - 61 = 10T + 49, \end{aligned} \tag{7.4.17}$$

$$\begin{aligned} H(\zeta^6) &= 6(\alpha\beta^5 + \beta\gamma^5 + \gamma\alpha^5) - 60(\alpha^2\beta^3 + \beta^2\gamma^3 + \gamma^2\alpha^3) \\ &\qquad + 60(\alpha^3\beta + \beta^3\gamma + \gamma^3\alpha) + 211 \\ &= 6 \times 3 - 60(-T-5) + 60(-T-8) + 211 = 49. \end{aligned} \tag{7.4.18}$$

Now let

$$\zeta_i = \zeta(\eta^i q), \quad i = 0, \ \ldots \ ,6. \tag{7.4.19}$$

Then (7.4.13)–(7.4.18) become

$$\sum_i \zeta_i = -7, \tag{7.4.20}$$

$$\sum_i \zeta_1^2 = 7, \tag{7.4.21}$$

$$\sum_i \zeta_i^3 = -49, \tag{7.4.22}$$

$$\sum_i \zeta_i^4 = -28T + 49, \tag{7.4.23}$$

$$\sum_i \zeta_i^5 = 70T + 343, \tag{7.4.24}$$

$$\sum_i \zeta_i^6 = 343, \tag{7.4.25}$$

We now employ the Newton–Girard formulas.

$$2\sum_{i<j} \zeta_i\zeta_j = \sum_i \zeta_i \sum_i \zeta_i - \sum_i \zeta_i^2$$

gives

$$\sum_{i<j} \zeta_i\zeta_j = 21, \tag{7.4.26}$$

$$3\sum_{i<j<k} \zeta_i\zeta_j\zeta_k = \sum_{i<j} \zeta_i\zeta_j \sum_i \zeta_i - \sum_i \zeta_i \sum_i \zeta_i^2 + \sum_i \zeta_i^3$$

gives

$$\sum_{i<j<k} \zeta_i\zeta_j\zeta_k = -49, \tag{7.4.27}$$

$$4\sum_{i<j<k<l} \zeta_i\zeta_j\zeta_k\zeta_l = \sum_{i<j<k} \zeta_i\zeta_j\zeta_l \sum_i \zeta_i - \sum_{i<j} \zeta_i\zeta_j \sum_i \zeta_i^2 + \sum_i \zeta_i \sum_i \zeta_i^3 - \sum_i \zeta_i^4$$

gives

$$\sum_{i<j<k<l} \zeta_i\zeta_j\zeta_k\zeta_l = 7T + 147. \tag{7.4.28}$$

In the same way, we find

$$\sum_{i<j<k<l<m} \zeta_i\zeta_j\zeta_k\zeta_l\zeta_m = -35T - 343, \tag{7.4.29}$$

$$\sum_{i<j<k<l<m<n} \zeta_i\zeta_j\zeta_k\zeta_l\zeta_m\zeta_n = 49T + 343, \tag{7.4.30}$$

and, from (7.3.3) and (7.4.1),

$$\zeta_0\zeta_1\zeta_2\zeta_3\zeta_4\zeta_5\zeta_6 = \frac{D(q)}{q^{14}E(q^{49})^7} = T^2. \tag{7.4.31}$$

It follows that $\zeta_0, \zeta_1, \ldots, \zeta_6$ are the roots of the equation

$$\begin{aligned} &X^7 + 7X^6 + 21X^5 + 49X^4 + (7T + 147)X^3 + (35T + 343)X^2 \\ &\quad + (49T + 343)X - T^2 = 0. \end{aligned} \tag{7.4.32}$$

In particular, $\zeta = \zeta_0$ is a root of (7.4.32), which yields (7.4.2).

7.5 Ramanujan's Partition Congruences for Powers of 7

We start by obtaining the 7-analogue of (6.4.1), namely

$$\sum_{n\geq 0} p(7^\beta n + \delta_\beta)q^n = \begin{cases} \sum_{i\geq 1} x_{\beta,i}q^{i-1}\dfrac{E(q^7)^{4i-1}}{E(q)^{4i}}, & \beta \text{ odd}, \\ \sum_{i\geq 1} x_{\beta,i}q^{i-1}\dfrac{E(q^7)^{4i}}{E(q)^{4i+1}}, & \beta \text{ even}, \end{cases} \tag{7.5.1}$$

where the coefficient vectors $\mathbf{x}_\beta = (x_{\beta,i})_{i\geq 1}$ are given recursively by

$$\mathbf{x}_1 = (7, 49, 0, \ldots) \tag{7.5.2}$$

and for $\beta \geq 1$,

$$\mathbf{x}_{\beta+1} = \begin{cases} \mathbf{x}_\beta A, & \beta \text{ odd}, \\ \mathbf{x}_\beta B, & \beta \text{ even}, \end{cases} \tag{7.5.3}$$

where $A = (a_{i,j})_{i,j\geq 1}$ and $B = (b_{i,j})_{i,j\geq 1}$ are defined by

$$a_{i,j} = m_{4i,i+j}, \quad b_{i,j} = m_{4i+1,i+j}, \tag{7.5.4}$$

where $M = (m_{i,j})_{i,j \geq 1}$ is defined as follows. The first seven rows of M are

$$\begin{aligned}
&(7,\ 7^2,\ 0,\ \dots),\\
&(10,\ 9\times 7^2,\ 2\times 7^4,\ 7^5,\ 0,\ \dots),\\
&(3,\ 114\times 7,\ 85\times 7^3,\ 24\times 7^5,\ 3\times 7^7,\ 7^8,\ 0,\ \dots),\\
&(0,\ 82\times 7,\ 176\times 7^3,\ 845\times 7^4,\ 272\times 7^6,\ 46\times 7^8,\ 4\times 7^{10},\ 7^{11},\ 0,\ \dots),\\
&(0,\ 190,\ 1265\times 7^2,\ 1895\times 7^4,\ 1233\times 7^6,\ 3025\times 7^7,\ 620\times 7^9,\ 75\times 7^{11},\\
&\qquad 5\times 7^{13},\ 7^{14},\ 0,\ \dots),\\
&(0,\ 27,\ 736\times 7^2,\ 16782\times 7^3,\ 20424\times 7^5,\ 12825\times 7^7,\ 4770\times 7^9,\ 7830\times 7^{10},\\
&\qquad 1178\times 7^{12},\ 111\times 7^{14},\ 6\times 7^{16},\ 7^{17},\ 0,\ \dots),\\
&(0,\ 1,\ 253\times 7^2,\ 1902\times 7^4,\ 4246\times 7^6,\ 31540\times 7^7,\ 19302\times 7^9,\ 7501\times 7^{11},\\
&\qquad 1944\times 7^{13},\ 2397\times 7^{14},\ 285\times 7^{16},\ 22\times 7^{18},\ 7^{20},\ 7^{20},\ 0,\ \dots),
\end{aligned} \tag{7.5.5}$$

and for $i \geq 4$, $m_{i,1} = 0$, for $i \geq 8$, $m_{i,2} = 0$, and for $i \geq 8$, $j \geq 3$,

$$\begin{aligned}
m_{i,j} &= 49m_{i-1,j-1} + 35m_{i-2,j-1} + 7m_{i-3,j-1}\\
&\quad + 343m_{i-1,j-2} + 343m_{i-2,j-2} + 147m_{i-3,j-2} + 49m_{i-4,j-2} + 21m_{i-5,j-2}\\
&\qquad + 7m_{i-6,j-2} + m_{i-7,j-2}.
\end{aligned} \tag{7.5.6}$$

Analogously to the case of powers of 5 in the previous chapter, the modular equation (7.4.2) can be written

$$\begin{aligned}
\frac{1}{\zeta^i} &= \frac{1}{T}\left(\frac{49}{\zeta^{i-1}} + \frac{35}{\zeta^{i-2}} + \frac{7}{\zeta^{i-3}}\right)\\
&\quad + \frac{1}{T^2}\left(\frac{343}{\zeta^{i-1}} + \frac{343}{\zeta^{i-2}} + \frac{147}{\zeta^{i-3}} + \frac{49}{\zeta^{i-4}} + \frac{21}{\zeta^{i-5}} + \frac{7}{\zeta^{i-6}} + \frac{1}{\zeta^{i-7}}\right).
\end{aligned} \tag{7.5.7}$$

From this it follows that

$$H\left(\frac{1}{\zeta^i}\right) = \sum_{j\geq 1} \frac{m_{i,j}}{T^j}, \tag{7.5.8}$$

where the $m_{i,j}$ satisfy (7.5.5) and (7.5.6).

We can check that $m_{4i,j} = 0$ and $m_{4i+1,j} = 0$ for $1 \leq j \leq i$, so we can write

$$H\left(\frac{1}{\zeta^{4i}}\right) = \sum_{j\geq 1} \frac{a_{i,j}}{T^{i+j}} \tag{7.5.9}$$

and

$$H\left(\frac{1}{\zeta^{4i+1}}\right) = \sum_{j\geq 1} \frac{b_{i,j}}{T^{i+j}}, \tag{7.5.10}$$

where the $a_{i,j}$ and $b_{i,j}$ are given by (7.5.4).

We can write (7.5.9)

$$H\left(\left(\frac{q^2E(q^{49})}{E(q)}\right)^{4i}\right) = \sum_{j\geq 1} a_{i,j}\left(\frac{q^7E(q^{49})^4}{E(q^7)^4}\right)^{i+j},$$

or,

$$H\left(q^{i-7}\frac{E(q^7)^{4i-1}}{E(q)^{4i}}\right) = \sum_{j\geq 1} a_{i,j}q^{7i-7}\frac{E(q^{49})^{4j}}{E(q^7)^{4j+1}}. \tag{7.5.11}$$

Similarly, we can write (7.5.10)

$$H\left(\left(\frac{q^2E(q^{49})}{E(q)}\right)^{4i+1}\right) = \sum_{j\geq 1} b_{i,j}\left(\frac{q^7E(q^{49})^4}{E(q^7)^4}\right)^{i+j}$$

or,

$$H\left(q^{i-5}\frac{E(q^7)^{4i}}{E(q)^{4i+1}}\right) = \sum_{j\geq 1} b_{i,j}q^{7j-7}\frac{E(q^{49})^{4j-1}}{E(q^7)^{4j}}. \tag{7.5.12}$$

We note also that if $\delta_1 = 5$ and

$$\delta_{\beta+1} = \begin{cases} 6\times 7^\beta + \delta_\beta, \ \beta \text{ odd}, \\ 4\times 7^\beta + \delta_\beta, \ \beta \text{ even}, \end{cases} \tag{7.5.13}$$

then

$$\delta_\beta = \begin{cases} \dfrac{17\times 7^\beta + 1}{24}, \ \beta \text{ odd}, \\ \dfrac{23\times 7^\beta + 1}{24}, \ \beta \text{ even} \end{cases} \tag{7.5.14}$$

and

$$1 \leq \delta_\beta < 7^\beta \text{ and } 24\delta_\beta \equiv 1 \pmod{7^\beta}. \tag{7.5.15}$$

That is, δ_β is the reciprocal of 24 modulo 7^β.

From this point on, the proof of (7.5.1) is very similar to that of (6.4.1).

We have, in (7.3.1), already verified the case $\beta = 1$,

$$\sum_{n\geq 0} p(7n+5)q^n = 7\frac{E(q^7)^3}{E(q)^4} + 49q\frac{E(q^7)^7}{E(q)^8}. \tag{7.5.16}$$

Now suppose β is odd and

$$\sum_{n\geq 0} p(7^{\beta}n+\delta_{\beta})q^{n} = \sum_{i\geq 1} x_{\beta,i}q^{i-1}\frac{E(q^7)^{4i-1}}{E(q)^{4i}}. \tag{7.5.17}$$

Then

$$\sum_{n\geq 0} p(7^{\beta}n+\delta_{\beta})q^{n-6} = \sum_{i\geq 1} x_{\beta,i}q^{i-7}\frac{E(q^7)^{4i-1}}{E(q)^{4i}}. \tag{7.5.18}$$

If we apply the operator H, then replace q by $q^{\frac{1}{7}}$, and make use of (7.5.13), we find that

$$\sum_{n\geq 0} p(7^{\beta+1}n+\delta_{\beta+1})q^{n} = \sum_{i\geq 1} x_{\beta+1,i}q^{i-1}\frac{E(q^7)^{4i}}{E(q)^{4i+1}}. \tag{7.5.19}$$

Now suppose β is even and

$$\sum_{n\geq 0} p(7^{\beta}n+\delta_{\beta})q^{n} = \sum_{i\geq 1} x_{\beta,i}q^{i-1}\frac{E(q^7)^{4i}}{E(q)^{4i+1}}. \tag{7.5.20}$$

Then

$$\sum_{n\geq 0} p(7^{\beta}n+\delta_{\beta})q^{n-4} = \sum_{i\geq 1} x_{\beta,i}q^{i-5}\frac{E(q^7)^{4i}}{E(q)^{4i+1}}. \tag{7.5.21}$$

If we apply the operator H, then replace q by $q^{\frac{1}{7}}$, and make use of (7.5.13), we find that

$$\sum_{n\geq 0} p(7^{\beta+1}n+\delta_{\beta+1})q^{n} = \sum_{i\geq 1} x_{\beta+1,i}q^{i-1}\frac{E(q^7)^{4i-1}}{E(q)^{4i}}, \tag{7.5.22}$$

and the proof of (7.5.1) is complete!

If $\nu(n)$ denotes the highest power of 7 that divides n, then it can be shown that

$$\nu(m_{i,j}) \geq \lfloor \tfrac{1}{4}(7j-2i-1) \rfloor,\ \ \nu(a_{i,j}) \geq \lfloor \tfrac{1}{4}(7j-i-1) \rfloor,\ \ \nu(b_{i,j}) \geq \lfloor \tfrac{1}{4}(7j-i-3) \rfloor \tag{7.5.23}$$

and that for $\beta \geq 1$,

$$\nu(x_{2\beta,i}) \geq \beta+1+\lfloor \tfrac{1}{4}(7i-6) \rfloor,\ \ \nu(x_{2\beta+1,i}) \geq \beta+1+\lfloor \tfrac{1}{4}(7i-4) \rfloor \tag{7.5.24}$$

while for $\beta \geq 1$,

$$\nu(x_{2\beta,1}) = \nu(x_{2\beta+1,1}) = \beta+1. \tag{7.5.25}$$

The result (7.1.1) follows directly.

Indeed, since $a_{1,1} = m_{4,2} = 82 \times 7$, $b_{1,1} = m_{5,2} = 190$, $82 \equiv -2 \pmod 7$ and $190 \equiv 1 \pmod 7$, it follows that for $\beta \geq 1$, modulo $7^{\beta+2}$,

$$\sum_{n\geq 0} p(7^{2\beta}n+\delta_{2\beta})q^n \equiv (-2)^{\beta}7^{\beta+1}\frac{E(q^7)^4}{E(q)^5} \equiv (-2)^{\beta}7^{\beta+1}E(q)^{23} \tag{7.5.26}$$

and for $\beta \geq 0$,

$$\sum_{n\geq 0} p(7^{2\beta+1}n+\delta_{2\beta+1})q^n \equiv (-2)^{\beta}7^{\beta+1}\frac{E(q^7)^3}{E(q)^4} \equiv (-2)^{\beta}7^{\beta+1}E(q)^{17}. \tag{7.5.27}$$

These should be compared with (6.6.7).

The number of terms in the sum on the right of (7.5.1) is

$$\frac{7^{\beta+1}-1}{24} \text{ if } \beta \text{ is odd, } \frac{7^{\beta+1}-7}{24} \text{ if } \beta \text{ is even.}$$

Exercise

We know that for $\beta \geq 0$,

$$\sum_{n\geq 0} p(7^{2\beta+1}n+\delta_{2\beta+1})q^n \equiv (-2)^{\beta}7^{\beta+1}E(q)^{17} \pmod{7^{\beta+1}}.$$

Show that for $\beta \geq 1$,

$$\begin{aligned} p(7^{2\beta}n+\delta_{2\beta}-7^{2\beta-1}) &\equiv 0 \pmod{7^{\beta+1}}, \\ p(7^{2\beta}n+\delta_{2\beta}-2\times 7^{2\beta-1}) &\equiv 0 \pmod{7^{\beta+1}} \\ \text{and}\quad p(7^{2\beta}n+\delta_{2\beta}-4\times 7^{2\beta-1}) &\equiv 0 \pmod{7^{\beta+1}}. \end{aligned} \tag{7.5.28}$$

So $p(n)$ is divisible by 49 at least $\frac{4}{49}$ of the time!

7.6 Two More Formulas

While writing the book, I discovered that

$$\sum_{n\geq 0} p(7n)q^n \sum_{n\geq 0} p(7n+2)q^n \sum_{n\geq 0} p(7n+6)q^n$$
$$= 22\frac{E(q^7)^5}{E(q)^8} + 2\times 7^3 q\frac{E(q^7)^9}{E(q)^{12}} + 25\times 7^3 q^2\frac{E(q^7)^{13}}{E(q)^{16}}$$

$$+3\times 7^5 q^3 \frac{E(q^7)^{17}}{E(q)^{20}} + 7^6 q^4 \frac{E(q^7)^{21}}{E(q)^{24}} \tag{7.6.1}$$

and

$$\sum_{n\geq 0} p(7n+1)q^n \sum_{n\geq 0} p(7n+3)q^n \sum_{n\geq 0} p(7n+4)q^n$$
$$= 15\frac{E(q^7)^5}{E(q)^8} + 12\times 7^2 q \frac{E(q^7)^9}{E(q)^{12}} + 24\times 7^3 q^2 \frac{E(q^7)^{13}}{E(q)^{16}}$$
$$+3\times 7^5 q^3 \frac{E(q^7)^{17}}{E(q)^{20}} + 7^6 q^4 \frac{E(q^7)^{21}}{E(q)^{24}}. \tag{7.6.2}$$

These are the counterparts of (6.7.10) and (6.7.11), and imply (3.7.5) and (3.7.6).

Endnotes.

§7.5 Garvan (1984) [47] calculated $\mathbf{x}_2$, as had Zuckerman(1939) [129]. Andrews–Berndt (AB III, p. 140) [7] discovered a fragment (LNB [112], p. 240), in which Ramanujan had written the first few $x_{2,i}$, but has $x_{2,4} = 335400 \times 7^8$, whereas the correct value is 2488800×7^7.

§7.5 The congruences (7.5.28) are due to Watson (1938) [123]. However, the case $\beta = 1$ appears in LNB [112], p. 144.

§7.5 The results (7.5.26) and (7.5.27) may be new.

§7.6 The results (7.6.1) and (7.6.2) are not new. They are proved in Kolberg (1957) [101].

References

7. G.E. Andrews, B.C. Berndt, *Ramanujan's Lost Notebook, Part III* (Springer, Berlin, 2012)
47. F.G. Garvan, A simple proof of Watson's partition congruences for powers of 7. J. Aust. Math. Soc. Ser. A **36**, 316–334 (1984)
101. O. Kolberg, Some identities involving the partition function. Math. Scand. **5**, 77–92 (1957)
112. S. Ramanujan, *The Lost Notebook and Other Unpublished Papers* (Springer, Berlin, 1988)
123. G.N. Watson, Ramanujans Vermutung über Zerfällungsanzahlen. J. Reine Angew. Math. **179**, 97–128 (1938)
129. H.S. Zuckerman, Identities analogous to Ramanujan's identities involving the partition function. Duke Math. J. **5**, 88–110 (1939)

Chapter 8
Ramanujan's 5-Dissection of Euler's Product

8.1 Introduction

Up to now, we have managed to obtain not only Ramanujan's most beautiful identity but also his partition congruences for powers of 5, using nothing more than the fact that

$$E(q) = qE(q^{25})\,(\alpha - 1 + \beta)$$

where α, β, defined in (6.2.5), satisfy

$$\alpha\beta = -1.$$

Ramanujan [59], p. 212, declares "It can be shewn that"

$$E(q) = E(q^{25})(R(q^5)^{-1} - q - q^2R(q^5)), \tag{8.1.1}$$

where

$$R(q) = \left(\begin{matrix} q, q^4 \\ q^2, q^3 \end{matrix}; q^5\right)_\infty. \tag{8.1.2}$$

Ramanujan then proceeds to prove (5.1.1), but because he never gives a proof of (8.1.1), Hardy says "Ramanujan never gave a complete proof [of (5.1.1)]" [58], p. 90.

In §§8.2, 8.3, we give two, completely different, proofs of (8.1.1).

Of course, (8.1.1) reveals that our α, β of Chapter 6 are given by

$$\alpha = q^{-1}R(q^5)^{-1}, \quad \beta = -qR(q^5). \tag{8.1.3}$$

In the light of this, in §8.4, we review some of our earlier results, and in §§8.5 and 8.6, we examine some remarkable identities of Ramanujan.

M.D. Hirschhorn, *The Power of q*, Developments in Mathematics 49,
DOI 10.1007/978-3-319-57762-3_8

8.2 Our First Proof of Ramanujan's 5-Dissection of Euler's Product

Our first proof of (8.1.1) is Carlitzian (by which I mean that it involves multiplication and rearrangement of series).

We have

$$\begin{aligned}E(q) &= \frac{(q,q^4,q^5;q^5)_\infty (q^2,q^3,q^5;q^5)_\infty}{(q^5;q^5)_\infty} \\ &= \frac{1}{(q^5;q^5)_\infty} \sum_{k,l=-\infty}^{\infty} (-1)^{k+l} q^{(5k^2-3k+5l^2-l)/2}. \qquad (8.2.1)\end{aligned}$$

We now split the sum according to the residue of $k+2l$ modulo 5.

If $k+2l \equiv 0 \pmod 5$, write $k+2l = 5m$, define $t = l-2m$, and then $k = m-2t$, $l = 2m+t$, and the contribution of these terms to the sum is

$$\begin{aligned}&\sum_{m,t=-\infty}^{\infty} (-1)^{3m-t} q^{(5(m-2t)^2-3(m-2t)+5(2m+t)^2-(2m+t))/2} \\ &\qquad = \sum_{m,t=-\infty}^{\infty} (-1)^{m+t} q^{(25m^2-5m+25t^2+5t)/2} \\ &\qquad = (q^{10},q^{15},q^{25};q^{25})_\infty^2. \qquad (8.2.2)\end{aligned}$$

If $k+2l \equiv 1 \pmod 5$, write $k+2l = 5m+1$, define $t = l-2m$, and then $k = m+1-2t$, $l = 2m+t$, and the contribution is

$$\begin{aligned}&\sum_{m,t=-\infty}^{\infty} (-1)^{3m+1-t} q^{(5(m+1-2t)^2-3(m+1-2t)+5(2m+t)^2-(2m+t))/2} \\ &\qquad = -\sum_{m,t=-\infty}^{\infty} (-1)^{m+t} q^{(25m^2+5m+25t^2-15t+2)/2} \\ &\qquad = -q(q^5;q^5)_\infty (q^{25};q^{25})_\infty. \qquad (8.2.3)\end{aligned}$$

If $k+2l \equiv 2 \pmod 5$, write $k+2l = 5m+2$, define $t = l-2m-1$, and then $k = m-2t$, $l = 2m+1+t$, and the contribution is

$$\begin{aligned}&\sum_{m,t=-\infty}^{\infty} (-1)^{3m+1-t} q^{(5(m-2t)^2-3(m-2t)+5(2m+1+t)^2-(2m+1+t))/2} \\ &\qquad = -\sum_{m,t=-\infty}^{\infty} (-1)^{m+t} q^{(25m^2+15m+25t^2+15t+4)/2} \\ &\qquad = -q^2(q^5,q^{20},q^{25};q^{25})_\infty^2. \qquad (8.2.4)\end{aligned}$$

If $k+2l \equiv 3 \pmod 5$, write $k+2l = 5m+3$, define $t = l-2m-1$, and then $k = m-2t+1$, $l = 2m+t+1$, and the contribution is

$$\sum_{m,t=-\infty}^{\infty} (-1)^{3m-t+2} q^{(5(m-2t+1)^2-3(m-2t+1)+5(2m+t+1)^2-(2m+t+1))/2}$$
$$= q^3 \sum_{m,t=-\infty}^{\infty} (-1)^{m+t} q^{(25m^2+25m+25t^2-5t)/2}$$
$$= 0. \tag{8.2.5}$$

Finally, if $k+2l \equiv 4 \pmod 5$, write $k+2l = 5m-1$, define $t = l-2m+1$, $k = m-2t+1$, $l = 2m+t-1$, and the contribution is

$$\sum_{m,t=-\infty}^{\infty} (-1)^{3m+t} q^{(5(m-2t+1)^2-3(m-2t+1)+5(2m+t-1)^2-(2m+t-1))/2}$$
$$= q^4 \sum_{m,t=-\infty}^{\infty} (-1)^{m+t} q^{(25m^2-5m+25t^2-25t)/2}$$
$$= 0. \tag{8.2.6}$$

(8.1.1) now follows from (8.2.1)–(8.2.6).

8.3 Our Second Proof of Ramanujan's 5-Dissection of Euler's Product

Our second proof of (8.1.1) is by a factorisation. That is, we will prove a pair of identities, the product of which is the desired identity.

We will prove the following two identities.

$$\prod_{k\geq 1}(1+\alpha q^k+q^{2k})(1-q^k) = (q^{10},q^{15},q^{25};q^{25})_\infty - \beta q(q^5,q^{20},q^{25};q^{25})_\infty \tag{8.3.1}$$

and

$$\prod_{k\geq 1}(1+\beta q^k+q^{2k})(1-q^k) = (q^{10},q^{15},q^{25};q^{25})_\infty - \alpha q(q^5,q^{20},q^{25};q^{25})_\infty \tag{8.3.2}$$

where $\alpha = \dfrac{1+\sqrt{5}}{2}$ and $\beta = \dfrac{1-\sqrt{5}}{2}$.

Warning: This notation is standard for Fibonacci aficionados. It is different from Ramanujan's and others'.

If we multiply (8.3.1) by (8.3.2), we find

$$(q;q)_\infty(q^5;q^5)_\infty = (q^{10},q^{15},q^{25};q^{25})_\infty^2 - q(q^5;q^5)_\infty(q^{25};q^{25})_\infty \\ - q^2(q^5,q^{20},q^{25};q^{25})_\infty^2, \tag{8.3.3}$$

and if we now divide by $(q^5;q^5)_\infty$, we obtain (8.1.1)

It is easy to establish (8.3.1) and (8.3.2).
We have (1.7.5),

$$(a^{-1}q,aq,q;q)_\infty = \sum_{k\geq 0}(-1)^k\left(a^k+a^{k-1}+\cdots+a^{-k}\right)q^{(k^2+k)/2}. \tag{8.3.4}$$

If we set $a=\eta=e^{\frac{2\pi i}{5}}$ and use the fact that

$$\eta^k+\eta^{k-1}+\cdots+\eta^{-k} = \begin{cases} 1 & \text{if } k\equiv 0 \pmod 5,\\ \alpha & \text{if } k\equiv 1 \pmod 5,\\ 0 & \text{if } k\equiv 2 \pmod 5,\\ -\alpha & \text{if } k\equiv 3 \pmod 5,\\ -1 & \text{if } k\equiv 4 \pmod 5,\end{cases} \tag{8.3.5}$$

(8.3.4) becomes

$$(\eta^{-1}q,\eta q,q;q)_\infty = (q^{10},q^{15},q^{25};q^{25})_\infty - \alpha q(q^5,q^{20},q^{25};q^{25})_\infty. \tag{8.3.6}$$

If we set $a=\eta^2$ in (8.3.4), we obtain

$$(\eta^{-2}q,\eta^2 q,q;q)_\infty = (q^{10},q^{15},q^{25};q^{25})_\infty - \beta q(q^5,q^{20},q^{25};q^{25})_\infty. \tag{8.3.7}$$

If we now use the facts that

$$\eta+\eta^{-1}=-\beta,\quad \eta^2+\eta^{-2}=-\alpha, \tag{8.3.8}$$

we see that (8.3.6) and (8.3.7) are (8.3.2) and (8.3.1), respectively.

8.4 Review of Earlier Results

We see that (6.2.4) can be written

$$R(q^5)^{-1}-q-q^2R(q^5)=\frac{E(q)}{E(q^{25})}, \tag{8.4.1}$$

while (6.2.7) can be written

$$R(q^5)^{-5} - 11q^5 - q^{10}R(q^5)^5 = \frac{E(q^5)^6}{E(q^{25})^6}, \tag{8.4.2}$$

or, if we replace q by $q^{\frac{1}{5}}$,

$$R(q)^{-5} - 11q - q^2R(q)^5 = \frac{E(q)^6}{E(q^5)^6}. \tag{8.4.3}$$

In §8.5, we will give a factorisation of (8.4.3).

Exercise: Show that in view of (8.1.3), (6.7.4) becomes

$$\frac{1}{E(q)} = \frac{E(q^{25})^5}{E(q^5)^6}(R(q^5)^{-4} + qR(q^5)^{-3} + 2q^2R(q^5)^{-2} + 3q^3R(q^5)^{-1} + 5q^4 \\ - 3q^5R(q^5) + 2q^6R(q^5)^2 - q^7R(q^5)^3 + q^8R(q^5)^4) \tag{8.4.4}$$

while (6.7.5)–(6.7.8) become

$$\sum_{n\geq 0} p(5n)q^n = \frac{E(q^5)^5}{E(q)^6}\left(R(q)^{-4} - 3qR(q)\right), \tag{8.4.5}$$

$$\sum_{n\geq 0} p(5n+1)q^n = \frac{E(q^5)^5}{E(q)^6}\left(R(q)^{-3} + 2qR(q)^2\right), \tag{8.4.6}$$

$$\sum_{n\geq 0} p(5n+2)q^n = \frac{E(q^5)^5}{E(q)^6}\left(2R(q)^{-2} - qR(q)^3\right), \tag{8.4.7}$$

$$\sum_{n\geq 0} p(5n+3)q^n = \frac{E(q^5)^5}{E(q)^6}\left(3R(q)^{-1} + qR(q)^4\right). \tag{8.4.8}$$

8.5 A Factorisation of Identity (8.4.3)

If we set

$$A = (q^{10}, q^{15}, q^{25}; q^{25})_\infty, \quad B = (q^5, q^{20}, q^{25}; q^{25})_\infty,$$

then (8.3.1) and (8.3.2) become

$$A - \beta qB = E(q)\prod_{k\geq 1}(1 + \alpha q^k + q^{2k}) \tag{8.5.1}$$

and

$$A - \alpha q B = E(q) \prod_{k \geq 1} (1 + \beta q^k + q^{2k}). \tag{8.5.2}$$

Suppose η is a fifth root of unity other than 1. Substitute q, ηq, $\eta^2 q$, $\eta^3 q$ and $\eta^4 q$ for q in (8.5.2), and multiply the five results, and we obtain

$$\begin{aligned} A^5 - \alpha^5 q^5 B^5 &= E(q)E(\eta q) \cdots E(\eta^4 q) \\ &\quad \times \prod_{k \geq 1} (1 + \beta q^k + q^{2k})(1 + \beta \eta^k q^k + \eta^{2k} q^{2k}) \cdots (1 + \beta \eta^{4k} q^k + \eta^{8k} q^{2k}) \\ &= \frac{E(q^5)^6}{E(q^{25})} \prod_{k \equiv 0 \pmod 5} (1 + \beta q^k + q^{2k})^5 \prod_{k \not\equiv 0 \pmod 5} (1 - q^{5k})^2 \\ &= \frac{E(q^5)^8}{E(q^{25})^3} \prod_{k \geq 1} (1 + \beta q^{5k} + q^{10k})^5. \end{aligned} \tag{8.5.3}$$

If in (8.5.3) we replace q by $q^{\frac{1}{5}}$, we obtain

$$(q^2, q^3, q^5; q^5)_\infty^5 - \alpha^5 q (q, q^4, q^5; q^5)_\infty^5 = \frac{E(q)^8}{E(q^5)^3} \prod_{k \geq 1} (1 + \beta q^k + q^{2k})^5. \tag{8.5.4}$$

Similarly, from (8.5.1) we obtain

$$(q^2, q^3, q^5; q^5)_\infty^5 - \beta^5 q (q, q^4, q^5; q^5)_\infty^5 = \frac{E(q)^8}{E(q^5)^3} \prod_{k \geq 1} (1 + \alpha q^k + q^{2k})^5. \tag{8.5.5}$$

If we multiply (8.5.4) by (8.5.5), we find

$$\begin{aligned} &(q^2, q^3, q^5; q^5)_\infty^{10} - 11 q (q, q^2, q^3, q^4, q^5, q^5; q^5)_\infty^5 - q^2 (q, q^4, q^5; q^5)_\infty^{10} \\ &= \frac{E(q)^{16}}{E(q^5)^6} \prod_{k \geq 1} (1 + \alpha q^k + q^{2k})^5 (1 + \beta q^k + q^{2k})^5 \\ &= \frac{E(q)^{16}}{E(q^5)^6} \prod_{k \geq 1} (1 + q^k + q^{2k} + q^{3k} + q^{4k})^5 \\ &= \frac{E(q)^{11}}{E(q^5)}. \end{aligned} \tag{8.5.6}$$

If we divide (8.5.6) by $(q;q)_\infty^5(q^5;q^5)_\infty^5 = E(q)^5E(q^5)^5$, we obtain

$$\left(\begin{matrix} q^2,q^3 \\ q,q^4 \end{matrix};q^5\right)_\infty^5 - 11q - q^2\left(\begin{matrix} q,q^4 \\ q^2,q^3 \end{matrix};q^5\right)_\infty^5 = \frac{E(q)^6}{E(q^5)^6},$$

which is (8.4.3). So the pair (8.5.4) and (8.5.5) constitute a factorisation of (8.4.3).

8.6 Ramanujan's Versions of (8.3.1) and (8.3.2), (8.5.4) and (8.5.5)

We start by noting that (8.3.1) and (8.3.2) can be written as follows.

$$(q^{10},q^{15},q^{25};q^{25})_\infty - \alpha(q^5,q^{20},q^{25};q^{25})_\infty = E(q^5)\Big/\prod_{k\geq 1}(1+\alpha q^k+q^{2k}) \quad (8.6.1)$$

and

$$(q^{10},q^{15},q^{25};q^{25})_\infty - \beta(q^5,q^{20},q^{25};q^{25})_\infty = E(q^5)\Big/\prod_{k\geq 1}(1+\beta q^k+q^{2k}). \quad (8.6.2)$$

If we divide (8.6.1) and (8.6.2) by $\sqrt{qE(q^5)E(q^{25})}$, we obtain

$$\sqrt{\frac{1}{q}\left(\begin{matrix} q^{10},q^{15} \\ q^5,q^{20} \end{matrix};q^{25}\right)_\infty} - \alpha\sqrt{q\left(\begin{matrix} q^5,q^{20} \\ q^{10},q^{15} \end{matrix};q^{25}\right)_\infty}$$
$$= \frac{1}{\sqrt{q}}\sqrt{\frac{E(q^5)}{E(q^{25})}}\Big/\prod_{k\geq 1}(1+\alpha q^k+q^{2k}) \quad (8.6.3)$$

and

$$\sqrt{\frac{1}{q}\left(\begin{matrix} q^{10},q^{15} \\ q^5,q^{20} \end{matrix};q^{25}\right)_\infty} - \beta\sqrt{q\left(\begin{matrix} q^5,q^{20} \\ q^{10},q^{15} \end{matrix};q^{25}\right)_\infty}$$
$$= \frac{1}{\sqrt{q}}\sqrt{\frac{E(q^5)}{E(q^{25})}}\Big/\prod_{k\geq 1}(1+\beta q^k+q^{2k}). \quad (8.6.4)$$

If we now replace q by $q^{\frac{1}{5}}$ and define

$$r(q) = q^{\frac{1}{5}}R(q), \quad (8.6.5)$$

(8.6.3) and (8.6.4) become

$$\frac{1}{\sqrt{r}} - \alpha\sqrt{r} = \frac{1}{q^{\frac{1}{10}}}\sqrt{\frac{E(q)}{E(q^5)}} \Big/ \prod_{k\geq 1}(1+\alpha q^{\frac{k}{5}} + q^{\frac{2k}{5}}) \tag{8.6.6}$$

and

$$\frac{1}{\sqrt{r}} - \beta\sqrt{r} = \frac{1}{q^{\frac{1}{10}}}\sqrt{\frac{E(q)}{E(q^5)}} \Big/ \prod_{k\geq 1}(1+\beta q^{\frac{k}{5}} + q^{\frac{2k}{5}}). \tag{8.6.7}$$

In the same way, (8.5.4) and (8.5.5) become

$$\left(\frac{1}{\sqrt{r}}\right)^5 - \left(\alpha\sqrt{r}\right)^5 = \frac{1}{\sqrt{q}}\sqrt{\frac{E(q)}{E(q^5)}} \Big/ \prod_{k\geq 1}(1+\alpha q^k + q^{2k})^5 \tag{8.6.8}$$

and

$$\left(\frac{1}{\sqrt{r}}\right)^5 - \left(\beta\sqrt{r}\right)^5 = \frac{1}{\sqrt{q}}\sqrt{\frac{E(q)}{E(q^5)}} \Big/ \prod_{k\geq 1}(1+\beta q^k + q^{2k})^5. \tag{8.6.9}$$

Endnotes.

§8.2 This proof is based on that in Hirschhorn (2000) [72].

§8.3 This proof appears in Hirschhorn (2011) [84].

§8.6 The identities (8.6.6)–(8.6.9) are given by Ramanujan in the LNB [112], p. 206, and are described by Andrews and Berndt [6], p. 24, as "remarkable".

References

6. G.E. Andrews, B.C. Berndt, *Ramanujan's Lost Notebook, Part I* (Springer, Berlin, 2005)
72. M.D. Hirschhorn, *An Identity of Ramanujan, and Applications, in q-Series from a Contemporary Perspective*, vol. 254, Contemporary Mathematics (AMS, Providence, 2000), pp. 229–234
84. M.D. Hirschhorn, Ramanujan's "most beautiful identity". Am. Math. Mon. **118**(9), 839–845 (2011)
112. S. Ramanujan, *The Lost Notebook and Other Unpublished Papers* (Springer, Berlin, 1988)

Chapter 9
A "Difficult and Deep" Identity of Ramanujan

9.1 Introduction

In this chapter, we make a start on obtaining an identity that Ramanujan included in his first letter to Hardy, which Hardy described as "both difficult and deep", and which helped to convince Hardy that Ramanujan was "a mathematician of the highest class". (Hardy (1940) [58], p. 9). We will complete the derivation of the identity in Chapter 15.

9.2 Proof Inspired by Chadwick Gugg

We have (8.5.2) and (8.5.1) ,

$$A-\alpha qB=E(q)\prod_{k\geq 1}(1+\beta q^k+q^{2k}) \tag{9.2.1}$$

and

$$A-\beta qB=E(q)\prod_{k\geq 1}(1+\alpha q^k+q^{2k}). \tag{9.2.2}$$

Now suppose $\eta=e^{\frac{2\pi i}{5}}$. Then

$$\eta+\eta^{-1}=-\beta,\quad \eta^2+\eta^{-2}=-\alpha. \tag{9.2.3}$$

Consider the quantity

$$Q=\frac{(A-\alpha\eta qB)(A-\alpha\eta^{-1}qB)(A-\beta\eta^2 qB)(A-\beta\eta^{-2}qB)}{(A-\alpha\eta^2 qB)(A-\alpha^2\eta^{-2}qB)(A-\beta\eta qB)(A-\beta\eta^{-1}qB)}. \tag{9.2.4}$$

M.D. Hirschhorn, *The Power of q*, Developments in Mathematics 49,
DOI 10.1007/978-3-319-57762-3_9

First, Q is an infinite product,

$$Q = \prod_{k\geq 1} t_k, \tag{9.2.5}$$

where

$$t_k = \frac{(1+\alpha\eta^{2k}q^k+\eta^{4k}q^{2k})(1+\alpha\eta^{-2k}q^k+\eta^{-4k}q^{2k})}{(1+\alpha\eta^{k}q^k+\eta^{2k}q^{2k})(1+\alpha\eta^{-k}q^k+\eta^{-2k}q^{2k})} \times \frac{(1+\beta\eta^{k}q^k+\eta^{2k}q^{2k})(1+\beta\eta^{-k}q^k+\eta^{-2k}q^{2k})}{(1+\beta\eta^{2k}q^k+\eta^{4k}q^{2k})(1+\beta\eta^{-2k}q^k+\eta^{-4k}q^{2k})}. \tag{9.2.6}$$

Now, if $k \equiv 0 \pmod 5$,

$$t_k = 1. \tag{9.2.7}$$

If $k \equiv 1$ or $4 \pmod 5$,

$$\begin{aligned} t_k &= \frac{(1+\alpha\eta^{2}q^k+\eta^{4}q^{2k})(1+\alpha\eta^{-2}q^k+\eta^{-4}q^{2k})}{(1+\alpha\eta q^k+\eta^{2}q^{2k})(1+\alpha\eta^{-1}q^k+\eta^{-2}q^{2k})} \times \frac{(1+\beta\eta q^k+\eta^{2}q^{2k})(1+\beta\eta^{-1}q^k+\eta^{-2}q^{2k})}{(1+\beta\eta^{2}q^k+\eta^{4}q^{2k})(1+\beta\eta^{-2}q^k+\eta^{-4}q^{2k})} \\ &= \frac{1-(1+\alpha)q^k+2\alpha q^{2k}-(1+\alpha)q^{3k}+q^{4k}}{1+q^k+q^{2k}+q^{3k}+q^{4k}} \times \frac{1-(1+\beta)q^k+2\beta q^{2k}-(1+\beta)q^{3k}+q^{4k}}{1+q^k+q^{2k}+q^{3k}+q^{4k}} \\ &= \frac{(1-q^k)^2(1+\beta q^k+q^{2k})(1-q^k)^2(1+\alpha q^k+q^{2k})}{(1+q^k+q^{2k}+q^{3k}+q^{4k})^2} \\ &= \frac{(1-q^k)^4(1+q^k+q^{2k}+q^{3k}+q^{4k})}{(1+q^k+q^{2k}+q^{3k}+q^{4k})^2} \\ &= \frac{(1-q^k)^5}{1-q^{5k}}. \end{aligned} \tag{9.2.8}$$

If $k \equiv 2$ or $3 \pmod 5$,

$$t_k = \frac{1-q^{5k}}{(1-q^k)^5}. \tag{9.2.9}$$

It follows from (9.2.5)–(9.2.9) that

$$Q = \left(\begin{matrix} q, q^4 \\ q^2, q^3 \end{matrix}; q^5\right)^5 \Big/ \left(\begin{matrix} q^5, q^{20} \\ q^{10}, q^{15} \end{matrix}; q^{25}\right)_\infty = R(q)^5/R(q^5) \tag{9.2.10}$$

where $R(q)$ is defined in (8.1.2).

On the other hand,

$$Q = \frac{(A-\alpha\eta qB)(A-\alpha\eta^{-1}qB)(A-\beta\eta^2 qB)(A-\beta\eta^{-2}qB)}{(A-\alpha\eta^2 qB)(A-\alpha\eta^{-2}qB)(A-\beta\eta qB)(A-\beta\eta^{-1}qB)}$$

$$= \frac{(A^2-qAB+\alpha^2q^2B^2)(A^2-qAB+\beta^2q^2B^2)}{(A^2+\alpha^2qAB+\alpha^2q^2B^2)(A^2+\beta^2qAB+\beta^2q^2B^2)}$$

$$= \frac{A^4-2qA^3B+4q^2A^2B^2-3q^3AB^3+q^4B^4}{A^4+3qA^3B+4q^2A^2B^2+2q^3AB^3+q^4B^4}. \tag{9.2.11}$$

If we divide (9.2.11) top and bottom by A^4, and make use of the fact that

$$\frac{B}{A} = R(q^5), \tag{9.2.12}$$

we find that (9.2.11) can be written

$$R(q)^5/R(q^5) = \frac{1-2qR(q^5)+4q^2R(q^5)^2-3q^3R(q^5)^3+R(q^5)^4}{1+3qR(q^5)+4q^2R(q^5)^2+2q^3R(q^5)^3+R(q^5)^4}. \tag{9.2.13}$$

If we now write

$$u = qR(q^5), \quad v = q^{\frac{1}{5}}R(q), \tag{9.2.14}$$

then (9.2.13) becomes

$$v^5 = u \cdot \frac{1-2u+4u^2-3u^3+u^4}{1+3u+4u^2+2u^3+u^4}, \tag{9.2.15}$$

which is, as we shall see in Chapter 15, essentially the “difficult and deep” identity of Ramanujan.

9.3 Proof Inspired by G.N. Watson

We can write (8.4.1), (8.4.2) and (8.4.3) as follows.

$$u^{-1}-1-u=\frac{E(q)}{qE(q^{25})}, \tag{9.3.1}$$

$$u^{-5}-11-u^{5}=\frac{E(q^5)^6}{q^5E(q^{25})^6}, \tag{9.3.2}$$

and

$$v^{-5}-11-v^{5}=\frac{E(q)^6}{qE(q^5)^6}. \tag{9.3.3}$$

It follows that

$$v^{-5}-11-v^{5}=\frac{(u^{-1}-1-u)^6}{u^{-5}-11-u^5}. \tag{9.3.4}$$

This is a quadratic for v^5, which we solve as follows.

We have

$$v^{-5}-11-v^{5}=\frac{1}{u^{-5}-11-u^5}(u^{-6}-6u^{-5}+9u^{-4}+10u^{-3}-30u^{-2}-6u^{-1} \\ +41+6u-30u^2-10u^3+9u^4+6u^5+u^6) \tag{9.3.5}$$

If we add 11, we obtain

$$v^{-5}-v^{5}=\frac{1}{u^{-5}-11-u^5}(u^{-6}+5u^{-5}+9u^{-4}+10u^{-3}-30u^{-2}-6u^{-1} \\ -80+6u-30u^2-10u^3+9u^4-5u^5+u^6) \tag{9.3.6}$$

If we square (9.3.6), we obtain

$$v^{-10}-2+v^{10}=\frac{1}{(u^{-5}-11-u^5)^2}(u^{-12}+10u^{-11}+43u^{-10}+110u^{-9}+121u^{-8} \\ -132u^{-7}-660u^{-6}-1496u^{-5}-660u^{-4}-1452u^{-3}+4334u^{-2} \\ -100u^{-1}+8042+100u+4334u^2+1452u^3-660u^4+1496u^5 \\ -660u^6+132u^7+121u^8-110u^9+43u^{10}-10u^{11}+u^{12}). \tag{9.3.7}$$

If we add 4, we obtain, somewhat miraculously,

$$\begin{aligned} v^{-10}+2+v^{10} &= \frac{1}{(u^{-5}-11-u^5)^2}\Big(u^{-12}+10u^{-11}+47u^{-10}+110u^{-9}+121u^{-8} \\ &-132u^{-7}-660u^{-6}-1584u^{-5}-660u^{-4}-1452u^{-3}+4334u^{-2} \\ &-100u^{-1}+8518+100u+4334u^2+1452u^3-660u^4+1584u^5 \\ &-660u^6+132u^7+121u^8-110u^9+47u^{10}-10u^{11}+u^{12}\Big) \\ &= \left(\frac{u^{-6}+5u^{-5}+11u^{-4}-66u^{-1}-66u-11u^4+5u^5-u^6}{u^{-5}-11-u^5}\right)^2. \end{aligned} \tag{9.3.8}$$

It follows that

$$v^{-5}+v^5 = \frac{u^{-6}+5u^{-5}+11u^{-4}-66u^{-1}-66u-11u^4+5u^5-u^6}{u^{-5}-11-u^5}. \tag{9.3.9}$$

If we subtract (9.3.6) from (9.3.9) and divide by 2, we obtain

$$v^5 = u\,\frac{u^{-5}-5u^{-4}+15u^{-3}-30u^{-2}+40u^{-1}-36+15u+5u^2-10u^3+5u^4-u^5}{u^{-5}-11-u^5}. \tag{9.3.10}$$

The numerator in (9.3.10) factorises as

$$\left(u^{-1}-1-u\right)\left(u^{-2}-2u^{-1}+4-3u+u^2\right)^2 \tag{9.3.11}$$

while the denominator factorises as

$$\left(u^{-1}-1-u\right)\left(u^{-2}-2u^{-1}+4-3u+u^2\right)\left(u^{-2}+3u^{-1}+4+2u+u^2\right). \tag{9.3.12}$$

So (9.3.10) becomes

$$v^5 = u\,\frac{u^{-2}-2u^{-1}+4-3u+u^2}{u^{-2}+3u^{-1}+4+2u+u^2},$$

which is (9.2.15).

Endnote.

The first proof was inspired by C. Gugg (2009) [55]. See also H.-C. Chan (2010) [25].
The second proof was inspired by that of G.N. Watson (1929) [122].

References

25. H.-C. Chan, On the Andrews-Schur's proof of the Rogers-Ramanujan identities. Ramanujan J. **23**, 417–431 (2010)
55. C. Gugg, A new proof of Ramanujan's modular equation relating $R(q)$ with $R(q^5)$. Ramanujan J. **20**, 163–177 (2009)
58. G.H. Hardy, *Ramanujan. Twelve Lectures on Subjects Suggested by His Life and Work* (AMS Chelsea, Providence, 1999)
90. M.D. Hirschhorn, An identity that may have changed the course of history, *Ramanujan 125*, vol. 627, Contemporary Mathematics (AMS, Providence, 2014), pp. 103–109
122. G.N. Watson, Theorems stated by Ramanujan (VII): theorems on continued fractions. J. Lond. Math. Soc. **4**, 39–48 (1929)

Chapter 10
The Quintuple Product Identity

10.1 Introduction

As Shaun Cooper tells us in his monumental paper, Cooper (2006) [33], the quintuple product identity has been discovered at least several times over about 100 years, and has been given at least 29 proofs in the literature!

In order to write the quintuple product identity in a concise and useful way, we need to introduce some more new notation. We saw in §1.2 and §3.6 how $(a;q)_\infty$ and $(a_1, \dots, a_n;q)_\infty$ are defined. We now define, for m, $n \geq 0$,

$$\left(\begin{matrix} a_1, \dots, a_m \\ b_1, \dots, b_n \end{matrix}; q\right)_\infty = \frac{(a_1, \dots, a_m;q)_\infty}{(b_1, \dots, b_n;q)_\infty}. \tag{10.1.1}$$

The quintuple product identity can be written

$$(a^{-1}, aq, q;q)_\infty (a^{-2}q, a^2q;q^2)_\infty = \sum_{-\infty}^{\infty} a^{3n} q^{(3n^2+n)/2} - \sum_{-\infty}^{\infty} a^{3n-1} q^{(3n^2-n)/2} \tag{10.1.2}$$

or, if we replace a by $-a$,

$$\left(\begin{matrix} a^{-2}, a^2q, q \\ a^{-1}, aq \end{matrix}; q\right)_\infty = \sum_{-\infty}^{\infty} (-1)^n a^{3n} q^{(3n^2+n)/2} + \sum_{-\infty}^{\infty} (-1)^n a^{3n-1} q^{(3n^2-n)/2}, \tag{10.1.3}$$

$$= (a^{-3}q, a^3q^2, q^3;q^3)_\infty + a^{-1}(a^{-3}q^2, a^3q, q^3;q^3)_\infty. \tag{10.1.4}$$

or, as Cooper details, in about a dozen other ways, some quite different in appearance from these two.

The original version of this chapter has been revised. The erratum to this chapter is available at https://doi.org/10.1007/978-3-319-57762-3_44.

M.D. Hirschhorn, *The Power of q*, Developments in Mathematics 49,
DOI 10.1007/978-3-319-57762-3_10

The quintuple product identity can be read, and used, in two quite different ways. It gives us the series expansion of the left side, but it also shows how in some circumstances two triple products can be summed to a single product.

In §10.2, I will present as direct a proof as I can, while in §10.3, I will present a generalisation, which is essentially no harder to obtain. Both proofs are Carlitzian.

In §10.4 we show how the quintuple product identity may be used to obtain Ramanujan's 5-dissection of Euler's product .

In §§10.5 and 10.6 we find the 7- and 11-dissections of Euler's product.

In §10.7, we obtain from the quintuple product identity two identities known to Ramanujan.

In §§10.8 and 10.9 we obtain further interesting results.

10.2 Proof of the Quintuple Product Identity

We have, by JTP,

$$\begin{aligned}\left(\begin{matrix}a^{-2},a^2q,q\\ a^{-1},aq\end{matrix};q\right)_\infty &= \frac{(-a^{-1},-aq,q;q)_\infty(a^{-2}q,a^2q,q^2;q^2)_\infty}{(q^2;q^2)_\infty}\\ &= \frac{1}{(q^2;q^2)_\infty}\sum_{k,l=-\infty}^{\infty}(-1)^l a^{k+2l}q^{(k^2+k)/2+l^2} \qquad (10.2.1)\end{aligned}$$

We now split this sum into three, according to the residue of $k+2l$ modulo 3.

In the case $k+2l\equiv 0 \pmod 3$, let $k+2l=3m$, and define $t=m-l$. Then $k=m+2t$, $l=m-t$, and the contribution of these terms to the sum on the right of (10.2.1) is

$$\begin{aligned}&\sum_{m,t=-\infty}^{\infty}(-1)^{m-t}a^{3m}q^{((m+2t)^2+(m+2t))/2+(m-t)^2}\\ &\qquad = \sum_{m,t=-\infty}^{\infty}(-1)^{m+t}a^{3m}q^{(3m^2+m)/2+3t^2+t}\\ &\qquad = (q^2;q^2)_\infty(a^{-3}q,a^3q^2,q^3;q^3)_\infty. \qquad (10.2.2)\end{aligned}$$

In the case $k+2l\equiv -1 \pmod 3$, let $k+2l=3m-1$ and $t=m-l$. Then $k=m+2t-1$, $l=m-t$, and the contribution of these terms is

$$\sum_{m,t=-\infty}^{\infty} (-1)^{m-t} a^{3m-1} q^{((m+2t-1)^2+m+2t-1))/2+(m-t)^2}$$
$$= \sum_{m,t=-\infty}^{\infty} (-1)^{m+t} a^{3m-1} q^{(3m^2-m)/2+3t^2+t}$$
$$= a^{-1}(q^2;q^2)_\infty (a^{-3}q^2, a^3q, q^3;q^3)_\infty. \qquad (10.2.3)$$

Finally, in the case $k+2l \equiv 1 \pmod 3$, let $k+2l = 3m+1$ and $t = m-l$. Then $k = m+2t+1$, $l = m-t$, and the contribution of these terms is

$$\sum_{m,t=-\infty}^{\infty} (-1)^{m-t} a^{3m+1} q^{((m+2t+1)^2+(m+2t+1))/2+(m-t)^2}$$
$$= \sum_{m,t=-\infty}^{\infty} (-1)^{m+t} a^{3m+1} q^{(3m^2+3m)/2+3t^2+3t+1}$$
$$= 0. \qquad (10.2.4)$$

If we substitute (10.2.2), (10.2.3) and (10.2.4) into (10.2.1), we obtain the quintuple product identity, in the form (10.1.4).

10.3 A Generalisation of the Quintuple Product Identity

In almost exactly the same way that we established (10.1.3), we can prove that

$$\begin{aligned}(-a^{-1}q, -aq, q^2;q^2)_\infty(-b^{-1}q^2, -bq^2, q^4;q^4)_\infty \\ = (-a^{-1}b^{-1}q^3, -abq^3, q^6;q^6)_\infty(-a^{-2}bq^6, -a^2b^{-1}q^6, q^{12};q^{12})_\infty \\ + aq(-a^{-1}b^{-1}q, -abq^5, q^6;q^6)_\infty(-a^{-2}bq^2, -a^2b^{-1}q^{10}, q^{12};q^{12})_\infty \\ + a^{-1}q(-a^{-1}b^{-1}q^5, -abq, q^6;q^6)_\infty(-a^{-2}bq^{10}, -a^2b^{-1}q^2, q^{12};q^{12})_\infty.\end{aligned} \qquad (10.3.1)$$

If in (10.3.1) we replace a by aq, b by $-a^2$, divide by $(q^4;q^4)_\infty$ then replace q by $q^{\frac{1}{2}}$, we retrieve (10.1.4).

Alternatively, if in (10.3.1) we replace a by aq, b by $-a^{-1}$, divide by $(q^2;q^2)_\infty$, then replace q by $q^{\frac{1}{4}}$, we retrieve (10.1.4).

Our proof of (10.3.1) goes as follows.

The left side of (10.3.1) is, by JTP,

$$\sum_{k,l=-\infty}^{\infty} a^k b^l q^{k^2+2l^2}. \qquad (10.3.2)$$

We split this sum into three, according to the residue modulo 3 of $k+2l$, and make the same substitutions we made in §10.2.

10.4 Ramanujan's 5-Dissection of Euler's Product Again

We have (1.6.1),

$$E(q)=\sum_{-\infty}^{\infty}(-1)^k q^{(3k^2-k)/2}. \tag{10.4.1}$$

We consider the terms corresponding to $k\equiv 0,\ 1,\ 2,\ 3,\ 4 \pmod 5$, by setting $k=5l,\ 5l+1, 5l+2,\ 5l-1$ and $5l-2$, and we obtain

$$\begin{aligned}
E(q)&=\sum_{-\infty}^{\infty}(-1)^l q^{(3(5l)^2-(5l))/2}-\sum_{-\infty}^{\infty}(-1)^l q^{(3(5l+1)^2-(5l+1))/2}\\
&\quad+\sum_{-\infty}^{\infty}(-1)^l q^{(3(5l+2)^2-(5l+2))/2}-\sum_{-\infty}^{\infty}(-1)^l q^{(3(5l-1)^2-(5l-1))/2}\\
&\quad+\sum_{-\infty}^{\infty}(-1)^l q^{(3(5l-2)^2-(5l-2))/2}\\
&=\left(\sum_{-\infty}^{\infty}(-1)^l q^{(75l^2-5l)/2}+q^5\sum_{-\infty}^{\infty}(-1)^l q^{(75l^2+55l)/2}\right)\\
&\quad-q\sum_{-\infty}^{\infty}(-1)^l q^{(75l^2+25l)/2}\\
&\quad-q^2\left(\sum_{-\infty}^{\infty}(-1)^l q^{(75l^2-35l)/2}-q^5\sum_{-\infty}^{\infty}(-1)^l q^{(75l^2-65l)/2}\right)\\
&=\left(\begin{matrix}q^{10},q^{15},q^{25}\\ q^5,q^{20}\end{matrix};q^{25}\right)_\infty-q(q^{25};q^{25})_\infty-q^2\left(\begin{matrix}q^5,q^{20},q^{25}\\ q^{10},q^{15}\end{matrix};q^{25}\right)_\infty\\
&=E(q^{25})(R(q^5)^{-1}-q-q^2R(q^5)),
\end{aligned}$$

as desired, where we have used the quintuple product (10.1.4) with q^{25} for q and $a=q^{-5}$ and $a=q^{-10}$, respectively to sum the pairs of triple products, as described in §10.1.

10.5 The 7-Dissection of Euler's Product

Exercise: Show, using the quintuple product identity, that the 7-dissection of $E(q)$ is

$$E(q) = E(q^{49})\left(\left(\begin{matrix} q^{14}, q^{35} \\ q^{7}, q^{42} \end{matrix}; q^{49}\right)_\infty - q\left(\begin{matrix} q^{21}, q^{28} \\ q^{14}, q^{35} \end{matrix}; q^{49}\right)_\infty \right.$$
$$\left. -q^2 + q^5\left(\begin{matrix} q^{7}, q^{42} \\ q^{21}, q^{28} \end{matrix}; q^{49}\right)_\infty \right). \tag{10.5.1}$$

Thus the α, β and γ of Chapter 7 are given by

$$\alpha = q^{-2}\left(\begin{matrix} q^{14}, q^{35} \\ q^{7}, q^{42} \end{matrix}; q^{49}\right)_\infty, \quad \beta = -q^{-1}\left(\begin{matrix} q^{21}, q^{28} \\ q^{14}, q^{35} \end{matrix}; q^{49}\right)_\infty, \quad \gamma = q^{3}\left(\begin{matrix} q^{7}, q^{42} \\ q^{21}, q^{28} \end{matrix}; q^{49}\right). \tag{10.5.2}$$

and it is immediate that $\alpha\beta\gamma = -1$.

10.6 The 11-Dissection of Euler's Product

Exercise: Show that the 11-dissection of $E(q)$ is

$$E(q) = E(q^{121})\left(\left(\begin{matrix} q^{44}, q^{77} \\ q^{22}, q^{99} \end{matrix}; q^{121}\right)_\infty - q\left(\begin{matrix} q^{22}, q^{99} \\ q^{11}, q^{110} \end{matrix}; q^{121}\right)_\infty - q^2\left(\begin{matrix} q^{55}, q^{66} \\ q^{33}, q^{88} \end{matrix}; q^{121}\right)_\infty \right.$$
$$\left. +q^5 + q^7\left(\begin{matrix} q^{33}, q^{88} \\ q^{44}, q^{77} \end{matrix}; q^{121}\right)_\infty - q^{15}\left(\begin{matrix} q^{11}, q^{110} \\ q^{55}, q^{66} \end{matrix}; q^{121}\right)_\infty \right). \tag{10.6.1}$$

If we set

$$\alpha = q^{-5}\left(\begin{matrix} q^{44}, q^{77} \\ q^{22}, q^{99} \end{matrix}; q^{121}\right)_\infty, \quad \beta = -q^{-4}\left(\begin{matrix} q^{22}, q^{99} \\ q^{11}, q^{110} \end{matrix}; q^{121}\right)_\infty,$$
$$\gamma = -q^{-3}\left(\begin{matrix} q^{55}, q^{66} \\ q^{33}, q^{88} \end{matrix}; q^{121}\right)_\infty, \quad \delta = q^{2}\left(\begin{matrix} q^{33}, q^{88} \\ q^{44}, q^{77} \end{matrix}; q^{121}\right)_\infty,$$
$$\varepsilon = -q^{10}\left(\begin{matrix} q^{11}, q^{110} \\ q^{55}, q^{66} \end{matrix}; q^{121}\right)_\infty, \tag{10.6.2}$$

then (10.6.1) becomes

$$E(q) = q^5 E(q^{121})(\alpha + \beta + \gamma + 1 + \delta + \varepsilon), \tag{10.6.3}$$

from (10.6.2) we deduce

$$\alpha\beta\gamma\delta\varepsilon = -1 \tag{10.6.4}$$

and Jacobi's identity (1.7.1) yields

$$\begin{aligned}
&\alpha^2\gamma+\alpha\beta^2+2\gamma\delta\varepsilon+2\beta\delta+\varepsilon^2=0,\\
&\alpha^2\delta+2\beta\gamma\varepsilon+\delta^2\varepsilon+2\alpha\gamma+\beta^2=0,\\
&2\alpha\beta\delta+\alpha\varepsilon^2+\gamma^2\varepsilon+2\beta\gamma+\delta^2=0,\\
&2\alpha\beta\varepsilon+\beta\gamma^2+\gamma\delta^2+\alpha^2+2\delta\varepsilon=0,\\
&2\alpha\gamma\delta+\beta^2\delta+\beta\varepsilon^2+2\alpha\varepsilon+\gamma^2=0
\end{aligned}$$

(invariant under $\alpha\to\gamma\to\beta\to\delta\to\varepsilon\to\alpha$)
and

$$\alpha\gamma^2+\beta\delta^2+\gamma\beta^2+\delta\varepsilon^2+\varepsilon\alpha^2=-4. \tag{10.6.5}$$

10.7 Two Identities of Ramanujan

We can write (10.1.2)

$$\begin{aligned}
(1-a^{-1})(a^{-1}q,aq,q;q)_\infty(a^{-2}q,a^2q;q^2)_\infty &= \sum_{n\geq 0}(a^{3n}-a^{-3n-1})q^{(3n^2+n)/2}\\
&\quad -\sum_{n\geq 1}(a^{3n-1}-a^{-3n})q^{(3n^2-n)/2}.
\end{aligned} \tag{10.7.1}$$

If $a\neq 1$ and we divide (10.7.1) by $1-a^{-1}$, we obtain

$$\begin{aligned}
(a^{-1}q,aq,q;q)_\infty(a^{-2}q,a^2q;q^2)_\infty &= \sum_{n\geq 0}(a^{3n}+a^{3n-1}+\cdots+a^{-3n})q^{(3n^2+n)/2}\\
&\quad -\sum_{n\geq 1}(a^{3n-1}+a^{3n-2}+\cdots+a^{-3n+1})q^{(3n^2-n)/2}.
\end{aligned} \tag{10.7.2}$$

If in (10.7.2) we let $a\to 1$, we obtain

$$\begin{aligned}
\frac{(q;q)_\infty^5}{(q^2;q^2)_\infty^2} &= \sum_{n\geq 0}(6n+1)q^{(3n^2+n)/2}-\sum_{n\geq 1}(6n-1)q^{(3n^2-n)/2}\\
&= \sum_{-\infty}^{\infty}(6n+1)q^{(3n^2+n)/2},
\end{aligned} \tag{10.7.3}$$

the first of our two identities.

If in (10.1.2) we replace q by q^2, then a by $a^{-1}q^{-1}$, we obtain

$$(1-a^{-2})(a^{-1}q,aq,q^2;q^2)_\infty(a^{-2}q^4,a^2q^4;q^4)_\infty = \sum_{-\infty}^{\infty} a^{3n}q^{3n^2+2n} - \sum_{-\infty}^{\infty} a^{3n-2}q^{3n^2-2n}$$
$$= \sum_{n\geq 0}(a^{3n}-a^{-3n-2})q^{3n^2+2n} - \sum_{n\geq 1}(a^{3n-2}-a^{-3n})q^{3n^2-2n}. \qquad (10.7.4)$$

If $a^2 \neq 1$ and we divide (10.7.4) by $1-a^{-2}$, we obtain

$$(a^{-1}q,aq,q^2;q^2)_\infty(a^{-2}q^4,a^2q^4;q^4)_\infty = \sum_{n\geq 0}(a^{3n}+a^{3n-2}+\cdots+a^{-3n})q^{3n^2+2n}$$
$$- \sum_{n\geq 1}(a^{3n-2}+a^{3n-4}+\cdots+a^{-3n+2})q^{3n^2-2n}. \qquad (10.7.5)$$

If in (10.7.5) we let $a \to 1$, we obtain

$$\frac{(q;q)_\infty^2(q^4;q^4)_\infty^2}{(q^2;q^2)_\infty} = \sum_{n\geq 0}(3n+1)q^{3n^2+2n} - \sum_{n\geq 1}(3n-1)q^{3n^2-2n}$$
$$= \sum_{-\infty}^{\infty}(3n+1)q^{3n^2+2n}, \qquad (10.7.6)$$

the second of Ramanujan's two identities.

The identities (10.7.3) and (10.7.6) appear (of course without proof) in Ramanujan's Notebooks II [111], p. 209 and in Ramanujan (1916) [108] (see Berndt [14], pp. 114–115, Ramanujan [59], p. 147, Berndt "Spirit", [15], p. 25).

Exercise: Set $a=-1$ in (10.7.2), and retrieve (1.6.1).

Exercise: Let $a \to -1$ in (10.7.5) (alternatively, put $-q$ for q in (10.7.6)) and obtain

$$\frac{(q^2;q^2)_\infty^5}{(q;q)_\infty^2} = \sum_{-\infty}^{\infty}(-1)^n(3n+1)q^{3n^2+2n}. \qquad (10.7.7)$$

Compare this with (10.7.3)!

10.8 Continuation

From (10.7.3), we will deduce that

$$\frac{\phi(-q)^3}{\phi(-q^3)} = 1-6\sum_{n\geq 0}\left(\frac{q^{6n+1}}{1-q^{6n+1}}-\frac{q^{6n+5}}{1-q^{6n+5}}\right)+18\sum_{n\geq 0}\left(\frac{q^{6n+2}}{1-q^{6n+2}}-\frac{q^{6n+4}}{1-q^{6n+4}}\right) \qquad (10.8.1)$$

and from (10.7.7) we will deduce that

$$\frac{\psi(q)^3}{\psi(q^3)} = 1+3\sum_{n\geq 0}\left(\frac{q^{6n+1}}{1-q^{6n+1}} - \frac{q^{6n+5}}{1-q^{6n+5}}\right). \tag{10.8.2}$$

We have (10.7.3)

$$\begin{aligned}
\frac{(q;q)_\infty^5}{(q^2;q^2)_\infty^2} &= \sum_{-\infty}^{\infty}(6n+1)q^{(3n^2+n)/2} \\
&= \left[\frac{d}{da}\left(\sum_{-\infty}^{\infty} a^{6n+1}q^{(3n^2+n)/2}\right)\right]_{a=1} \\
&= \left[\frac{d}{da}\left(a(-a^{-6}q,-a^6q^2,q^3;q^3)_\infty\right)\right]_{a=1} \\
&= \Bigg[(-a^{-6}q,-a^6q^2,q^3;q^3)_\infty\Bigg(1-6\sum_{n\geq 0}\Bigg(\frac{a^{-6}q^{3n+1}}{1+a^{-6}q^{3n+1}} \\
&\qquad\qquad - \frac{a^6q^{3n+2}}{1+a^6q^{3n+2}}\Bigg)\Bigg)\Bigg]_{a=1} \\
&= (-q,-q^2,q^3;q^3)_\infty\left(1-6\sum_{n\geq 0}\left(\frac{q^{3n+1}}{1+q^{3n+1}} - \frac{q^{3n+2}}{1+q^{3n+2}}\right)\right) \\
&= \frac{(q^2;q^2)_\infty(q^3;q^3)_\infty^2}{(q;q)_\infty(q^6;q^6)_\infty}\left(1-6\sum_{n\geq 0}\left(\frac{q^{3n+1}}{1+q^{3n+1}} - \frac{q^{3n+2}}{1+q^{3n+2}}\right)\right).
\end{aligned} \tag{10.8.3}$$

It follows that

$$\frac{(q;q)_\infty^6(q^6;q^6)_\infty}{(q^2;q^2)_\infty^3(q^3;q^3)_\infty^2} = 1-6\sum_{n\geq 0}\left(\frac{q^{3n+1}}{1+q^{3n+1}} - \frac{q^{3n+2}}{1+q^{3n+2}}\right),$$

or, by (1.5.8),

$$\begin{aligned}
\frac{\phi(-q)^3}{\phi(-q^3)} &= 1-6\sum_{n\geq 0}\left(\frac{q^{3n+1}}{1+q^{3n+1}} - \frac{q^{3n+2}}{1+q^{3n+2}}\right) \\
&= 1-6\sum_{n\geq 0}\left(\frac{q^{3n+1}}{1-q^{3n+1}} - \frac{q^{3n+2}}{1-q^{3n+2}}\right) \\
&\quad + 6\sum_{n\geq 0}\left(\frac{q^{3n+1}}{1-q^{3n+1}} - \frac{q^{3n+1}}{1+q^{3n+1}}\right) - 6\sum_{n\geq 0}\left(\frac{q^{3n+2}}{1-q^{3n+2}} - \frac{q^{3n+2}}{1+q^{3n+2}}\right) \\
&= 1-6\sum_{n\geq 0}\left(\frac{q^{3n+1}}{1-q^{3n+1}} - \frac{q^{3n+2}}{1-q^{3n+2}}\right) + 12\sum_{n\geq 0}\left(\frac{q^{6n+2}}{1-q^{6n+2}} - \frac{q^{6n+4}}{1-q^{6n+4}}\right)
\end{aligned}$$

$$= 1 - 6\sum_{n\ge 0}\left(\frac{q^{6n+1}}{1-q^{6n+1}} + \frac{q^{6n+4}}{1-q^{6n+4}} - \frac{q^{6n+2}}{1-q^{6n+2}} - \frac{q^{6n+5}}{1-q^{6n+5}}\right)$$
$$+ 12\sum_{n\ge 0}\left(\frac{q^{6n+2}}{1-q^{6n+2}} - \frac{q^{6n+4}}{1-q^{6n+4}}\right)$$
$$= 1 - 6\sum_{n\ge 0}\left(\frac{q^{6n+1}}{1-q^{6n+1}} - \frac{q^{6n+5}}{1-q^{6n+5}}\right) + 18\sum_{n\ge 0}\left(\frac{q^{6n+2}}{1-q^{6n+2}} - \frac{q^{6n+4}}{1-q^{6n+4}}\right),$$

which is (10.8.1).

Also (10.7.7),

$$\frac{(q^2;q^2)_\infty^5}{(q;q)_\infty^2} = \sum_{-\infty}^{\infty}(-1)^n(3n+1)q^{3n^2+2n}$$
$$= \left[\frac{d}{da}\left(\sum_{-\infty}^{\infty}(-1)^n a^{3n+1}q^{3n^2+2n}\right)\right]_{a=1}$$
$$= \left[\frac{d}{da}\left(a(a^{-3}q, a^3q^5, q^6;q^6)_\infty\right)\right]_{a=1}$$
$$= \left[(a^{-3}q, a^3q^5, q^6;q^6)_\infty\left(1 + 3\sum_{n\ge 0}\left(\frac{a^{-3}q^{6n+1}}{1-a^{-3}q^{6n+1}} - \frac{a^3q^{6n+5}}{1-a^3q^{6n+5}}\right)\right)\right]_{a=1}$$
$$= (q, q^5, q^6;q^6)_\infty\left(1 + 3\sum_{n\ge 0}\left(\frac{q^{6n+1}}{1-q^{6n+1}} - \frac{q^{6n+5}}{1-q^{6n+5}}\right)\right)$$
$$= \frac{(q;q)_\infty(q^6;q^6)_\infty^2}{(q^2;q^2)_\infty(q^3;q^3)_\infty}\left(1 + 3\sum_{n\ge 0}\left(\frac{q^{6n+1}}{1-q^{6n+1}} - \frac{q^{6n+5}}{1-q^{6n+5}}\right)\right). \quad (10.8.4)$$

It follows that

$$\frac{(q^2;q^2)_\infty^6(q^3;q^3)_\infty}{(q;q)_\infty^3(q^6;q^6)_\infty^2} = 1 + 3\sum_{n\ge 0}\left(\frac{q^{6n+1}}{1-q^{6n+1}} - \frac{q^{6n+5}}{1-q^{6n+5}}\right),$$

or, by (1.5.7),

$$\frac{\psi(q)^3}{\psi(q^3)} = 1 + 3\sum_{n\ge 0}\left(\frac{q^{6n+1}}{1-q^{6n+1}} - \frac{q^{6n+5}}{1-q^{6n+5}}\right),$$

which is (10.8.2).

10.9 Some Interesting Results

From (10.8.1) and (10.8.2) we obtain the eye–catching result (the right side is even),

$$\frac{\phi(-q)^3}{\phi(-q^3)}+2\frac{\psi(q)^3}{\psi(q^3)}=3\left(1+6\sum_{n\geq 0}\left(\frac{q^{6n+2}}{1-q^{6n+2}}-\frac{q^{6n+4}}{1-q^{6n+4}}\right)\right). \tag{10.9.1}$$

We also obtain

$$4\frac{\psi(q)^3}{\psi(q^3)}-\frac{\phi(-q)^3}{\phi(-q^3)}=3\left(1+6\sum_{n\geq 0}\left(\frac{q^{3n+1}}{1-q^{3n+1}}-\frac{q^{3n+2}}{1-q^{3n+2}}\right)\right). \tag{10.9.2}$$

If we cast our eye ahead to Chapter 21, we will see that these become

$$\frac{\phi(-q)^3}{\phi(-q^3)}+2\frac{\psi(q)^3}{\psi(q^3)}=3a(q^2) \tag{10.9.3}$$

and

$$4\frac{\psi(q)^3}{\psi(q^3)}-\frac{\phi(-q)^3}{\phi(-q^3)}=3a(q), \tag{10.9.4}$$

where

$$a(q)=\sum_{m,n=-\infty}^{\infty}q^{m^2+mn+n^2}. \tag{10.9.5}$$

Endnotes.

§10.2 This proof is essentially that given in Hirschhorn (1998) [71].

§10.3 A slightly different proof was given in Hirschhorn (1988) [68].

References

14. B.C. Berndt, *Ramanujan's Notebooks, Part III* (Springer, Berlin, 1991)
15. B.C. Berndt, *Number Theory in the Spirit of Ramanujan*, vol. 34, Student Mathematical Library (AMS, Providence, 2006)
33. S. Cooper, The quintuple product identity. Int. J. Number Theory **2**, 115–161 (2006)
59. G.H. Hardy, P.V. Seshu Aiyar, B.M. Wilson, *Collected Papers of Srinivasa Ramanujan* (AMS Chelsea, Providence, 2000)
68. M.D. Hirschhorn, A generalisation of the quintuple product identity. J. Aust. Math. Soc. Ser. A **44**, 42–45 (1988)
71. M.D. Hirschhorn, On the expansion of Ramanujan's continued fraction. Ramanujan J. **2**, 521–527 (1998)
108. S. Ramanujan, On certain arithmetical functions. Proc. Camb. Philos. Soc. **XXII**, 159–184 (1916)

Chapter 11
Winquist's Identity

11.1 Introduction

This chapter is devoted to proving the identity known as Winquist's identity,

$$\begin{aligned}
&(a^{-1},aq,b^{-1},bq,a^{-1}b^{-1},abq,a^{-1}b,ab^{-1}q,q,q;q)_\infty \\
&= \sum_{k,l=-\infty}^{\infty} (-1)^{k+l}(a^{3k}b^{3l} - a^{3k}b^{-3l-1} - a^{3l-1}b^{3k+1} + a^{-3l-2}b^{3k+1}) \\
&\qquad\qquad \times q^{(3k^2+3k+3l^2+l)/2} \qquad (11.1.1)
\end{aligned}$$

$$\begin{aligned}
&= (a^{-3},a^3q^3,b^{-3}q,b^3q^2,q^3,q^3;q^3)_\infty - b^{-1}(a^{-3},a^3q^3,b^{-3}q^2,b^3q,q^3,q^3;q^3)_\infty \\
&-a^{-1}b(a^{-3}q,a^3q^2,b^{-3},b^3q^3,q^3,q^3;q^3)_\infty + a^{-2}b(a^{-3}q^2,a^3q,b^{-3},b^3q^3,q^3,q^3;q^3)_\infty. \qquad (11.1.2)
\end{aligned}$$

Note that there are many alternative ways of writing (11.1.1).

As you can see, (11.1.1) can be regarded as giving the expansion of the product on the left, but can also be used to sum the double series on the right to a product. We will use it in both directions.

We will employ (11.1.1) to give three quite different proofs of Ramanujan's partition congruence

$$p(11n+6) \equiv 0 \pmod{11} \qquad (11.1.3)$$

in the next two chapters.

In §11.2, we will prove (11.1.1) and (11.1.2).

M.D. Hirschhorn, *The Power of q*, Developments in Mathematics 49,
DOI 10.1007/978-3-319-57762-3_11

11.2 A Proof of Winquist's Identity

We have, by JTP,

$$\begin{aligned}&(a^{-1},aq,b^{-1},bq,a^{-1}b^{-1},abq,a^{-1}b,ab^{-1}q,q,q,q,q,q;q)_\infty\\&=\sum_{r,s,t,u=-\infty}^{\infty}(-1)^{r+s+t+u}a^{r+t+u}b^{s+t-u}q^{(r^2+r+s^2+s+t^2+t+u^2+u)/2}.\end{aligned}\tag{11.2.1}$$

We split this sum into nine, according to the residue modulo 3 of each of $r+t+u$ and $s+t-u$.

Thus, for example, if $r+t+u\equiv 0$ and $s+t-u\equiv 0$, we can suppose that

$$r+t+u=3k,\quad s+t-u=3l,$$

we can define v and w by

$$v=k+l-t,\quad w=u-(k-l),\tag{11.2.2}$$

and then we have

$$r=k+v-w,\quad s=l+v+w,\quad t=k+l-v,\quad u=k-l+w.\tag{11.2.3}$$

The contribution of these terms to the right side of (11.2.1) is

$$\begin{aligned}&\sum_{k,l,v,w=-\infty}^{\infty}(-1)^{k+l+v+w}a^{3k}b^{3l}q^{(3k^2+3k+3l^2+l+3v^2+v+3w^2+w)/2}\\&=\sum_{k,l=-\infty}^{\infty}(-1)^{k+l}a^{3k}b^{3l}q^{(3k^2+3k+3l^2+l)/2}(q,q;q)_\infty.\end{aligned}\tag{11.2.4}$$

Let us now consider the remaining eight cases, using the parametrisation (11.2.2). If $r+t+u\equiv 0$ and $s+t-u\equiv -1$, let

$$r+t+u=3k,\quad s+t-u=3l-1,$$

and then

$$r=k+v-w,\quad s=l+v+w-1,\quad t=k+l-v,\quad u=k-l+w$$

and the contribution of these terms to the right side of (11.2.1) is

$$\sum_{k,l,v,w=-\infty}^{\infty} (-1)^{k+l+v+w+1} a^{3k} b^{3l-1} q^{(3k^2+3k+3l^2-l+3v^2-v+3w^2-w)/2}$$
$$= \sum_{k,l=-\infty}^{\infty} (-1)^{k+l+1} a^{3k} b^{3l-1} q^{3k^2+3k+3l^2-l)/2} (q,q;q)_\infty$$
$$= \sum_{k,l=-\infty}^{\infty} (-1)^{k+l+1} a^{3k} b^{-3l-1} q^{(3k^2+3k+3l^2+l)/2} (q,q;q)_\infty. \quad (11.2.5)$$

If $r+t+u \equiv -1$ and $s+t-u \equiv 1$, let

$$r+t+u = 3k-1, \quad s+t-u = 3l+1,$$

and then

$$r = k+v-w-1, \quad s = l+v+w+1, \quad t = k+l-v, \quad u = k-l+w$$

and the contribution of these terms to the right side of (11.2.1) is

$$\sum_{k,l,v,w=-\infty}^{\infty} (-1)^{k+l+v+w} a^{3k-1} b^{3l+1} q^{(3k^3+k+3l^2+3l+3v^2+v+3w^2+5w+2)/2}$$
$$= \sum_{k,l,v,w=-\infty}^{\infty} (-1)^{k+l+v+w+1} a^{3k-1} b^{3l+1} q^{(3k^2+k+3l^2+3l+3v^2+v+3w^2-w)/2}$$
$$= \sum_{k,l=-\infty}^{\infty} (-1)^{k+l+1} a^{3l-1} b^{3k+1} q^{(3k^2+3k+3l^2+l)/2} (q,q;q)_\infty. \quad (11.2.6)$$

If $r+t+u \equiv 1$ and $s+t-u \equiv 1$, let

$$r+t+u = 3k+1, \quad s+t-u = 3l+1,$$

and then

$$r = k+v-w+1, \quad s = l+v+w+1, \quad t = k+l-v, \quad u = k-l+w$$

and the contribution of these terms to the right side of (11.2.1) is

$$\begin{aligned}
&\sum_{k,l,v,w=-\infty}^{\infty} (-1)^{k+l+v+w} a^{3k+1} b^{3l+1} q^{(3k^2+5k+3l^2+3l+3v^2+5v+3w^2+w+4)/2} \\
&\quad = \sum_{k,l,v,w=-\infty}^{\infty} (-1)^{k+l+v+w} a^{3k-2} b^{3l+1} q^{(3k^2-k+3l^2+3l+3v^2-v+3w^2+w)/2} \\
&\quad = \sum_{k,l=-\infty}^{\infty} (-1)^{k+l} a^{-3k-2} b^{3l+1} q^{(3k^2+k+3l^2+3l)/2} (q,q;q)_\infty \\
&\quad = \sum_{k,l=-\infty}^{\infty} (-1)^{k+l} a^{-3l-2} b^{3k+1} q^{(3k^2+3k+3l^2+l)/2} (q,q;q)_\infty. \qquad (11.2.7)
\end{aligned}$$

The other five cases contribute nothing, since

$$\sum_{-\infty}^{\infty} (-1)^v q^{(3v^2+3v)/2} = 0. \qquad (11.2.8)$$

If we add the contributions to (11.2.1) in (11.2.4)–(11.2.7) and divide by $(q,q;q)_\infty$, we obtain (11.1.1); (11.1.2) follows by JTP.

Endnote.

§11.1 Winquist's identity appears in Winquist (1969) [124].

References

67. M.D. Hirschhorn, A generalisation of Winquist's identity and a conjecture of Ramanujan. J. Indian Math. Soc. **51**, 49–55 (1987)

124. L. Winquist, An elementary proof of $p(11n+6) \equiv 0 \pmod{11}$. J. Comb. Theory **6**, 56–59 (1969)

Chapter 12
The Crank of a Partition

12.1 Introduction

In this chapter, we will meet the concept of the crank of a partition, and obtain new proofs of Ramanujan's partition congruences

$$p(5n+4) \equiv 0 \pmod 5, \quad p(7n+5) \equiv 0 \pmod 7 \text{ and } p(11n+6) \equiv 0 \pmod{11}. \tag{12.1.1}$$

In 1944, while still an undergraduate, Freeman Dyson [42] defined the rank of a partition as the largest part minus the number of parts, and conjectured that the partitions of $5n+4$ and of $7n+5$ fall into five (respectively seven) **equinumerous** classes, according to the residues of their ranks modulo 5 (respectively 7), thus providing explanations of the congruences $p(5n+4) \equiv 0 \pmod 5$ and $p(7n+5) \equiv 0 \pmod 7$. This does not, however, happen for the partitions of 6.

Dyson's conjection was proved by Atkin–Swinnerton-Dyer [12]. Garvan later discovered that Ramanujan had an identity in the lost notebook [112] that implied Dyson's conjecture for 5.

At the same time in 1944, Dyson had conjectured that there might be a partition statistic, which he dubbed the "crank" that would, in the same way as the rank did for 5 and 7, divide the partitions of $11n+6$ into 11 equinumerous classes according to the residues of their cranks modulo 11, thus giving an explanation of the congruence $p(11n+6) \equiv 0 \pmod{11}$.

As will be revealed in this chapter, George E. Andrews and Frank Garvan did indeed discover such a crank, remarkably on the day after the Ramanujan Centenary Conference in 1987, and it works to explain not only the mod 11 congruence, but also the mod 5 and mod 7 congruences!

The original version of this chapter has been revised. The erratum to this chapter is available at https://doi.org/10.1007/978-3-319-57762-3_44.

M.D. Hirschhorn, *The Power of q*, Developments in Mathematics 49,
DOI 10.1007/978-3-319-57762-3_12

12.2 Introducing the Crank

Prompted by a line in the Lost Notebook ([112], p. 20) equivalent to (12.3.4) and another ([112], p. 19) which begins with only the left side of (12.3.8), Frank Garvan was led to consider the product

$$\frac{(q;q)_\infty}{(a^{-1}q;q)_\infty(aq;q)_\infty}. \tag{12.2.1}$$

We can expand this product as a series in ascending powers of q with coefficients functions of a.

$$\frac{(q;q)_\infty}{(a^{-1}q;q)_\infty(aq;q)_\infty} = \sum_{n\geq 0} c_n(a)q^n. \tag{12.2.2}$$

Indeed, since the product is invariant under $a \leftrightarrow a^{-1}$, the coefficients are reciprocal functions, that is,

$$c_n(a^{-1}) = c_n(a). \tag{12.2.3}$$

The expansion begins

$$\frac{(q;q)_\infty}{(a^{-1}q;q)_\infty(aq;q)_\infty} = 1+(a^{-1}-1+a)q+(a^{-2}+a^2)q^2+(a^{-3}+1+a^3)q^3$$
$$+\,(a^{-4}+a^{-2}+1+a^2+a^4)q^4+(a^{-5}+a^{-3}+a^{-1}+1+a+a^3+a^5)q^5+\cdots. \tag{12.2.4}$$

If in (12.2.2) we set $a=1$, we see that

$$c_n(1) = p(n), \tag{12.2.5}$$

the number of partitions of n.

Let us write

$$c_n(a) = \sum_{k=-n}^{n} p(n,k)a^k. \tag{12.2.6}$$

Note that the $p(n,k)$ are integers and that

$$\sum_{k=-n}^{n} p(n,k) = c_n(1) = p(n). \tag{12.2.7}$$

In 1986, Garvan conjectured that for $n>1$,

$$p(n,k) \geq 0, \tag{12.2.8}$$

and strongly suggested to George Andrews and me that for $n > 1$, $p(n,k)$ counts the number of partitions of n possessing some numerical–valued property with value k, and given what we will see in what follows, that property should be the crank, proposed by Dyson (1944) [42] to exist.

About a year later, Andrews and Garvan (1988) [9] not only proved (12.2.8) but in doing so found a description of the crank of a partition in terms of the parts, and we will give that description in §12.4.

In §12.3 we will show that if η is any 5th root of unity (other than 1),

$$c_{5n+4}(\eta) = 0, \tag{12.2.9}$$

if η is any 7th root of unity (other than 1),

$$c_{7n+5}(\eta) = 0, \tag{12.2.10}$$

while if η is any 11th root of unity (other than 1),

$$c_{11n+6}(\eta) = 0. \tag{12.2.11}$$

It follows that

$$\begin{aligned}\sum_{k\equiv -2 \pmod 5} p(5n+4,k) &= \sum_{k\equiv -1 \pmod 5} p(5n+4,k) = \sum_{k\equiv 0 \pmod 5} p(5n+4,k) \\ &= \sum_{k\equiv 1 \pmod 5} p(5n+4,k) = \sum_{k\equiv 2 \pmod 5} p(5n+4,k),\end{aligned} \tag{12.2.12}$$

$$\begin{aligned}\sum_{k\equiv -3 \pmod 7} p(7n+5,k) &= \sum_{k\equiv -2 \pmod 7} p(7n+5,k) = \cdots \\ &= \sum_{k\equiv 3 \pmod 7} p(7n+5,k),\end{aligned} \tag{12.2.13}$$

and

$$\sum_{k\equiv -5 \pmod{11}} p(11n+6,k) = \cdots = \sum_{k\equiv 5 \pmod{11}} p(11n+6,k). \tag{12.2.14}$$

These are precisely the properties required by Dyson of the crank, so $p(n,k)$ is the number of partitions of n with crank k.

Of course, (12.2.12)–(12.2.14) imply the congruences (12.1.1).

12.3 Proofs of (12.2.9)–(12.2.11)

Suppose η is a 5th root of unity other than 1. Then

$$\frac{(q;q)_\infty}{(\eta^{-1}q;q)_\infty(\eta q;q)_\infty} = \frac{(q;q)_\infty(\eta^{-2}q,\eta^2 q,q;q)_\infty}{(q^5;q^5)_\infty}. \tag{12.3.1}$$

Now we have (8.1.1),

$$(q;q)_\infty = (q^{25};q^{25})_\infty\left(\left(\begin{matrix} q^{10},q^{15} \\ q^5,q^{20} \end{matrix};q^{25}\right)_\infty - q - q^2\left(\begin{matrix} q^5,q^{20} \\ q^{10},q^{15} \end{matrix};q^{25}\right)_\infty\right) \tag{12.3.2}$$

and (8.3.7),

$$(\eta^2 q,\eta^{-2}q,q;q)_\infty = (q^{10},q^{15},q^{25};q^{25})_\infty + (\eta+\eta^{-1})q(q^5,q^{20},q^{25};q^{25})_\infty \tag{12.3.3}$$

If we substitute (12.3.2) and (12.3.3) into (12.3.1), we obtain

$$\begin{aligned}\frac{(q;q)_\infty}{(\eta^{-1}q;q)_\infty(\eta q;q)_\infty} &= \frac{(q^{25};q^{25})^2_\infty}{(q^5;q^5)_\infty}\left(\frac{(q^{10},q^{15};q^{25})^2_\infty}{(q^5,q^{20};q^{25})_\infty} + (\eta-1+\eta^{-1})q(q^{10},q^{15};q^{25})_\infty\right. \\ &\left. + (\eta^2+\eta^{-2})q^2(q^5,q^{20};q^{25})_\infty - (\eta+\eta^{-1})q^3\frac{(q^5,q^{20};q^{25})^2_\infty}{(q^{10},q^{15};q^{25})_\infty}\right).\end{aligned} \tag{12.3.4}$$

This does not contain any terms in which the power of q is 4 modulo 5, so

$$c_{5n+4}(\eta) = 0,$$

which is (12.2.9).

Suppose η is a 7th root of unity other than 1. Then

$$\begin{aligned}\frac{(q;q)_\infty}{(\eta^{-1}q;q)_\infty(\eta q;q)_\infty} &= \frac{(q;q)_\infty(\eta^{-3}q,\eta^{-2}q,\eta^2 q,\eta^3 q,q;q)_\infty}{(q^7;q^7)_\infty} \\ &= \frac{(\eta^{-3}q,\eta^3 q,q;q)_\infty(\eta^{-2}q,\eta^2 q,q;q)_\infty}{(q^7;q^7)_\infty}\end{aligned} \tag{12.3.5}$$

Now,

$$\eta^k+\eta^{k-1}+\cdots+\eta^{-k}=\begin{cases}1 & \text{if } k\equiv 0 \pmod 7,\\ \eta+1+\eta^{-1} & \text{if } k\equiv 1 \pmod 7,\\ -(\eta^3+\eta^{-3}) & \text{if } k\equiv 2 \pmod 7,\\ 0 & \text{if } k\equiv 3 \pmod 7,\\ \eta^3+\eta^{-3} & \text{if } k\equiv 4 \pmod 7,\\ -(\eta+1+\eta^{-1}) & \text{if } k\equiv 5 \pmod 7,\\ -1 & \text{if } k\equiv 6 \pmod 7,\end{cases} \tag{12.3.6}$$

so (1.7.5) yields

$$\begin{aligned}(\eta^{-1}q,\eta q,q;q)_\infty &= (q^{21},q^{28},q^{49};q^{49})_\infty-(\eta+1+\eta^{-1})q(q^{14},q^{35},q^{49};q^{49})_\infty\\ &\quad-(\eta^3+\eta^{-3})q^3(q^7,q^{42},q^{49};q^{49})_\infty.\end{aligned} \tag{12.3.7}$$

If we set η^2 and η^3 for η in (12.3.7) and substitute the results into (12.3.5), we obtain

$$\begin{aligned}&\frac{(q;q)_\infty}{(\eta^{-1}q;q)_\infty(\eta q;q)_\infty}\\ &=\frac{(q^{49};q^{49})_\infty^2}{(q^7;q^7)_\infty}\big((q^{21},q^{28};q^{49})_\infty^2+(\eta-1+\eta^{-1})q(q^{14},q^{21},q^{28},q^{35};q^{49})_\infty\\ &\quad+(\eta^2+\eta^{-2})q^2(q^{14},q^{35};q^{49})_\infty^2+(\eta^3+1+\eta^{-3})q^3(q^7,q^{21},q^{28},q^{42};q^{49})_\infty\\ &\quad-(\eta+\eta^{-1})q^4(q^7,q^{14},q^{35},q^{42};q^{49})_\infty-(\eta^2+1+\eta^{-2})q^6(q^7,q^{42};q^{49})_\infty^2\big).\end{aligned} \tag{12.3.8}$$

This does not contain any terms in which the power of q is 5 modulo 7, so

$$c_{7n+5}(\eta)=0,$$

which is (12.2.10).

Suppose η is an 11th root of unity other than 1. Then

$$\begin{aligned}\frac{(q;q)_\infty}{(\eta^{-1}q,q)_\infty(\eta q,q)_\infty}&=\frac{(q;q)_\infty(\eta^{-5}q,\eta^{-4}q,\eta^{-3}q,\eta^{-2}q,\eta^2q,\eta^3q,\eta^4q,\eta^5q,q;q)_\infty}{(q^{11};q^{11})_\infty}\\ &=\frac{(\eta^{-5}q,\eta^{-4}q,\eta^{-3}q,\eta^{-2}q,\eta^2q,\eta^3q,\eta^4q,\eta^5q,q,q;q)_\infty}{(q^{11};q^{11})_\infty}.\end{aligned} \tag{12.3.9}$$

If in Winquist's identity (11.1.1) we set $a=\eta^6$, $b=\eta^2$ and divide by $(1-\eta^{-2})(1-\eta^{-4})(1-\eta^{-6})(1-\eta^{-8})$, we find that the numerator of the right side of (12.3.9) is

$$
\begin{aligned}
&(\eta^{-5}q,\eta^{-4}q,\eta^{-3}q,\eta^{-2}q,\eta^{2}q,\eta^{3}q,\eta^{4}q,\eta^{5}q,q,q;q)_\infty \\
&= \frac{1}{(\eta-\eta^{-1})(\eta^2-\eta^{-2})(\eta^3-\eta^{-3})(\eta^4-\eta^{-4})} \\
&\times \sum_{k,l=-\infty}^{\infty} (-1)^{k+l}(\eta^{18k+6l+10}-\eta^{18k-6l+8}-\eta^{6k+18l+6}+\eta^{6k-18l}) \\
&\qquad\qquad \times q^{(3k^2+3k+3l^2+l)/2}. \qquad (12.3.10)
\end{aligned}
$$

The terms in which the power of q is 6 modulo 11 are given by $k \equiv 5 \pmod{11}$, $l \equiv -2 \pmod{11}$, since the congruence

$$(3k^2+3k+3l^2+l)/2 \equiv 6 \pmod{11} \qquad (12.3.11)$$

is equivalent to

$$(6k+3)^2+(6l+1)^2 \equiv 0 \pmod{11}, \qquad (12.3.12)$$

of which the only solution is

$$k \equiv 5, \ \ l \equiv -2 \pmod{11}. \qquad (12.3.13)$$

If we replace k by $11k+5$, l by $11l-2$, we find that the coefficient becomes

$$\eta^{198k+66l+88}-\eta^{198k-66l+110}-\eta^{66k+198l}+\eta^{66k-198l+66}=0. \qquad (12.3.14)$$

So

$$c_{11n+6}(\eta)=0,$$

which is (12.2.11), and we are done.

With some work we can obtain the 11–analogue of (12.3.4) and (12.3.8), namely

$$
\begin{aligned}
&\frac{(q;q)_\infty}{(\eta^{-1}q;q)_\infty(\eta q;q)_\infty} \\
&= \frac{(q^{121};q^{121})_\infty^2}{(q^{11};q^{11})_\infty}((q^{22},q^{33},q^{44},q^{55},q^{66},q^{77},q^{88},q^{99};q^{121})_\infty \\
&+(\eta-1+\eta^{-1})q(q^{11},q^{44},q^{55},q^{55},q^{66},q^{66},q^{77},q^{110};q^{121})_\infty \\
&+(\eta^2+\eta^{-2})q^2(q^{22},q^{33},q^{33},q^{55},q^{66},q^{88},q^{88},q^{99};q^{121})_\infty \\
&+(\eta^3+1+\eta^{-3})q^3(q^{22},q^{22},q^{44},q^{55},q^{66},q^{77},q^{99},q^{99};q^{121})_\infty \\
&+(\eta^4+\eta^2+1+\eta^{-2}+\eta^{-4})q^4(q^{11},q^{33},q^{44},q^{55},q^{66},q^{77},q^{88},q^{110};q^{121})_\infty \\
&-(\eta^4+\eta^2+\eta^{-2}+\eta^{-4})q^5(q^{11},q^{33},q^{44},q^{44},q^{77},q^{77},q^{88},q^{110};q^{121})_\infty \\
&+(\eta^4+\eta+\eta^{-1}+\eta^{-4})q^7(q^{11},q^{22},q^{44},q^{55},q^{66},q^{77},q^{99},q^{110};q^{121})_\infty
\end{aligned}
$$

$$
\begin{aligned}
&+ (\eta^4+\eta^3+\eta+\eta^{-1}+\eta^{-3}+\eta^{-4})q^{19}(q^{11},q^{11},q^{22},q^{33},q^{88},q^{99},q^{110},q^{110};q^{121})_\infty \\
&- (\eta^4+1+\eta^{-4})q^{9}(q^{11},q^{22},q^{33},q^{55},q^{66},q^{88},q^{99},q^{110};q^{121})_\infty \\
&- (\eta^3+\eta^{-3})q^{10}(q^{11},q^{22},q^{33},q^{44},q^{77},q^{88},q^{99},q^{110};q^{121})_\infty). \qquad (12.3.15)
\end{aligned}
$$

The terms in (12.3.10) in which the power of q is congruent to r modulo 11 are given by

$$
3\frac{k^2+k}{2}+\frac{3l^2+l}{2}\equiv r \pmod{11}. \qquad (12.3.16)
$$

The solutions of (12.3.16) can be found from the following table.

k \ l	$3\Delta_k$	4	5	−5	−4	−3	−2	−1	0	1	2	3
P_l		4	−4	2	0	1	5	1	0	2	−4	4
0	0	4	−4	2	0	1	5	1	0	2	−4	4
1	3	−4	−1	5	3	4	−3	4	3	5	−1	−4
2	−2	2	5	0	−2	−1	3	−1	−2	0	5	2
3	−4	0	3	−2	−4	−3	1	−3	−4	−2	3	0
4	−3	1	4	−1	−3	−2	2	−2	−3	−1	4	1
5	1	5	−3	3	1	2	**−5**	2	1	3	−3	5
−5	−3	1	4	−1	−3	−2	2	−2	−3	−1	4	1
−4	−4	0	3	−2	−4	−3	1	−3	−4	−2	3	0
−3	−2	2	5	0	−2	−1	3	−1	−2	0	5	2
−2	3	−4	−1	5	3	4	−3	4	3	5	−1	−4
−1	0	4	−4	2	0	1	5	1	0	2	−4	4

(12.3.17)

Note that with $r\equiv -5$ at the centre of the table, the only solution is $k\equiv 5,\ l\equiv -2$, as established in (12.3.13). For each r other than -5, there are 12 solutions.

For example, if $r\equiv 0$, the 12 solutions are

$$
\begin{aligned}
(k,l)\equiv&(0,0),\ (-1,0),\ (0,-4),\ (-1,-4),\ (2,1),\ (-3,1), \\
&(2,-5),\ (-3,-5),\ (3,3),\ (-4,3),\ (3,4),\ (-4,4). \qquad (12.3.18)
\end{aligned}
$$

If we replace (k,l) by $(11k+0,11l+0)$, $(11k-1,11l+0)$ and so on, the component sum in (12.3.10) collapses to

$$
\begin{aligned}
&(-\eta^5-\eta^3+\eta+2+\eta^{-1}-\eta^{-3}-\eta^{-5}) \\
&\times ((q^{165},q^{176},q^{187},q^{198},q^{363},q^{363};q^{363})_\infty - q^{22}(q^{55},q^{165},q^{198},q^{308},q^{363},q^{363};q^{363})_\infty \\
&- q^{33}(q^{66},q^{77},q^{286},q^{297},q^{363},q^{363};q^{363})_\infty - q^{44}(q^{44},q^{66},q^{297},q^{319},q^{363},q^{363};q^{363})_\infty).
\end{aligned}
$$

(12.3.19)

The numerical factor cancels with

$$(\eta-\eta^{-1})(\eta^2-\eta^{-2})(\eta^3-\eta^{-3})(\eta^4-\eta^{-4}) = -\eta^5-\eta^3+\eta+2+\eta^{-1} -\eta^{-3}-\eta^{-5} \tag{12.3.20}$$

and the other factor becomes

$$(q^{22},q^{33},q^{44},q^{55},q^{66},q^{77},q^{88},q^{99},q^{121},q^{121};q^{121})_\infty \tag{12.3.21}$$

by (11.1.2) with q replaced by q^{121}, $a=q^{-55}$ and $b=q^{-22}$.

We can go through a similar process for each residue r, and obtain (12.3.15).

12.4 The Definition of the Crank

The Andrews–Garvan crank of a partition is defined in terms of the parts as follows.

For $n>1$, if π is a partition of n, let $l(\pi)$ denote the largest part in π, $\omega(\pi)$ denote the number of 1's in π, and $\mu(\pi)$ denote the number of parts greater than $\omega(\pi)$. Then the crank of π, $c(\pi)$, is given by

$$c(\pi)=\begin{cases} l(\pi) & \text{if } \omega(\pi)=0,\\ \mu(\pi)-\omega(\pi) & \text{if } \omega(\pi)>0.\end{cases}$$

This definition was motivated by the inspired rewriting by Andrews and Garvan [9] of the generating function, making use of the q-binomial theorem (see the Appendix),

$$\begin{aligned}
\frac{(q;q)_\infty}{(a^{-1}q;q)_\infty(aq;q)_\infty} &= \frac{1-q}{(aq;q)_\infty}\cdot\frac{(q^2;q)_\infty}{(a^{-1}q;q)_\infty}\\
&= \frac{1-q}{(aq;q)_\infty}\cdot\sum_{\omega\geq 0}\frac{(aq;q)_\omega}{(q;q)_\omega}(a^{-1}q)^\omega\\
&= \frac{1-q}{(aq;q)_\infty}+\sum_{\omega\geq 1}\frac{a^{-\omega}q^\omega}{(q^2;q)_{\omega-1}(aq^{\omega+1};q)_\infty}\\
&= (1-q)\sum_{l\geq 0}\frac{a^lq^l}{(q;q)_l}+\sum_{\omega\geq 1}\frac{a^{-\omega}q^\omega}{(q^2;q)_{\omega-1}(aq^{\omega+1};q)_\infty}\\
&= 1-q+\sum_{l\geq 1}\frac{a^lq^l}{(q^2;q)_{l-1}}+\sum_{\omega\geq 1}\frac{a^{-\omega}q^\omega}{(q^2;q)_{\omega-1}(aq^{\omega+1};q)_\infty}\\
&= 1-q+\sum_{l\geq 1}a^lq^l(1+q^2+q^{2+2}+\cdots)(1+q^3+q^{3+3}+\cdots)\cdots(1+q^l+\cdots)
\end{aligned}$$

$$+\sum_{\omega\geq 1} a^{-\omega} q^{\omega\times 1}(1+q^2+q^{2+2}+\cdots)\cdots(1+q^{\omega}+q^{\omega+\omega}+\cdots)$$
$$\times\,(1+aq^{\omega+1}+a^2q^{(\omega+1)+(\omega+1)}+\cdots)$$
$$\times\,(1+aq^{\omega+2}+a^2q^{(\omega+2)+(\omega+2)}+\cdots)$$
$$\times\cdots, \tag{12.4.1}$$

and interpreting the right side.

Endnotes.

§§12.2, 12.3 This all follows Garvan (1988) [48], with the exception of (12.3.15). which was stated without proof in Hirschhorn (1990) [69]. Ekin (1998) [43] published a proof.

References

9. G.E. Andrews, F.G. Garvan, Dyson's crank of a partition. Bull. Am. Math. Soc. **18**, 167–171 (1988)
12. A.O.L. Atkin, H.P.F. Swinnerton-Dyer, Some properties of partitions, Proc. Lond. Math. Soc. **4**, 84–106 (1954)
42. F.J. Dyson, Some guesses in the theory of partitions. Eureka (Cambridge) **8**, 10–15 (1944)
43. A.B. Ekin, Inequalities for the crank. J. Comb. Theory Ser. A **83**, 283–289 (1998)
48. F.G. Garvan, New combinatorial interpretations of Ramanujan's partition congruences mod 5, 7 and 11. Trans. Am. Math. Soc. **305**, 47–77 (1988)
69. M.D. Hirschhorn, A birthday present for Ramanujan. Am. Math. Mon. **97**, 398–400 (1990)
112. S. Ramanujan, *The Lost Notebook and Other Unpublished Papers* (Springer, Berlin, 1988)

Chapter 13
Two More Proofs of $p(11n+6)\equiv 0 \pmod{11}$, and More

13.1 Introduction

The purpose of this chapter is to present another proof of the mod 11 partition congruence and to derive the Atkin–Swinnerton-Dyer congruences for the modulus 11.

We will start by proving, in §13.2, that

$$48E(q)^{10} = \sum_{m,n=-\infty}^{\infty}(-1)^{m+n}((6m+3)^3(6n+1)-(6m+3)(6n+1)^3)q^{(3m^2+3m+3n^2+n)/2}. \tag{13.1.1}$$

In §13.3, we will use (13.1.1) to give another proof of

$$p(11n+6)\equiv 0 \pmod{11}$$

and in §13.4 we will use (13.1.1) to obtain the Atkin–Swinnerton-Dyer partition congruences for the modulus 11,

$$\sum_{n\geq 0} p(11n)q^n \equiv \frac{(q^2,q^3,q^4,q^5,q^6,q^7,q^8,q^9,q^{11},q^{11};q^{11})_\infty}{(q;q)_\infty},$$

$$\sum_{n\geq 0} p(11n+1)q^n \equiv \frac{(q,q^4,q^5,q^5,q^6,q^6,q^7,q^{10},q^{11},q^{11};q^{11})_\infty}{(q;q)_\infty},$$

$$\sum_{n\geq 0} p(11n+2)q^n \equiv 2\frac{(q^2,q^3,q^3,q^5,q^6,q^8,q^8,q^9,q^{11},q^{11};q^{11})_\infty}{(q;q)_\infty},$$

$$\sum_{n\geq 0} p(11n+3)q^n \equiv 3\frac{(q^2,q^2,q^4,q^5,q^6,q^7,q^9,q^9,q^{11},q^{11};q^{11})_\infty}{(q;q)_\infty},$$

M.D. Hirschhorn, *The Power of q*, Developments in Mathematics 49,
DOI 10.1007/978-3-319-57762-3_13

$$\sum_{n\geq 0} p(11n+4)q^n \equiv 5\frac{(q,q^3,q^4,q^5,q^6,q^7,q^8,q^{10},q^{11},q^{11};q^{11})_\infty}{(q;q)_\infty},$$

$$\sum_{n\geq 0} p(11n+5)q^n \equiv 7\frac{(q,q^3,q^4,q^4,q^7,q^7,q^8,q^{10},q^{11},q^{11};q^{11})_\infty}{(q;q)_\infty},$$

$$\sum_{n\geq 0} p(11n+6)q^n \equiv 0,$$

$$\sum_{n\geq 0} p(11n+7)q^n \equiv 4\frac{(q,q^2,q^4,q^5,q^6,q^7,q^9,q^{10},q^{11},q^{11};q^{11})_\infty}{(q;q)_\infty},$$

$$\sum_{n\geq 0} p(11n+8)q^n \equiv 6q\frac{(q,q,q^2,q^3,q^8,q^9,q^{10},q^{10},q^{11},q^{11};q^{11})_\infty}{(q;q)_\infty},$$

$$\sum_{n\geq 0} p(11n+9)q^n \equiv 8\frac{(q,q^2,q^3,q^5,q^6,q^8,q^9,q^{10},q^{11},q^{11};q^{11})_\infty}{(q;q)_\infty}$$

and

$$\sum_{n\geq 0} p(11n+10)q^n \equiv 9\frac{(q,q^2,q^3,q^4,q^7,q^8,q^9,q^{10},q^{11},q^{11};q^{11})_\infty}{(q;q)_\infty}. \tag{13.1.2}$$

The counterparts for the modulus 5 are in §3.6, and for modulus 7 in §3.7.

13.2 The Expansion of $E(q)^{10}$

In Winquist's identity (11.1.1), put a^2 for a, b^2 for b, and multiply by ab, and we obtain

$$\begin{aligned}&(a-a^{-1})(b-b^{-1})(a^{-2}q,a^2q,b^{-2}q,b^2q,a^{-2}b^{-2},a^2b^2q,a^{-2}b^2,a^2b^{-2}q,q,q;q)_\infty\\ &= \sum_{m,n=-\infty}^{\infty} (-1)^{m+n}(a^{6m+1}b^{6n+1} - a^{6m+1}b^{-6n-1} - a^{6n-1}b^{6m+3} + a^{-6n-3}b^{6m+3})\\ &\qquad \times q^{(3m^2+3m+3n^2+n)/2}.\end{aligned} \tag{13.2.1}$$

We now differentiate with respect to b, put $b=1$ and multiply by a^2, and we obtain

$$\begin{aligned}&2(a-a^{-1})^3(a^{-2}q,a^{-2}q,a^{-2}q,a^2q,a^2q,a^2q,q,q,q,q;q)_\infty\\ &= \sum_{m,n=-\infty}^{\infty} (-1)^{m+n}(2(6n+1)a^{6m+3} - (6m+3)a^{6n+1} + (6m+3)a^{-6n-1})\\ &\qquad \times q^{(3m^2+3m+3n^2+n)/2}.\end{aligned} \tag{13.2.2}$$

We next do the following three times: differentiate with respect to a then multiply by a. Then put $a = 1$, divide by 2, and we obtain

$$48(q;q)_\infty^{10} = \sum_{m,n=-\infty}^{\infty} (-1)^{m+n}((6m+3)^3(6n+1) - (6m+3)(6n+1)^3)q^{(3m^3+3m+3n^2+n)/2}, \tag{13.2.3}$$

which is (13.1.1).

13.3 $p(11n+6) \equiv 0 \pmod{11}$ Again

We can write (13.1.1) as follows.

$$E(q)^{10} = \sum_{n\geq 0} c(n)q^n, \tag{13.3.1}$$

where

$$c(n) = \frac{1}{48} \sum_{\substack{\alpha\equiv 3 \pmod 6 \\ \beta\equiv 1 \pmod 6 \\ (\alpha^2+\beta^2-10)/24=n}} (-1)^{(\alpha+\beta-4)/6}\alpha\beta(\alpha^2-\beta^2). \tag{13.3.2}$$

In particular,

$$c(11n+6) = \frac{1}{48} \sum_{\substack{\alpha\equiv 3 \pmod 6 \\ \beta\equiv 1 \pmod 6 \\ (\alpha^2+\beta^2-10)/24=11n+6}} (-1)^{(\alpha+\beta-4)/6}\alpha\beta(\alpha^2-\beta^2). \tag{13.3.3}$$

Now, the equation

$$\frac{\alpha^2+\beta^2-10}{24} = 11n+6 \tag{13.3.4}$$

entails

$$\alpha^2+\beta^2 = 264n+154,$$

$$\alpha^2+\beta^2 \equiv 0 \pmod{11}, \tag{13.3.5}$$

so

$$\alpha \equiv 0 \pmod{11} \text{ and } \beta \equiv 0 \pmod{11}. \tag{13.3.6}$$

Set

$$\alpha = -11A, \;\; \beta = -11B. \tag{13.3.7}$$

Then

$$A \equiv 3 \pmod{6}, \;\; B \equiv 1 \pmod{6} \tag{13.3.8}$$

and

$$264n + 154 = 121(A^2 + B^2). \tag{13.3.9}$$

It follows that

$$264n + 154 \equiv 0 \pmod{121},$$

or,

$$n \equiv 4 \pmod{11}. \tag{13.3.10}$$

So $c(11n+6) = 0$ unless $n = 11m+4$, in which case

$$264(11m+4) + 154 = 121(A^2 + B^2),$$

$$A^2 + B^2 = 24m + 10,$$

$$\frac{A^2 + B^2 - 10}{24} = m$$

and then

$$\begin{aligned} c(11n+6) &= c(121m+50) \\ &= \frac{11^4}{48} \sum_{\substack{A \equiv 3 \pmod{6} \\ B \equiv 1 \pmod{6} \\ (A^2+B^2-10)/24 = m}} (-1)^{(-11A-11B-4)/6} AB(A^2 - B^2) \\ &= \frac{11^4}{48} \sum_{\substack{A \equiv 3 \pmod{6} \\ B \equiv 1 \pmod{6} \\ (A^2+B^2-10)/24 = m}} (-1)^{(A+B-4)/6} AB(A^2 - B^2) \\ &= 11^4 c(m). \end{aligned} \tag{13.3.11}$$

It follows that

$$
\begin{aligned}
\sum_{n\geq 0} c(11n+6)q^n &= \sum_{m\geq 0} c(121m+50)q^{11m+4} \\
&= 11^4 \sum_{m\geq 0} c(m)q^{11m+4} \\
&= 11^4 q^4 E(q^{11})^{10}, \qquad (13.3.12)
\end{aligned}
$$

$$
\sum_{n\geq 0} c(11n+6)q^{11n} = 11^4 q^{44} E(q^{121})^{10}, \qquad (13.3.13)
$$

or,

$$
H(q^{-6}E(q)^{10}) = 11^4 q^{44} E(q^{121})^{10}, \qquad (13.3.14)
$$

$$
H\left(\left(\frac{E(q)}{q^5 E(q^{121})}\right)^{10}\right) = 11^4, \qquad (13.3.15)
$$

and if we write

$$
\zeta = \frac{E(q)}{q^5 E(q^{121})}, \qquad (13.3.16)
$$

(13.3.15) becomes

$$
H\left(\zeta^{10}\right) = 11^4. \qquad (13.3.17)
$$

Isn't that neat?!

Then we have, modulo 11,

$$
\begin{aligned}
\sum_{n\geq 0} p(n)q^n &= \frac{1}{E(q)} = \frac{E(q)^{10}}{E(q)^{11}} \equiv \frac{E(q)^{10}}{E(q^{11})} \\
&= q^{50} \frac{E(q^{121})^{10}}{E(q^{11})} \left(\frac{E(q)}{q^5 E(q^{121})}\right)^{10} \\
&= q^{50} \frac{E(q^{121})^{10}}{E(q^{11})} \zeta^{10}, \qquad (13.3.18)
\end{aligned}
$$

$$
\sum_{n\geq 0} p(n)q^{n-6} \equiv q^{44} \frac{E(q^{121})^{10}}{E(q^{11})} \zeta^{10}, \qquad (13.3.19)
$$

$$
H\left(\sum_{n\geq 0} p(n)q^{n-6}\right) \equiv q^{44} \frac{E(q^{121})^{10}}{E(q^{11})} H\left(\zeta^{10}\right) \qquad (13.3.20)
$$

and finally,

$$
\sum_{n\geq 0} p(11n+6)q^{11n} \equiv 0.
$$

Exercise: Start with $H(\zeta^4) = -5$ (6.3.3) and $H(\zeta^6) = 49$ (7.4.18) and devise proofs of $p(5n+4)\equiv 0 \pmod 5$ and $p(7n+5)\equiv 0 \pmod 7$ similar to the proof above.

13.4 The Atkin–Swinnerton-Dyer Congruences for Modulus 11

Modulo 11, (13.1.1) becomes

$$E(q)^{10} \equiv 3\left(\sum_{-\infty}^{\infty}(-1)^n(6n+3)^3 q^{(3n^2+3n)/2}\sum_{-\infty}^{\infty}(-1)^n(6n+1)q^{(3n^2+n)/2}\right.$$
$$\left. -\sum_{-\infty}^{\infty}(-1)^n(6n+3)q^{(3n^2+3n)/2}\sum_{-\infty}^{\infty}(-1)^n(6n+1)^3 q^{(3n^2+n)/2}\right). \tag{13.4.1}$$

It is easy to show by considering $n \equiv -5, -4, \ldots, 5 \pmod{11}$ that

$$\sum_{-\infty}^{\infty}(-1)^n(6n+3)q^{(3n^2+3n)/2} \equiv -5A_1+4q^3A_3-3q^9A_5+2q^{18}A_7-q^{30}A_9, \tag{13.4.2}$$

$$\sum_{-\infty}^{\infty}(-1)^n(6n+3)^3q^{(3n^2+3n)/2} \equiv -A_1+5q^3A_3-4q^9A_5+2q^{18}A_7-3q^{30}A_9, \tag{13.4.3}$$

$$\sum_{-\infty}^{\infty}(-1)^n(6n+1)q^{(3n^2+n)/2} \equiv B_1+5qB_5+4q^2B_7+2q^7B_{13}-5q^{12}B_{17}+3q^{15}B_{19}$$
$$-q^{22}B_{23}+3q^{26}B_{25}-4q^{35}B_{29}+2q^{40}B_{31} \tag{13.4.4}$$

and

$$\sum_{-\infty}^{\infty}(-1)^n(6n+1)^3q^{(3n^2+n)/2} \equiv B_1+4qB_5-2q^2B_7-3q^7B_{13}-4q^{12}B_{17}+5q^{15}B_{19}$$
$$-q^{22}B_{23}+5q^{26}B_{25}+2q^{35}B_{29}-3q^{40}B_{31} \tag{13.4.5}$$

where

$$A_k=\sum_{-\infty}^{\infty}(-1)^n q^{(363n^2+33kn)/2}=(q^{(363-33k)/2},q^{(363+33k)/2},q^{363};q^{363})_\infty \tag{13.4.6}$$

and

$$B_k = \sum_{-\infty}^{\infty}(-1)^n q^{(363n^2+11kn)/2} = (q^{(363-11k)/2}, q^{(363+11k)/2}, q^{363}; q^{363})_\infty. \tag{13.4.7}$$

If we substitute (13.4.2)–(13.4.5) into (13.4.1) and rearrange, we obtain

$$\begin{aligned}E(q)^{10} \equiv\ & (A_1B_1 - q^{22}A_1B_{23} - q^{33}A_7B_{19} - q^{44}A_7B_{25})\\ &+ q(A_1B_5 - q^{11}A_1B_{17} - q^{44}A_9B_{19} - q^{55}A_9B_{25})\\ &+ 2q^2(A_1B_7 - q^{22}A_5B_{19} - q^{33}A_5B_{25} - q^{33}A_1B_{29})\\ &+ 3q^3(A_3B_1 - q^{22}A_3B_{23} - q^{22}A_7B_{13} - q^{55}A_7B_{31})\\ &+ 5q^4(A_3B_5 - q^{11}A_3B_{17} - q^{33}A_9B_{13} - q^{66}A_9B_{31})\\ &+ 7q^5(A_3B_7 - q^{11}A_5B_{13} - q^{33}A_3B_{29} - q^{44}A_5B_{31})\\ &+ 4q^7(A_1B_{13} - q^{11}A_3B_{19} - q^{22}A_3B_{25} + q^{33}A_1B_{31})\\ &+ 6q^{19}(A_7B_5 - q^{11}A_9B_1 - q^{11}A_7B_{17} + q^{33}A_9B_{23})\\ &+ 8q^9(A_5B_1 - q^{11}A_7B_7 - q^{22}A_5B_{23} + q^{44}A_7B_{29})\\ &+ 9q^{10}(A_5B_5 - q^{11}A_5B_{17} - q^{22}A_9B_7 + q^{55}A_9B_{29}).\end{aligned} \tag{13.4.8}$$

Every bracketed term on the right of (13.4.8) can, remarkably, be expressed as a single product. For example, the first term is

$$\begin{aligned}&A_1B_1 - q^{22}A_1B_{23} - q^{33}A_7B_{19} - q^{44}A_7B_{25}\\ &= (q^{22}, q^{33}, q^{44}, q^{55}, q^{66}, q^{77}, q^{88}, q^{99}, q^{121}, q^{121}; q^{121})_\infty\end{aligned} \tag{13.4.9}$$

where we have used Winquist's identity (11.1.2) with q replaced by q^{121} and $a = q^{-55}, b = q^{-22}$.

For the remaining terms, the relevant substitutions into (11.1.2) are respectively

$$\begin{aligned}(a,b) = &(q^{-55}, q^{-11}),\ (q^{-55}, q^{-33}),\ (q^{-44}, q^{-22}),\ (q^{-44}, q^{-11}),\ (q^{-44}, q^{-33}),\\ &(q^{-55}, q^{-44}),\ (q^{-22}, q^{-11}),\ (q^{-33}, q^{-22}) \text{ and } (q^{-33}, q^{-11}).\end{aligned} \tag{13.4.10}$$

If we substitute these ten results into (13.4.8), substitute the resulting congruence into

$$\sum_{n\geq 0} p(n)q^n \equiv \frac{E(q)^{10}}{E(q^{11})} \pmod{11},$$

and 11-dissect the result, we obtain (13.1.2).

As noted in §3.8, it follows from (13.1.2) that, modulo 11,

$$\sum_{n\geq 0} p(11n)q^n \sum_{n\geq 0} p(11n+4)q^n \sum_{n\geq 0} p(11n+7)q^n \sum_{n\geq 0} p(11n+9)q^n \sum_{n\geq 0} p(11n+10)q^n \equiv -(q;q)_\infty^{65} \tag{13.4.11}$$

and

$$\sum_{n\geq 0} p(11n+1)q^n \sum_{n\geq 0} p(11n+2)q^n \sum_{n\geq 0} p(11n+3)q^n \sum_{n\geq 0} p(11n+5)q^n \sum_{n\geq 0} p(11n+19)q^n \equiv -(q;q)_\infty^{65}. \tag{13.4.12}$$

Endnotes.

§§13.2, 13.4 appear in Hirschhorn (2000) [73].

This chapter owes a lot to the input of Shaun Cooper (especially §13.3) and Frank Garvan.

References

39. S. Cooper, M.D. Hirschhorn, R.P. Lewis, Powers of Euler's product and related identities. Ramanujan J. **3**, 153–158 (1999)
73. M.D. Hirschhorn, Winquist and the Atkin–Swinnerton-Dyer congruences for modulus 11. Australas. J. Comb. **22**, 101–104 (2000)

Chapter 14
Partitions Where Even Parts Come in Two Colours

14.1 Introduction

Recently, Hei-Chi Chan (Chan (2010) [26, 27]) studied partitions in which the even parts come in two colours. If $p^*(n)$ denotes the number of such partitions of n, he showed that $p^*(n)$ possesses similar properties modulo powers of 3 to those the partition function $p(n)$ possesses modulo powers of 5, 7 and 11. Specifically, he showed that if $\alpha \geq 1$ and δ_α is the reciprocal of 8 modulo 3^α, then

$$p^*(3^\alpha n + \delta_\alpha) \equiv 0 \pmod{3^{1+2\lfloor \frac{\alpha}{2} \rfloor}}. \tag{14.1.1}$$

The purpose of this chapter is to provide a sketch of a proof of (14.1.1), and close by obtaining some new results. All proofs are left as exercises for the reader.

14.2 The Generating Function for the $p^*(n)$

The $p^*(n)$ are defined by

$$\sum_{n\geq 0} p^*(n)q^n = \frac{1}{F(q)}, \tag{14.2.1}$$

where

$$F(q) = (q;q^2)_\infty (q^2;q^2)_\infty^2 = (q;q)_\infty (q^2;q^2)_\infty = \phi(-q)\psi(q). \tag{14.2.2}$$

M.D. Hirschhorn, *The Power of q*, Developments in Mathematics 49,
DOI 10.1007/978-3-319-57762-3_14

14.3 The 3-Dissection of $F(q)$

In this section, we derive the 3-dissection of $F(q)$,

$$F(q) = F(q^9)\left(X(q^3)^{-1} - q - 2q^2X(q^3)\right), \tag{14.3.1}$$

where

$$X(q) = \frac{(q;q)_\infty(q^6;q^6)^3_\infty}{(q^2;q^2)_\infty(q^3;q^3)^3_\infty} = \left(\begin{matrix} q, q^5 \\ q^3, q^3 \end{matrix}; q^6\right)_\infty.$$

Begin by showing that the 3-dissections of $\phi(-q)$ and $\psi(q)$ are

$$\phi(-q) = \frac{(q^9;q^9)^2_\infty}{(q^{18};q^{18})_\infty} - 2q\frac{(q^3;q^3)_\infty(q^{18};q^{18})^2_\infty}{(q^6;q^6)_\infty(q^9;q^9)_\infty} \tag{14.3.2}$$

and

$$\psi(q) = \frac{(q^6;q^6)_\infty(q^9;q^9)^2_\infty}{(q^3;q^3)_\infty(q^{18};q^{18})_\infty} + q\frac{(q^{18};q^{18})^2_\infty}{(q^9;q^9)_\infty}. \tag{14.3.3}$$

Now, (14.3.2) can be written

$$\phi(-q) = \phi(-q^9)\left(1 - 2qX(q^3)\right) \tag{14.3.4}$$

and (14.3.3) can be written

$$\psi(q) = \psi(q^9)\left(X(q^3)^{-1} + q\right). \tag{14.3.5}$$

If we multiply (14.3.4) and (14.3.5), we obtain (14.3.1).

14.4 Hei-Chi Chan's Most Beautiful Identity?

In this section we will prove

$$\sum_{n\geq 0} p^*(3n+2)q^n = 3\frac{F(q^3)^3}{F(q)^4}, \tag{14.4.1}$$

Hei-Chi Chan's analogue of Ramanujan's most beautiful identity.

We have, with ω a cube root of unity other than 1,

$$\sum_{n\geq 0} p^*(n)q^n = \frac{1}{F(q)} = \frac{F(\omega q)F(\omega^2 q)}{F(q)F(\omega q)F(\omega^2 q)}. \tag{14.4.2}$$

Now, (see §5.2),

$$F(q)F(\omega q)F(\omega^2 q) = \frac{F(q^3)^4}{F(q^9)}, \tag{14.4.3}$$

and by (14.3.1),

$$\begin{aligned} F(\omega q)F(\omega^2 q) &= F(q^9)^2(X(q^3)^{-1} - \omega q - 2\omega^2 q^2 X(q^3)) \\ &\quad \times (X(q^3)^{-1} - \omega^2 q - 2\omega q^2 X(q^3)) \\ &= F(q^9)^2(X(q^3)^{-2} + qX(q^3)^{-1} \\ &\quad + 3q^2 - 2q^3 X(q^3) + 4q^4 X(q^3)^2). \end{aligned} \tag{14.4.4}$$

It follows from (14.4.2), (14.4.3) and (14.4.4) that

$$\sum_{n\geq 0} p^*(n)q^n = \frac{F(q^9)^3}{F(q^3)^4}(X(q^3)^{-2} + qX(q^3)^{-1} + 3q^2 - 2q^3X(q^3) + 4q^4X(q^3)^2). \tag{14.4.5}$$

From (14.4.5) we obtain (14.4.1).

14.5 The Modular Equation of Degree 3

Let

$$\zeta = \frac{F(q)}{qF(q^9)}, \quad \alpha = \frac{1}{qX(q^3)}, \quad \beta = qX(q^3), \quad T = \frac{F(q^3)^4}{q^3F(q^9)^4}. \tag{14.5.1}$$

Then (14.3.1) gives

$$\zeta = \alpha - 1 - 2\beta. \tag{14.5.2}$$

It follows that

$$\zeta^2 = \alpha^2 - 2\alpha - 3 + 4\beta + 4\beta^2 \tag{14.5.3}$$

and

$$\zeta^3 = \alpha^3 - 3\alpha^2 - 3\alpha + 11 + 6\beta - 12\beta^2 - 8\beta^3. \tag{14.5.4}$$

It follows from (14.5.2)–(14.5.4), (14.5.1) and (14.4.3) that

$$\begin{aligned} \zeta^3 + 3\zeta^2 + 9\zeta &= \alpha^3 - 7 - 8\beta^3 \\ &= \zeta(q)\zeta(\omega q)\zeta(\omega^2 q) \end{aligned}$$

$$\begin{aligned}&= \frac{F(q)F(\omega q)F(\omega^2 q)}{q^3 F(q^9)^3}\\ &= \frac{F(q^3)^4}{q^3 F(q^9)^4}\\ &= T. \end{aligned} \tag{14.5.5}$$

This can be written

$$\frac{1}{\zeta^i} = \frac{1}{T}\left(\frac{9}{\zeta^{i-1}} + \frac{3}{\zeta^{i-2}} + \frac{1}{\zeta^{i-3}}\right). \tag{14.5.6}$$

If H is the huffing operator modulo 3,

$$H\left(\frac{1}{\zeta^i}\right) = \frac{1}{T}\left(9H\left(\frac{1}{\zeta^{i-1}}\right) + 3H\left(\frac{1}{\zeta^{i-2}}\right) + H\left(\frac{1}{\zeta^{i-3}}\right)\right). \tag{14.5.7}$$

It is easy to verify that

$$H(\zeta) = -1, \quad H(\zeta^2) = -3, \tag{14.5.8}$$

and hence that for $i \geq 1$,

$$H\left(\frac{1}{\zeta^i}\right) = \sum_{j\geq 1} \frac{m_{i,j}}{T^j}, \tag{14.5.9}$$

where the $m_{i,j}$, $i, j \geq 1$ form a matrix M, the first three rows of which are

$$\begin{pmatrix} 3 & 0 & 0 & 0 \ldots \\ 2 & 27 & 0 & 0 \ldots \\ 1 & 27 & 243 & 0 \ldots \end{pmatrix} \tag{14.5.10}$$

and for $i \geq 4$, $m_{i,1} = 0$ and for $j \geq 2$, the $m_{i,j}$ satisfy

$$m_{i,j} = 9m_{i-1,j-1} + 3m_{i-2,j-1} + m_{i-3,j-1}. \tag{14.5.11}$$

Indeed, it is easy to check that $m_{i,j} = 0$ for $i \geq 3j+1$ and for $j \geq i+1$. It follows that $m_{4i,j} = 0$ and $m_{4i+1,j} = 0$ for $1 \leq j \leq i$.

We can write

$$H\left(\frac{1}{\zeta^{4i}}\right) = \sum_{j=1}^{4i} \frac{m_{i,j}}{T^j} = \sum_{j=i+1}^{4i} \frac{m_{i,j}}{T^j} = \sum_{j=1}^{3i} \frac{a_{i,j}}{T^{i+j}} \tag{14.5.12}$$

and

$$H\left(\frac{1}{\zeta^{4i+1}}\right) = \sum_{j=1}^{4i+1} \frac{m_{i,j}}{T^j} = \sum_{j=i+1}^{4i+1} \frac{m_{i,j}}{T^j} = \sum_{j=1}^{3i+1} \frac{b_{i,j}}{T^{i+j}} \tag{14.5.13}$$

where

$$a_{i,j} = m_{4i,i+j} \text{ and } b_{i,j} = m_{4i+1,i+j}. \tag{14.5.14}$$

We can write (14.5.12)

$$H\left(\left(\frac{qF(q^9)}{F(q)}\right)^{4i}\right) = \sum_{j=1}^{3i} a_{i,j} \left(\frac{q^3F(q^9)^4}{F(q^3)^4}\right)^{i+j}$$

or,

$$H\left(q^{i-3}\frac{F(q^3)^{4i-1}}{F(q)^{4i}}\right) = \sum_{j=1}^{3i} a_{i,j}q^{3j-3}\frac{F(q^9)^{4j}}{F(q^3)^{4j+1}}. \tag{14.5.15}$$

Similarly, we can write (14.5.13)

$$H\left(\left(\frac{qF(q^9)}{F(q)}\right)^{4i+1}\right) = \sum_{j=1}^{3i+1} b_{i,j} \left(\frac{q^3F(q^9)^4}{F(q^3)^4}\right)^{i+j}$$

or,

$$H\left(q^{i-2}\frac{F(q^3)^{4i}}{F(q)^{4i+1}}\right) = \sum_{j=1}^{3i+1} b_{i,j}q^{3j-3}\frac{F(q^9)^{4j-1}}{F(q^3)^{4j}}. \tag{14.5.16}$$

14.6 The Generating Function $\sum_{n\geq 0} p^*(3^\alpha n + \delta_\alpha)q^n$

Matrices A and B are defined by (14.5.10), (14.5.11) and (14.5.14). Define coefficient vectors $\mathbf{x}_\alpha = (x_{\alpha,i})_{i\geq 1}$ recursively by

$$\mathbf{x}_1 = (3,0,0,0, \dots) \tag{14.6.1}$$

and for $\alpha \geq 1$,

$$\mathbf{x}_{\alpha+1} = \begin{cases} \mathbf{x}_\alpha A, & \alpha \text{ odd,} \\ \mathbf{x}_\alpha B. & \alpha \text{ even.} \end{cases} \tag{14.6.2}$$

Note that if $\delta_1 = 2$ and

$$\delta_{\alpha+1} = \begin{cases} 2\times 3^\alpha + \delta_\alpha & \text{if } \alpha \text{ is odd,} \\ 3^\alpha + \delta_\alpha & \text{if } \alpha \text{ is even} \end{cases} \tag{14.6.3}$$

then

$$\delta_\alpha = \begin{cases} \dfrac{5\times 3^\alpha + 1}{8} & \text{if } \alpha \text{ is odd,} \\ \dfrac{7\times 3^\alpha + 1}{8} & \text{if } \alpha \text{ is even} \end{cases} \tag{14.6.4}$$

and

$$1 \leq \delta_\alpha < 3^\alpha \text{ and } 8\delta_\alpha \equiv 1 \pmod{3^\alpha}. \tag{14.6.5}$$

That is, δ_α is the reciprocal modulo 3^α of 8.

It is now straightforward to prove that for $\alpha \geq 0$,

$$\sum_{n\geq 0} p^*(3^{2\alpha+1}n + \delta_{2\alpha+1})q^n = \sum_{i=1}^{(3^{2\alpha+2}-1)/8} x_{2\alpha+1,i}q^{i-1}\frac{F(q^3)^{4i-1}}{F(q)^{4i}},$$
$$\sum_{n\geq 0} p^*(3^{2\alpha+2}n + \delta_{2\alpha+2})q^n = \sum_{i=1}^{(3^{2\alpha+3}-3)/8} x_{2\alpha+2,i}q^{i-1}\frac{F(q^3)^{4i}}{F(q)^{4i+1}}. \tag{14.6.6}$$

14.7 Hei-Chi Chan's Partition Congruences

It is not hard to prove that, if $\nu(n)$ denotes the highest power of 3 that divides n, then

$$\nu(m_{i,j}) \geq 3j - i - 1,\ \ \nu(a_{i,j}) \geq 3j - i - 1,\ \nu(b_{i,j}) \geq 3j - i - 2, \tag{14.7.1}$$

but note that

$$\nu(a_{1,1}) = \nu(m_{4,2}) = \nu(18) = 2. \tag{14.7.2}$$

It follows that for $\alpha \geq 0$,

$$\begin{aligned} &\nu(x_{2\alpha+1,1}) = 2\alpha + 1, \\ &\nu(x_{2\alpha+1,i}) \geq 2\alpha + 1 + 3(i-1) \text{ for } i > 1, \\ &\nu(x_{2\alpha+2,1}) = 2\alpha + 3, \\ &\nu(x_{2\alpha+2,i}) \geq 2\alpha + 2 + 3(i-1) \text{ for } i > 1. \end{aligned} \tag{14.7.3}$$

It follows from (14.6.6) and (14.7.3) that for $\alpha \geq 1$,

$$\sum_{n\geq 0} p^*(3^\alpha n+\delta_\alpha)q^n \equiv 0 \pmod{3^{1+2\lfloor \frac{\alpha}{2}\rfloor}}, \tag{14.7.4}$$

and (14.1.1) follows.

14.8 Some New Results

Since $a_{1,1}/9 = 2 \equiv -1 \pmod 3$ and $b_{1,1} = 5 \equiv -1 \pmod 3$, we obtain the following result, analogous to (6.6.7), or to (7.5.26) and (7.5.27).

For $\alpha \geq 0$,

$$\sum_{n\geq 0} p^*(3^{2\alpha+1}n+\delta_{2\alpha+1})q^n \equiv 3^{2\alpha+1}F(q)^5 \pmod{3^{2\alpha+2}}, \tag{14.8.1}$$

$$\sum_{n\geq 0} p^*(3^{2\alpha+2}n+\delta_{2\alpha+2})q^n \equiv -3^{2\alpha+3}F(q)^7 \pmod{3^{2\alpha+4}}. \tag{14.8.2}$$

Indeed, we can strengthen (14.8.1) to

$$\sum_{n\geq 0} p^*(3^{2\alpha+1}n+\delta_{2\alpha+1})q^n \equiv 3^{2\alpha+1}F(q)^5 \pmod{3^{2\alpha+3}} \tag{14.8.3}$$

by using the fact that

$$F(q^3)^3 \equiv F(q)^9 \pmod 9. \tag{14.8.4}$$

Exercise: Make use of (1.5.11), (14.3.4) and (14.3.5) to obtain the 3-dissection of $(q;q)_\infty^3$ in the form

$$(q;q)_\infty^3 = (q^9;q^9)_\infty^3\left(\left(X(q^3)^{-1}+4q^3X(q^3)^2\right)-3q\right). \tag{14.8.5}$$

Exercise: Let $u = qX(q^3)$ and $v = q^{\frac{1}{3}}X(q)$. Then

$$u^{-1}-1-2u = \frac{F(q)}{qF(q^9)},\quad u^{-3}-7-8u^3 = \frac{F(q^3)^4}{q^3F(q^9)^4}\ \text{ and }\ v^{-3}-7-8v^3 = \frac{F(q)^4}{qF(q^3)^4}.$$

Hence show that

$$v^{-3}-7-8v^3 = \frac{(u^{-1}-1-2u)^4}{u^{-3}-7-8u^3}.$$

Solve this quadratic for v^3 and show that

$$v^3 = u \cdot \frac{1-u+u^2}{1+2u+4u^2}.$$

Endnotes.

It has recently been proved, independently by H. H. Chan and P. C. Toh (Chan–Toh (2010) [28]) and by X. Xiong (Xiong (2010)) [125] that for $\alpha \geq 2$,

$$p^*(5^\alpha n + \delta_\alpha) \equiv 0 \pmod{5^{\lfloor \frac{\alpha}{2} \rfloor}},$$

where δ_α is the reciprocal of 8 modulo 5^α.
I have not yet managed to find an elementary proof.

Using (8.1.1), (40.2.3), (40.2.4), (41.1.2) and (41.1.3), I have managed to show that the 5-dissection of $F(q)$ is

$$\begin{aligned} F(q) = F(q^{25}) \Bigg(& \left(\begin{matrix} q^{10}, q^{10}, q^{15}, q^{25}, q^{25}, q^{35}, q^{40}, q^{40} \\ q^5, q^5, q^{20}, q^{20}, q^{30}, q^{30}, q^{45}, q^{45} \end{matrix} ; q^{50} \right)_\infty \\ & - q \left(\begin{matrix} q^5, q^{20}, q^{20}, q^{25}, q^{25}, q^{30}, q^{30}, q^{45} \\ q^{10}, q^{10}, q^{15}, q^{15}, q^{35}, q^{35}, q^{40}, q^{40} \end{matrix} ; q^{50} \right)_\infty \\ & - 2q^2 \left(\begin{matrix} q^{15}, q^{20}, q^{30}, q^{35} \\ q^{10}, q^{25}, q^{25}, q^{40} \end{matrix} ; q^{50} \right)_\infty \\ & + q^3 \\ & - 2q^9 \left(\begin{matrix} q^5, q^{10}, q^{40}, q^{45} \\ q^{20}, q^{25}, q^{25}, q^{30} \end{matrix} ; q^{50} \right)_\infty \Bigg). \end{aligned}$$

This may be useful.

Note added in proof: I have succeeded! (May 5, 2017).

References

26. H-C. Chan, Ramanujan's cubic continued fraction and an analog of his "Most Beautiful Identity". Int. J. Number Theory **6**, 673–680 (2010)
27. H-C. Chan, Ramanujan's cubic continued fraction and Ramanujan type congruences for a certain partition function. Int. J. Number Theory **6**, 819–831 (2010)
28. H.H. Chan, P.C. Toh, New analogues of Ramanujan's partition identities. J. Number Theory **130**, 1898–1913 (2010)
125. X. Xiong, The number of cubic partitions modulo powers of 5. Scientia Sinica **41**(1), 1–15 (2011)

Chapter 15
The Rogers–Ramanujan Identities and the Rogers–Ramanujan Continued Fraction

15.1 Introduction

The Rogers–Ramanujan identities are the following.

$$\sum_{n\geq 0} \frac{q^{n^2}}{(q;q)_n} = \frac{1}{(q,q^4;q^5)_\infty}, \tag{15.1.1}$$

$$\sum_{n\geq 0} \frac{q^{n^2+n}}{(q;q)_n} = \frac{1}{(q^2,q^3;q^5)_\infty}. \tag{15.1.2}$$

These identities were first proved by Rogers (Rogers (1894) [114]), but were made famous when Ramanujan rediscovered them. He did not immediately have a proof of them, but in fact, as pointed out by Dick Askey, they can be derived by a limiting process from an (earlier) identity of his ([111], p. 194 Entry 7, [14], p.16). The identities were later independently discovered and proved by Schur (Schur (1917) [117]) and Baxter (Baxter (1980) [13]), leading Askey to say (perhaps tongue-in-cheek?) that the identities should properly be called the Rogers–Schur–Baxter identities.

Important historical note! According to notes taken by Marshall Hall Jr. of Hardy's talks at the Institute of Advanced Study in 1936 (Hall (1937) [57]), Ramanujan found a proof before he came across Rogers's proof! But Hardy doesn't make any such statement in the published version, his book "Ramanujan" (Hardy (1940) [58], esp. p. 91).

The identities are easy to discover, but somewhat difficult to prove. They have given rise to a huge body of work. The proof I will present in §15.2 is due to Ramanujan (Ramanujan (1919) [110]), while in §15.3, I will outline Andrews' proof (Andrews (1970) [3]).

We now turn to discussing the Rogers–Ramanujan continued fraction. There is some ambiguity in the literature as to the precise meaning of the term "the Rogers–Ramanujan continued fraction", but this ambiguity is of no great importance.

M.D. Hirschhorn, *The Power of q*, Developments in Mathematics 49,
DOI 10.1007/978-3-319-57762-3_15

Some writers (myself included) have, in the past, used the term to mean

$$1+\cfrac{q}{1+\cfrac{q^2}{1+\cfrac{q^3}{1+\cfrac{q^4}{1+\ddots}}}} \tag{15.1.3}$$

but nowadays, what is meant is

$$\cfrac{1}{1+\cfrac{q}{1+\cfrac{q^2}{1+\cfrac{q^3}{1+\cfrac{q^4}{1+\ddots}}}}} \tag{15.1.4}$$

or

$$\cfrac{q^{\frac{1}{5}}}{1+\cfrac{q}{1+\cfrac{q^2}{1+\cfrac{q^3}{1+\cfrac{q^4}{1+\ddots}}}}} \tag{15.1.5}$$

In this book, by "the Rogers–Ramanujan continued fraction", I will mean (15.1.4).

In §§15.4 and 15.5, we will give two demonstrations that

$$\cfrac{1}{1+\cfrac{q}{1+\cfrac{q^2}{1+\cfrac{q^3}{1+\cfrac{q^4}{1+\ddots}}}}} = R(q), \tag{15.1.6}$$

where $R(q)$ is defined in (8.1.2).

The first derivation is Ramanujan's original, which avoids the use of the Rogers–Ramanujan identities, the second makes use of Rogers–Ramanujan identities.

A finite version of (15.1.6) is given in §15.6.

In §15.7, we will complete the derivation of the "difficult and deep" identity of Ramanujan, begun in Chapter 9.

15.2 Ramanujan's Proof of the Rogers–Ramanujan Identities

We prove the following identity, from which we obtain (15.1.1) on setting $a = 1$, and (15.1.2) on setting $a = q$, each time making use of JTP.

$$(aq;q)_\infty \sum_{n\geq 0} \frac{a^n q^{n^2}}{(q;q)_n} = 1 + \sum_{k\geq 1} (-1)^k a^{2k} q^{(5k^2-k)/2}(1-aq^{2k})\frac{(aq;q)_{k-1}}{(q;q)_k}. \quad (15.2.1)$$

Let $G(a)$ denote the right side of (15.2.1),

$$G(a) = 1 + \sum_{k\geq 1} (-1)^k a^{2k} q^{(5k^2-k)/2}(1-aq^{2k})\frac{(aq;q)_{k-1}}{(q;q)_k}. \quad (15.2.2)$$

Then

$$\begin{aligned} G(a) &= 1 + \sum_{k\geq 1} (-1)^k a^{2k} q^{(5k^2-k)/2}((1-q^k)+q^k(1-aq^k))\frac{(aq;q)_{k-1}}{(q;q)_k} \\ &= 1 + \sum_{k\geq 1} (-1)^k a^{2k} \left(q^{(5k^2-k)/2}\frac{(aq;q)_{k-1}}{(q;q)_{k-1}} + q^{(5k^2+k)/2}\frac{(aq;q)_k}{(q;q)_k}\right) \\ &= 1 - a^2q^2 + \sum_{k\geq 1} (-1)^k a^{2k} q^{(5k^2+k)/2}(1-a^2q^{4k+2})\frac{(aq;q)_k}{(q;q)_k}. \quad (15.2.3) \end{aligned}$$

If we divide (15.2.3) by $1 - aq$, we obtain

$$\frac{G(a)}{1-aq} = 1 + aq + \sum_{k\geq 1} (-1)^k a^{2k} q^{(5k^2+k)/2}(1-a^2q^{4k+2})\frac{(aq^2;q)_{k-1}}{(q;q)_k}. \quad (15.2.4)$$

Also, from (15.2.2) we have

$$G(aq) = 1 + \sum_{k\geq 1} (-1)^k a^{2k} q^{(5k^2+3k)/2}(1-aq^{2k+1})\frac{(aq^2;q)_{k-1}}{(q;q)_k}. \quad (15.2.5)$$

If we subtract (15.2.5) from (15.2.4), we find

$$\begin{aligned} &\frac{G(a)}{1-aq} - G(aq) \\ &= aq + \sum_{k\geq 1} (-1)^k a^{2k} q^{(5k^2+k)/2}((1-a^2q^{4k+2}) - q^k(1-aq^{2k+1}))\frac{(aq^2;q)_{k-1}}{(q;q)_k} \\ &= aq + \sum_{k\geq 1} (-1)^k a^{2k} q^{(5k^2+k)/2}((1-q^k) + aq^{3k+1}(1-aq^{k+1}))\frac{(aq^2;q)_{k-1}}{(q;q)_k} \end{aligned}$$

$$= aq + \sum_{k\geq 1}(-1)^k a^{2k}\left(q^{(5k^2+k)/2}\frac{(aq^2;q)_{k-1}}{(q;q)_{k-1}} + aq^{(5k^2+7k+2)/2}\frac{(aq^2;q)_k}{(q;q)_k}\right)$$

$$= aq - a^2q^3 + \sum_{k\geq 1}(-1)^k a^{2k+1}q^{(5k^2+7k+2)/2}(1-aq^{2k+2})\frac{(aq^2;q)_k}{(q;q)_k}$$

$$= aq(1-aq^2) + aq(1-aq^2)\sum_{k\geq 1}(-1)^k a^{2k}q^{(5k^2+7k)/2}(1-aq^{2k+2})\frac{(aq^3;q)_{k-1}}{(q;q)_k}$$

$$= aq(1-aq^2)G(aq^2). \tag{15.2.6}$$

Thus we have

$$G(a) = (1-aq)G(aq) + aq(1-aq)(1-aq^2)G(aq^2). \tag{15.2.7}$$

If we now write

$$G(a) = (aq;q)_\infty F(a), \tag{15.2.8}$$

(15.2.7) becomes

$$F(a) = F(aq) + aqF(aq^2). \tag{15.2.9}$$

The solution of (15.2.9) with $F(0) = 1$ is

$$F(a) = \sum_{n\geq 0}\frac{a^n q^{n^2}}{(q;q)_n}. \tag{15.2.10}$$

It follows that

$$G(a) = (aq;q)_\infty \sum_{n\geq 0}\frac{a^n q^{n^2}}{(q;q)_n},$$

which is (15.2.1).

15.3 Andrews' Proof of the Rogers–Ramanujan Identities

Andrews (1970) [3] proved the following polynomial identity by induction on n.

If $\alpha = 0$ or 1,

$$\sum_{k\geq 0} q^{k^2+\alpha k}\begin{bmatrix} n+1-\alpha-k \\ k \end{bmatrix}_q = \sum_{k=-\infty}^{\infty}(-1)^k q^{(5k^2+k)/2-2\alpha k}\begin{bmatrix} n+1 \\ \lfloor (n+1-5k)/2\rfloor+\alpha \end{bmatrix}_q. \tag{15.3.1}$$

If in (15.3.1) we let $n \to \infty$ and use JTP, we obtain both (15.1.1) and (15.1.2) .

15.4 The Rogers–Ramanujan Continued Fraction (Ramanujan's Original Derivation)

The identity (15.1.6) is the special case $a = 1$ of the following result, which we obtain directly from (15.2.7).

$$\cfrac{1}{1+\cfrac{aq}{1+\cfrac{aq^2}{1+\cfrac{aq^3}{1+\cfrac{aq^4}{1+\ddots}}}}} = \frac{\sum_{k\geq 0}(-1)^k a^{2k} q^{(5k^2+3k)/2}(1-aq^{2k+1})\frac{(aq;q)_k}{(q;q)_k}}{\sum_{k\geq 0}(-1)^k a^{2k} q^{(5k^2+k)/2}(1-a^2q^{4k+2})\frac{(aq;q)_k}{(q;q)_k}}. \tag{15.4.1}$$

Thus (15.2.7) can be written

$$\frac{G(a)}{(1-aq)G(aq)} = 1 + \frac{aq(1-aq^2)G(aq^2)}{G(aq)}, \tag{15.4.2}$$

or,

$$\frac{(1-aq)G(aq)}{G(a)} = \frac{1}{1+aq\cdot\left(\frac{(1-aq^2)G(aq^2)}{G(aq)}\right)}. \tag{15.4.3}$$

If we iterate (15.4.3), we find

$$\frac{(1-aq)G(aq)}{G(a)} = \cfrac{1}{1+\cfrac{aq}{1+\cfrac{aq^2}{1+\ddots\cfrac{aq^{n-1}}{1+aq^n\cdot\left(\cfrac{(1-aq^{n+1})G(aq^{n+1})}{G(aq^n)}\right)}}}}. \tag{15.4.4}$$

If we take $|q| < 1$ and let $n \to \infty$, $\dfrac{(1-aq^{n+1})G(aq^{n+1})}{G(aq^n)} \to 1$, and we find

$$\frac{(1-aq)G(aq)}{G(a)} = \cfrac{1}{1+\cfrac{aq}{1+\cfrac{aq^2}{1+\cfrac{aq^3}{1+\cfrac{aq^4}{1+\ddots}}}}}. \tag{15.4.5}$$

If we now make use of (15.2.5) and (15.2.3), we obtain (15.4.1).

15.5 The Rogers–Ramanujan Continued Fraction (Second Derivation)

We can write (15.2.9)

$$\frac{F(a)}{F(aq)} = 1 + \frac{aq}{\left(\dfrac{F(aq)}{F(aq^2)}\right)}, \tag{15.5.1}$$

or,

$$\frac{F(aq)}{F(a)} = \frac{1}{1+aq\cdot\left(\dfrac{F(aq^2)}{F(aq)}\right)}. \tag{15.5.2}$$

If we iterate this, we find

$$\frac{F(aq)}{F(a)} = \cfrac{1}{1+\cfrac{aq}{1+\cfrac{aq^2}{1+\cfrac{aq^3}{1+\ddots\cfrac{aq^{n-1}}{1+aq^n\cdot\left(\cfrac{F(aq^{n+1})}{F(aq^n)}\right)}}}}}. \tag{15.5.3}$$

If we take $|q| < 1$ and let $n \to \infty$, $F(aq^n) \to F(0) = 1$, and we find

$$\frac{F(aq)}{F(a)} = \cfrac{1}{1+\cfrac{aq}{1+\cfrac{aq^2}{1+\cfrac{aq^3}{1+\cfrac{aq^4}{1+\ddots}}}}}. \tag{15.5.4}$$

That is,

$$\cfrac{1}{1+\cfrac{aq}{1+\cfrac{aq^2}{1+\cfrac{aq^3}{1+\cfrac{aq^4}{1+\ddots}}}}} = \sum_{n\geq 0}\frac{a^n q^{n^2+n}}{(q;q)_n} \Big/ \sum_{n\geq 0}\frac{a^n q^{n^2}}{(q;q)_n}. \tag{15.5.5}$$

If we now set $a = 1$ and make use of the Rogers–Ramanujan identities (15.1.1) and (15.1.2), we find

$$\cfrac{1}{1+\cfrac{q}{1+\cfrac{q^2}{1+\cfrac{q^3}{1+\cfrac{q^4}{1+\ddots}}}}} = \sum_{n\geq 0}\frac{q^{n^2+n}}{(q;q)_n} \Big/ \sum_{n\geq 0}\frac{q^{n^2}}{(q;q)_n} = \left(\begin{matrix} q, q^4 \\ q^2, q^3 \end{matrix}; q^5\right)_\infty,$$

which is (15.1.6).

15.6 A Finite Version of the Rogers–Ramanujan Continued Fraction

For $n \geq 0$, let

$$P_n(a) = \sum_{k \geq 0} a^k q^{k^2} \begin{bmatrix} n+1-k \\ k \end{bmatrix}_q. \tag{15.6.1}$$

Then

$$P_0(a) = 1 \text{ and } P_1(a) = 1 + aq, \tag{15.6.2}$$

and it follows directly from the recurrence for $\begin{bmatrix} n \\ k \end{bmatrix}_q$ (see §1.8) that

$$P_n(a) = P_{n-1}(aq) + aqP_{n-2}(aq^2). \tag{15.6.3}$$

This can be written

$$\frac{P_n(a)}{P_{n-1}(aq)} = 1 + \frac{aq}{\left(\dfrac{P_{n-1}(aq)}{P_{n-2}(aq^2)}\right)}, \tag{15.6.4}$$

or,

$$\frac{P_{n-1}(aq)}{P_n(a)} = \frac{1}{1 + aq \cdot \left(\dfrac{P_{n-2}(aq^2)}{P_{n-1}(aq)}\right)}. \tag{15.6.5}$$

By iteration (or by induction on n),

$$\frac{P_{n-1}(aq)}{P_n(a)} = \cfrac{1}{1 + \cfrac{aq}{1 + \cfrac{aq^2}{1 + \cfrac{aq^3}{1 + \ddots \cfrac{aq^{n-1}}{1 + \cfrac{aq^n}{1}}}}}}. \tag{15.6.6}$$

That is,

$$\cfrac{1}{1+\cfrac{aq}{1+\cfrac{aq^2}{1+\cfrac{aq^3}{1+\ddots\cfrac{aq^{n-1}}{1+\cfrac{aq^n}{1}}}}}} = \sum_{k\geq 0} a^k q^{k^2+k} \begin{bmatrix} n-k \\ k \end{bmatrix}_q \Big/ \sum_{k\geq 0} a^k q^{k^2} \begin{bmatrix} n+1-k \\ k \end{bmatrix}_q . \tag{15.6.7}$$

If in (15.6.7) we take $|q| < 1$ and let $n \longrightarrow \infty$ and we obtain (15.4.5). If we then set $a = 1$, we obtain (15.1.6).

Alternatively, if in (15.6.7) we set $a = 1$, use (15.3.1) and then let $n \longrightarrow \infty$, we obtain (15.1.6).

15.7 The "Difficult and Deep" Identity of Chapter 9

If we combine (9.2.14) and (9.2.15) with (15.1.6), we are able to say with Ramanujan that

If

$$u = \cfrac{q}{1+\cfrac{q^5}{1+\cfrac{q^{10}}{1+\cfrac{q^{15}}{1+\ddots}}}} \quad \text{and} \quad v = \cfrac{q^{\frac{1}{5}}}{1+\cfrac{q}{1+\cfrac{q^2}{1+\cfrac{q^3}{1+\ddots}}}}$$

then

$$v^5 = u\frac{1-2u+4u^2-3u^3+u^4}{1+3u+4u^2+2u^3+u^4}. \tag{15.7.1}$$

Endnotes.

§15.2 I gave a slightly different proof of (15.2.7) in Hirschhorn (2014) [90].

§15.6 I would like to be able to claim that I was the first to publish (15.6.7) (Hirschhorn (1972) [61]), but, in fact it appears in a paper of P. Kesava Menon (Menon (1965) [99]). And indeed, it is in Ramanujan's Notebooks ([111], p. 196) (See [14], p. 31 for a slightly different proof).

§15.6 My introduction to Ramanujan's work came in 1969, when I was studying under George Szekeres. A combinatorial problem he was working on gave rise to the

Rogers–Ramanujan continued fraction as a generating function. In trying to make sense of the infinite continued fraction, I was led to discover (15.6.7).

§§15.7, 9.1 My proofs of Ramanujan's result (15.7.1) were inspired by those of C. Gugg (2009) [55] and G.N. Watson (1929) [122]. A proof was given also by H.-C. Chan (2010) [25].

References

11. G.E. Andrews, A polynomial identity which implies the Rogers-Ramanujan identities. Scripta Math. **28**, 297–305 (1970)
13. R.J. Baxter, Hard hexagons: exact solution. J. Phys. A **13**, L61–L70 (1980)
14. B.C. Berndt, *Ramanujan's Notebooks, Part III* (Springer, Berlin, 1991)
25. H.-C. Chan, On the Andrews-Schur's proof of the Rogers-Ramanujan identities. Ramanujan J. **23**, 417–431 (2010)
55. C. Gugg, A new proof of Ramanujan's modular equation relating $R(q)$ with $R(q^5)$. Ramanujan J. **20**, 163–177 (2009)
57. M. Hall Jr., Lectures by Godfrey Harold Hardy on the Mathematical Work of Ramanujan, Institute of Advanced Study, Princeton (1936)
58. G.H. Hardy, *Ramanujan. Twelve Lectures on Subjects Suggested by His Life and Work* (AMS Chelsea, Providence, 1999)
61. M.D. Hirschhorn, Partitions and Ramanujan's continued fraction. Duke Math. J. **39**, 789–791 (1972)
90. M.D. Hirschhorn, An identity that may have changed the course of history, *Ramanujan 125*, vol. 627, Contemporary Mathematics (AMS, Providence, 2014), pp. 103–109
99. P. Kesava Menon, On Ramanujan's continued fraction and related identities. J. Lond. Math. Soc. **40**, 49–54 (1965)
110. S. Ramanujan, Proof of certain identities in combinatory analysis. Proc. Camb. Philos. Soc. **XIX**, 214–216 (1919)
111. S. Ramanujan, Notebooks, II, Tata Institute of Fundamental Research, Mumbai (2012)
114. L.J. Rogers, Second memoir on the expansion of certain infinite products. Proc. Lond. Math. Soc. (2) **XXV**, 318–343 (1894)
117. I. Schur, Ein Beitrag zur additiven Zahlentheorie und zur Theorie der Kettenbrüche. Berliner Sitzungsberichte **23**, 301–321 (1917)
122. G.N. Watson, Theorems stated by Ramanujan (VII): theorems on continued fractions. J. Lond. Math. Soc. **4**, 39–48 (1929)

Chapter 16
The Series Expansion of the Rogers–Ramanujan Continued Fraction and Its Reciprocal

16.1 Introduction

In 1968 or early 1969, George Szekeres expanded $R(q)$ and $R(q)^{-1}$ as series (we will see in §16.2 how he might have done this), and observed a surprising phenomenon. We will write*

$$R(q) = \sum_{n\geq 0} u_n q^n, \quad R(q)^{-1} = \sum_{n\geq 0} v_n q^n. \tag{16.1.1}$$

With some early exceptions, the sign of the u_n and of the v_n is periodic with period 5. Thus, apart from the facts that

$$u_3 = u_8 = u_{13} = u_{23} = 0, \quad v_2 = v_4 = v_9 = 0, \tag{16.1.2}$$

we have

$$u_n \begin{cases} > 0 \text{ if } n \equiv 0 \text{ or } 2 \pmod 5, \\ < 0 \text{ if } n \equiv 1,\ 3 \text{ or } 4 \pmod 5, \end{cases} \tag{16.1.3}$$

$$v_n \begin{cases} > 0 \text{ if } n \equiv 0 \text{ or } 1 \pmod 5, \\ < 0 \text{ if } n \equiv 2,\ 3 \text{ or } 4 \pmod 5. \end{cases} \tag{16.1.4}$$

The object of this chapter is to give straightforward proofs of (16.1.2)–(16.1.4).

Incidentally, it is interesting to note that when the Lost Notebook came to light some years after Szekeres made his observations, it was revealed that Ramanujan had calculated the u_n and v_n for n beyond 1000 ! ([112], p. 49). But he left us no conclusions. (*The definitions of u_n and v_n are Ramanujan's.)

M.D. Hirschhorn, *The Power of q*, Developments in Mathematics 49,
DOI 10.1007/978-3-319-57762-3_16

16.2 Expanding the Rogers–Ramanujan Continued Fraction and Its Reciprocal

We have (15.1.6)

$$\sum_{n\geq 0} u_n q^n = R(q) = \left(\begin{matrix} q, q^4 \\ q^2, q^3 \end{matrix}; q^5\right)_\infty \tag{16.2.1}$$

$$= \frac{(q,q^4,q^5;q^5)_\infty}{(q^2,q^3,q^5;q^5)_\infty}$$

$$= \sum_{k\in\mathbb{Z}} (-1)^k q^{(5k^2-3k)/2} \Big/ \sum_{k\in\mathbb{Z}} (-1)^k q^{(5k^2-k)/2} \tag{16.2.2}$$

so

$$\sum_{k\in\mathbb{Z}} (-1)^k q^{(5k^2-k)/2} \sum_{n\geq 0} u_n q^n = \sum_{k\in\mathbb{Z}} (-1)^k q^{(5k^2-3k)/2}, \tag{16.2.3}$$

from which it follows that

$$u_n - u_{n-2} - u_{n-3} + u_{n-9} + u_{n-11} - - + + \cdots$$
$$= \begin{cases} 0 \text{ unless } n \text{ is of the form } (5k^2-3k)/2,\ k\in\mathbb{Z}, \\ (-1)^k \text{ if } n \text{ is of the form } (5k^2-3k)/2,\ k\in\mathbb{Z}, \end{cases} \tag{16.2.4}$$

where 2, 3, 9, 11, ... are the numbers of the form $(5k^2-k)/2$, $k \in \mathbb{Z}$.

Similarly,

$$v_n - v_{n-1} - v_{n-4} + v_{n-7} + v_{n-13} - - + + \cdots$$
$$= \begin{cases} 0 \text{ unless } n \text{ is of the form } (5k^2-k)/2,\ k\in\mathbb{Z}, \\ (-1)^k \text{ if } n \text{ is of the form } (5k^2-k)/2,\ k\in\mathbb{Z}, \end{cases} \tag{16.2.5}$$

where 1, 4, 7, 13, ... are the numbers of the form $(5k^2-3k)/2$, $k \in \mathbb{Z}$.

Alternative recurrences may be found by differentiating (16.2.1) and its reciprocal. If I remember correctly, Szekeres used these recurrences to calculate the u_n and v_n.

Exercise: Obtain these recurrences.

16.3 Proofs of Szekeres's Observations

We have, using the quintuple product identity (10.1.3),

$$\begin{aligned}
\sum_{n\geq 0} u_n q^n &= \begin{pmatrix} q, q^4 \\ q^2, q^3 \end{pmatrix}; q^5 \Big)_\infty \\
&= \frac{1}{(q^5;q^5)_\infty} \begin{pmatrix} q, q^4, q^5 \\ q^2, q^3 \end{pmatrix}; q^5 \Big)_\infty \\
&= \frac{1}{(q^5;q^5)_\infty} \left(\sum_{-\infty}^{\infty} (-1)^n q^{(15n^2+7n)/2} - \sum_{-\infty}^{\infty} (-1)^n q^{(15n^2+13n+2)/2} \right).
\end{aligned} \tag{16.3.1}$$

We now split each sum into five, according to the residue modulo 5 of n. Thus we replace n by $5n$, $5n+1$, $5n+2$, $5n-1$ and $5n-2$ to obtain

$$\begin{aligned}
\sum_{n\geq 0} u_n q^n = \frac{1}{(q^5;q^5)_\infty} \Bigg(&\sum_{-\infty}^{\infty} (-1)^n q^{(375n^2+35n)/2} - \sum_{-\infty}^{\infty} (-1)^n q^{(375n^2+185n+22)/2} \\
&+ \sum_{-\infty}^{\infty} (-1)^n q^{(375n^2+335n+74)/2} - \sum_{-\infty}^{\infty} (-1)^n q^{(375n^2+115n+8)/2} \\
&+ \sum_{-\infty}^{\infty} (-1)^n q^{(375n^2+265n+46)/2} - \sum_{-\infty}^{\infty} (-1)^n q^{(375n^2+65n+2)/2} \\
&+ \sum_{-\infty}^{\infty} (-1)^n q^{(375n^2+215n+30)/2} - \sum_{-\infty}^{\infty} (-1)^n q^{(375n^2+365n+88)/2} \\
&+ \sum_{-\infty}^{\infty} (-1)^n q^{(375n^2+85n+4)/2} - \sum_{-\infty}^{\infty} (-1)^n q^{(375n^2+235n+36)/2} \Bigg).
\end{aligned} \tag{16.3.2}$$

We now group these ten sums into pairs according to the residue modulo 5 of the indices.

$$\begin{aligned}
\sum_{n\geq 0} u_n q^n = \frac{1}{(q^5;q^5)_\infty} \Bigg(&\left(\sum_{-\infty}^{\infty} (-1)^n q^{(375n^2+35n)/2} + q^{15} \sum_{-\infty}^{\infty} (-1)^n q^{(375n^2+215n)/2} \right) \\
&- q \left(\sum_{-\infty}^{\infty} (-1)^n q^{(375n^2+65n)/2} + q^{10} \sum_{-\infty}^{\infty} (-1)^n q^{(375n^2+185n)/2} \right) \\
&+ q^2 \left(\sum_{-\infty}^{\infty} (-1)^n q^{(375n^2+85n)/2} + q^{35} \sum_{-\infty}^{\infty} (-1)^n q^{(375n^2+335n)/2} \right) \\
&- q^{18} \left(\sum_{-\infty}^{\infty} (-1)^n q^{(375n^2+235n)/2} - q^5 \sum_{-\infty}^{\infty} (-1)^n q^{(375n^2+265n)/2} \right) \\
&- q^4 \left(\sum_{-\infty}^{\infty} (-1)^n q^{(375n^2+115n)/2} + q^{40} \sum_{-\infty}^{\infty} (-1)^n q^{(375n^2+365n)/2} \right) \Bigg).
\end{aligned} \tag{16.3.3}$$

Each pair can now be summed by the quintuple product identity (10.1.3) with q replaced by q^{125}, and $a = q^{-15}$, q^{-10}, q^{-35}, q^{-60}, q^{-40} respectively, to give

$$\begin{aligned}\sum_{n\geq 0} u_n q^n = \frac{1}{(q^5;q^5)_\infty}\Bigg(&\left(\begin{matrix} q^{30}, q^{95}, q^{125} \\ q^{15}, q^{110} \end{matrix} ; q^{125}\right)_\infty - q\left(\begin{matrix} q^{20}, q^{105}, q^{125} \\ q^{10}, q^{115} \end{matrix} ; q^{125}\right)_\infty \\ &+q^2\left(\begin{matrix} q^{55}, q^{70}, q^{125} \\ q^{35}, q^{90} \end{matrix} ; q^{125}\right)_\infty - q^{18}\left(\begin{matrix} q^{5}, q^{120}, q^{125} \\ q^{60}, q^{65} \end{matrix} ; q^{125}\right)_\infty \\ &-q^4\left(\begin{matrix} q^{45}, q^{80}, q^{125} \\ q^{40}, q^{85} \end{matrix} ; q^{125}\right)_\infty\Bigg). \end{aligned} \tag{16.3.4}$$

It follows that, for example, that

$$\begin{aligned}\sum_{n\geq 0} u_{5n} q^n &= \frac{1}{(q;q)_\infty}\left(\begin{matrix} q^{6}, q^{19}, q^{25} \\ q^{3}, q^{22} \end{matrix} ; q^{25}\right)_\infty \\ &= 1/(q, q^2, q^3, q^3, q^4, q^5, q^7, q^8, q^9, q^{10}, q^{11}, q^{12}, q^{13}, q^{14}, q^{15}, q^{16}, \\ &\qquad q^{17}, q^{18}, q^{20}, q^{21}, q^{22}, q^{22}, q^{23}, q^{24}; q^{25})_\infty, \end{aligned} \tag{16.3.5}$$

Similarly,

$$\begin{aligned}\sum_{n\geq 0} u_{5n+1} q^n = -1/(&q, q^2, q^2, q^3, q^5, q^6, q^7, q^8, q^9, q^{10}, q^{11}, q^{12}, q^{13}, q^{14}, q^{15}, q^{16}, \\ &q^{17}, q^{18}, q^{19}, q^{20}, q^{22}, q^{23}, q^{23}, q^{24}; q^{25})_\infty, \end{aligned} \tag{16.3.6}$$

$$\begin{aligned}\sum_{n\geq 0} u_{5n+2} q^n = 1/(&q, q^2, q^3, q^4, q^5, q^6, q^7, q^7, q^8, q^9, q^{10}, q^{12}, q^{13}, q^{15}, q^{16}, q^{17}, \\ &q^{18}, q^{18}, q^{19}, q^{20}, q^{21}, q^{22}, q^{23}, q^{24}; q^{25})_\infty, \end{aligned} \tag{16.3.7}$$

$$\begin{aligned}\sum_{n\geq 0} u_{5n+3} q^n = -q^3/(&q^2, q^3, q^4, q^5, q^6, q^7, q^8, q^9, q^{10}, q^{11}, q^{12}, q^{12}, q^{13}, q^{13}, \\ &q^{14}, q^{15}, q^{16}, q^{17}, q^{18}, q^{19}, q^{20}, q^{21}, q^{22}, q^{23}; q^{25})_\infty \end{aligned} \tag{16.3.8}$$

and

$$\begin{aligned}\sum_{n\geq 0} u_{5n+4} q^n = -1/(&q, q^2, q^3, q^4, q^5, q^6, q^7, q^8, q^8, q^{10}, q^{11}, q^{12}, q^{13}, q^{14}, q^{15}, q^{17}, \\ &q^{17}, q^{18}, q^{19}, q^{20}, q^{21}, q^{22}, q^{23}, q^{24}; q^{25})_\infty. \end{aligned} \tag{16.3.9}$$

In the same way, we find that

$$\sum_{n\geq 0} v_n q^n = \frac{1}{(q^5;q^5)_\infty}\Bigg(\left(\begin{matrix} q^{40},q^{85},q^{125} \\ q^{20},q^{105} \end{matrix};q^{125}\right)_\infty + q\left(\begin{matrix} q^{60},q^{65},q^{125} \\ q^{30},q^{95} \end{matrix};q^{125}\right)_\infty$$
$$-q^7\left(\begin{matrix} q^{35},q^{90},q^{125} \\ q^{45},q^{80} \end{matrix};q^{125}\right)_\infty - q^3\left(\begin{matrix} q^{10},q^{115},q^{125} \\ q^{5},q^{120} \end{matrix};q^{125}\right)_\infty$$
$$-q^{14}\left(\begin{matrix} q^{15},q^{110},q^{125} \\ q^{55},q^{70} \end{matrix};q^{125}\right)_\infty\Bigg), \tag{16.3.10}$$

$$\sum_{n\geq 0} v_{5n} q^n = 1/(q,q^2,q^3,q^4,q^4,q^5,q^6,q^7,q^9,q^{10},q^{11},q^{12},q^{13},q^{14},q^{15},q^{16},q^{18},$$
$$q^{19},q^{20},q^{21},q^{21},q^{22},q^{23},q^{24};q^{25})_\infty, \tag{16.3.11}$$
$$\sum_{n\geq 0} v_{5n+1} q^n = 1/(q,q^2,q^3,q^4,q^5,q^6,q^6,q^7,q^8,q^9,q^{10},q^{11},q^{14},q^{15},q^{16},q^{17},q^{18},$$
$$q^{19},q^{19},q^{20},q^{21},q^{22},q^{23},q^{24};q^{25})_\infty, \tag{16.3.12}$$
$$\sum_{n\geq 0} v_{5n+2} q^n = -q/(q,q^2,q^3,q^4,q^5,q^6,q^8,q^9,q^9,q^{10},q^{11},q^{12},q^{13},q^{14},q^{15},q^{16},$$
$$q^{16},q^{17},q^{19},q^{20},q^{21},q^{22},q^{23},q^{24};q^{25})_\infty, \tag{16.3.13}$$
$$\sum_{n\geq 0} v_{5n+3} q^n = -1/(q,q,q^3,q^4,q^5,q^6,q^7,q^8,q^9,q^{10},q^{11},q^{12},q^{13},q^{14},q^{15},q^{16},$$
$$q^{17},q^{18},q^{19},q^{20},q^{21},q^{22},q^{24},q^{24};q^{25})_\infty \tag{16.3.14}$$

and

$$\sum_{n\geq 0} v_{5n+4} q^n = -q^2/(q,q^2,q^4,q^5,q^6,q^7,q^8,q^9,q^{10},q^{11},q^{11},q^{12},q^{13},q^{14},q^{14},q^{15},$$
$$q^{16},q^{17},q^{18},q^{19},q^{20},q^{21},q^{23},q^{24};q^{25})_\infty. \tag{16.3.15}$$

It follows that

$$u_3 = u_8 = u_{13} = u_{23} = 0, \;\; v_2 = v_4 = v_9 = 0. \tag{16.3.16}$$

Now let us write $\sum_{n\geq 0} a_n q^n > \sum_{n\geq 0} b_n q^n$ if $a_n \geq b_n$ for all $n \geq 0$. Then all of

$$\sum_{n\geq 0} u_{5n}q^n,\; \sum_{n\geq 0} -u_{5n+1}q^n,\; \sum_{n\geq 0} u_{5n+2}q^n,\; \sum_{n\geq 0} -u_{5n+4}q^n,$$

$$\sum_{n\geq 0} v_{5n}q^n,\; \sum_{n\geq 0} v_{5n+1}q^n,\; \sum_{n\geq 0} -v_{5n+7}q^n,\; \sum_{n\geq 0} -v_{5n+3}q^n \text{ and } \sum_{n\geq 0} -v_{5n+14}q^n$$

are greater than

$$\frac{1}{1-q} = 1+q+q^2+\cdots, \tag{16.3.17}$$

and

$$\begin{aligned}\sum_{n\geq 0} -u_{5n+18}q^n &> \frac{1}{(1-q^2)(1-q^3)} = \frac{1+q^3+\cdots}{1-q^2} > \frac{1+q^3}{1-q^2} \\ &= \frac{1-q+q^2}{1-q} = 1+\frac{q^2}{1-q} \\ &= 1+q^2+q^3+q^4+\cdots. \end{aligned} \tag{16.3.18}$$

(16.1.2)–(16.1.4) follow from (16.3.16)–(16.3.18).

Endnotes.

The first proofs of Szekeres's observations were given by Andrews (1981) [4]. They are based on some formulas in [112], and are very much combinatorial. These proofs were later simplified by Dennis Eichhorn ([6], p. 111).

Earlier, Richmond and Szekeres (1978) [113] had used the circle method to show that

$$u_n \sim \frac{\sqrt{2}}{(5n)^{\frac{3}{4}}} \exp\left\{\frac{4\pi}{25}\sqrt{5n}\right\}\left(\cos\left(\frac{4\pi n}{5}+\frac{3\pi}{25}\right)+\mathrm{O}(n^{-\frac{1}{2}})\right)$$

and

$$v_n \sim \frac{\sqrt{2}}{(5n)^{\frac{3}{4}}} \exp\left\{\frac{4\pi}{25}\sqrt{5n}\right\}\left(\cos\left(\frac{2\pi n}{5}-\frac{4\pi}{25}\right)+\mathrm{O}(n^{-\frac{1}{2}})\right)$$

which imply that (16.1.3) and (16.1.4) hold for n sufficiently large.
(The dominant terms had been found in 1969 by Hirschhorn and Szekeres but they were not able to show the error term is small. As a consequence, that work was not published. Nor, disappointingly, was it referred to in Richmond–Szekeres [113].)

Somewhat different proofs of Szekeres's observations are given in Hirschhorn (1998) [71]. The proofs given in the text are new from (16.3.16) on.

Ramanujan ([112], p. 50) had a version of (16.3.3) and of the corresponding formula for the v_n, but failed to notice the step to (16.3.4). (Luckily for me!)

References

4. G.E. Andrews, Ramanujan's "Lost" notebook III the Rogers-Ramanujan continued fraction. Adv. Math. **41**, 186–208 (1981)
6. G.E. Andrews, B.C. Berndt, *Ramanujan's Lost Notebook, Part I* (Springer, Berlin, 2005)

71. M.D. Hirschhorn, On the expansion of Ramanujan's continued fraction. Ramanujan J. **2**, 521–527 (1998)
112. S. Ramanujan, *The Lost Notebook and Other Unpublished Papers* (Springer, Berlin, 1988)
113. L.B. Richmond, G. Szekeres, The Taylor coefficients of certain infinite products. Acta Scientiarum Mathematicarum **40**, 347–369 (1978)

Chapter 17
The 2- and 4-Dissections of the Rogers–Ramanujan Continued Fraction and Its Reciprocal

17.1 Introduction

In Chapter 16, we found the 5-dissections of the Rogers–Ramanujan continued fraction and its reciprocal.

In this chapter, we will give the 2-dissections,

$$R(q) = \left(\begin{matrix} q^4, q^4, q^{16}, q^{16} \\ q^2, q^{10}, q^{10}, q^{18} \end{matrix} ; q^{20} \right)_\infty - q \left(\begin{matrix} q^4, q^6, q^{14}, q^{16} \\ q^8, q^{10}, q^{10}, q^{12} \end{matrix} ; q^{20} \right)_\infty \tag{17.1.1}$$

and

$$R(q)^{-1} = \left(\begin{matrix} q^8, q^8, q^{12}, q^{12} \\ q^6, q^{10}, q^{10}, q^{14} \end{matrix} ; q^{20} \right)_\infty + q \left(\begin{matrix} q^2, q^8, q^{12}, q^{18} \\ q^4, q^{10}, q^{10}, q^{16} \end{matrix} ; q^{20} \right)_\infty \tag{17.1.2}$$

as well as the 4-dissections,

$$\begin{aligned} R(q) &= \left(\begin{matrix} q^4, q^{16}, q^{16}, q^{24}, q^{24}, q^{24}, q^{24}, q^{36}, q^{44}, q^{56}, q^{56}, q^{56}, q^{56}, q^{64}, q^{64}, q^{76} \\ q^{12}, q^{12}, q^{20}, q^{20}, q^{20}, q^{20}, q^{28}, q^{28}, q^{52}, q^{52}, q^{60}, q^{60}, q^{60}, q^{60}, q^{68}, q^{68} \end{matrix} ; q^{80} \right)_\infty \\ &- q \left(\begin{matrix} q^4, q^{16}, q^{16}, q^{16}, q^{24}, q^{24}, q^{24}, q^{36}, q^{44}, q^{56}, q^{56}, q^{56}, q^{64}, q^{64}, q^{64}, q^{76} \\ q^8, q^{12}, q^{20}, q^{20}, q^{20}, q^{20}, q^{28}, q^{32}, q^{48}, q^{52}, q^{60}, q^{60}, q^{60}, q^{60}, q^{68}, q^{72} \end{matrix} ; q^{80} \right)_\infty \\ &+ q^2 \left(\begin{matrix} q^4, q^{16}, q^{16}, q^{16}, q^{16}, q^{24}, q^{24}, q^{36}, q^{44}, q^{56}, q^{56}, q^{64}, q^{64}, q^{64}, q^{64}, q^{76} \\ q^8, q^8, q^{20}, q^{20}, q^{20}, q^{20}, q^{32}, q^{32}, q^{48}, q^{48}, q^{60}, q^{60}, q^{60}, q^{60}, q^{72}, q^{72} \end{matrix} ; q^{80} \right)_\infty \\ &+ q^7 \left(\begin{matrix} q^4, q^4, q^4, q^{16}, q^{24}, q^{36}, q^{36}, q^{36}, q^{44}, q^{44}, q^{44}, q^{56}, q^{64}, q^{76}, q^{76}, q^{76} \\ q^8, q^{12}, q^{20}, q^{20}, q^{20}, q^{20}, q^{28}, q^{32}, q^{48}, q^{52}, q^{60}, q^{60}, q^{60}, q^{60}, q^{68}, q^{72} \end{matrix} ; q^{80} \right)_\infty \end{aligned} \tag{17.1.3}$$

The original version of this chapter has been revised. The erratum to this chapter is available at https://doi.org/10.1007/978-3-319-57762-3_44.

M.D. Hirschhorn, *The Power of q*, Developments in Mathematics 49,
DOI 10.1007/978-3-319-57762-3_17

and

$$
\begin{aligned}
R(q)^{-1}
&= \left(\begin{matrix} q^8, q^8, q^{12}, q^{28}, q^{32}, q^{32}, q^{32}, q^{32}, q^{48}, q^{48}, q^{48}, q^{48}, q^{52}, q^{68}, q^{72}, q^{72} \\ q^{16}, q^{16}, q^{20}, q^{20}, q^{20}, q^{20}, q^{24}, q^{24}, q^{56}, q^{56}, q^{60}, q^{60}, q^{60}, q^{60}, q^{64}, q^{64} \end{matrix} ; q^{80} \right)_\infty \\
&+ q \left(\begin{matrix} q^8, q^{12}, q^{12}, q^{12}, q^{28}, q^{28}, q^{28}, q^{32}, q^{48}, q^{52}, q^{52}, q^{52}, q^{68}, q^{68}, q^{68}, q^{72} \\ q^4, q^{16}, q^{20}, q^{20}, q^{20}, q^{20}, q^{24}, q^{36}, q^{44}, q^{56}, q^{60}, q^{60}, q^{60}, q^{60}, q^{64}, q^{76} \end{matrix} ; q^{80} \right)_\infty \\
&+ q^6 \left(\begin{matrix} q^8, q^8, q^8, q^8, q^{12}, q^{28}, q^{32}, q^{32}, q^{48}, q^{48}, q^{52}, q^{68}, q^{72}, q^{72}, q^{72}, q^{72} \\ q^4, q^4, q^{20}, q^{20}, q^{20}, q^{20}, q^{36}, q^{36}, q^{44}, q^{44}, q^{60}, q^{60}, q^{60}, q^{60}, q^{76}, q^{76} \end{matrix} ; q^{80} \right)_\infty \\
&- q^3 \left(\begin{matrix} q^8, q^8, q^8, q^{12}, q^{28}, q^{32}, q^{32}, q^{32}, q^{48}, q^{48}, q^{48}, q^{52}, q^{68}, q^{72}, q^{72}, q^{72} \\ q^4, q^{16}, q^{20}, q^{20}, q^{20}, q^{20}, q^{24}, q^{36}, q^{44}, q^{56}, q^{60}, q^{60}, q^{60}, q^{60}, q^{64}, q^{76} \end{matrix} ; q^{80} \right)_\infty .
\end{aligned}
\tag{17.1.4}
$$

The 2-dissections were stated by Ramanujan ([112], p. 50) (though written differently) and first proved by Andrews (1981) [4], while the 4-dissections were conjectured in Hirschhorn (1998) [71], and first proved by Lewis and Liu (2000) [103]. The proofs presented here follow Hirschhorn (2011) [85].

In a final section, we will also briefly discuss a set of 40 identities to be found (in Watson's writing) on pages 236–7 of Ramanujan's Lost Notebook [112].

17.2 The 2-Dissections

We have

$$
\begin{aligned}
R(q)^{-1} &= \left(\begin{matrix} q^2, q^3 \\ q, q^4 \end{matrix} ; q^5 \right)_\infty = \left(\begin{matrix} q^2, q^3, q^7, q^8 \\ q, q^4, q^6, q^9 \end{matrix} ; q^{10} \right)_\infty \\
&= \left(\begin{matrix} q^2, q^8 \\ q^4, q^6 \end{matrix} ; q^{10} \right)_\infty \frac{(-q, q^3, q^7, -q^9; q^{10})_\infty}{(q^2, q^{18}; q^{20})_\infty} \\
&= \frac{(q^8, q^{12}; q^{20})_\infty}{(q^4, q^6, q^{10}, q^{10}; q^{10})_\infty} \sum_{k,l=-\infty}^{\infty} (-1)^l q^{5k^2-4k+5l^2-2l} \\
&= \frac{(q^8, q^{12}; q^{20})_\infty}{(q^4, q^6, q^{10}, q^{10}; q^{10})_\infty} \left(\sum_{r,s=-\infty}^{\infty} (-1)^{r+s} q^{5(r-s)^2-4(r-s)+5(r+s)^2-2(r+s)} \right. \\
&\qquad \left. + \sum_{r,s=-\infty}^{\infty} (-1)^{r+s} q^{5(r-s+1)^2-4(r-s+1)+5(r+s)^2-2(r+s)} \right) \\
&= \frac{(q^8, q^{12}; q^{20})_\infty}{(q^4, q^6, q^{10}, q^{10}; q^{10})_\infty} \left(\sum_{r,s=-\infty}^{\infty} (-1)^{r+s} q^{10r^2-6r+10s^2+2s} \right. \\
&\qquad \left. + q \sum_{r,s=-\infty}^{\infty} (-1)^{r+s} q^{10r^2+4r+10s^2-8s} \right)
\end{aligned}
$$

$$= \frac{(q^8,q^{12};q^{20})_\infty}{(q^4,q^6,q^{10},q^{10};q^{10})_\infty}((q^4,q^8,q^{12},q^{16},q^{20},q^{20};q^{20})_\infty$$

$$\cdot + q(q^2,q^6,q^{14},q^{18},q^{20},q^{20};q^{20})_\infty)$$

$$= \left(\begin{matrix} q^8,q^8,q^{12},q^{12} \\ q^6,q^{10},q^{10},q^{14} \end{matrix};q^{20}\right)_\infty + q\left(\begin{matrix} q^2,q^8,q^{12},q^{18} \\ q^4,q^{10},q^{10},q^{16} \end{matrix};q^{20}\right)_\infty,$$

which is (17.1.2). The proof of (17.1.1) is similar, so omitted.

17.3 The 4-Dissections

We have, from (17.1.2),

$$\sum_{n\geq 0} v_{2n}q^{2n} = \left(\begin{matrix} q^8,q^8,q^{12},q^{12} \\ q^6,q^{10},q^{10},q^{14} \end{matrix};q^{20}\right)_\infty$$

$$= \frac{(q^8,q^8,q^{12},q^{12};q^{20})_\infty}{(q^{12},q^{20},q^{20},q^{28};q^{40})_\infty}(-q^6,-q^{10},-q^{10},-q^{14};q^{20})_\infty$$

$$= \frac{(q^8,q^8,q^{12},q^{12};q^{20})_\infty}{(q^{12},q^{20},q^{20},q^{20},q^{20},q^{28},q^{40},q^{40};q^{40})_\infty}\sum_{k,l=-\infty}^{\infty} q^{10k^2-4k+10l^2}$$

$$= \frac{(q^8,q^8,q^{12},q^{12};q^{20})_\infty}{(q^{12},q^{20},q^{20},q^{20},q^{20},q^{28},q^{40},q^{40};q^{40})_\infty}$$
$$\times\left(\sum_{r,s=-\infty}^{\infty} q^{10(r+s)^2-4(r+s)+10(r-s)^2} + \sum_{r,s=-\infty}^{\infty} q^{10(r+s+1)^2-4(r+s+1)+10(r-s)^2}\right)$$

$$= \frac{(q^8,q^8,q^{12},q^{12};q^{20})_\infty}{(q^{12},q^{20},q^{20},q^{20},q^{20},q^{28},q^{40},q^{40};q^{40})_\infty}$$
$$\times\left(\sum_{r,s=-\infty}^{\infty} q^{20r^2-4r+20s^2-4s} + q^6\sum_{r,s=-\infty}^{\infty} q^{20r^2+16r+20s^2+16s}\right)$$

$$= \frac{(q^8,q^8,q^{12},q^{12};q^{20})_\infty}{(q^{12},q^{20},q^{20},q^{20},q^{20},q^{28},q^{40},q^{40};q^{40})_\infty}$$
$$\times((-q^{16},-q^{16},-q^{24},-q^{24},q^{40},q^{40};q^{40})_\infty$$
$$+q^6(-q^4,-q^4,-q^{36},-q^{36},q^{40},q^{40};q^{40})_\infty)$$

$$= \left(\begin{matrix} q^8,q^8,q^{12},q^{28},q^{32},q^{32} \\ q^{20},q^{20},q^{20},q^{20} \end{matrix};q^{40}\right)_\infty$$
$$\times\left(\frac{(q^{32},q^{32},q^{48},q^{48};q^{80})_\infty}{(q^{16},q^{16},q^{24},q^{24};q^{40})_\infty} + q^6\frac{(q^8,q^8,q^{72},q^{72};q^{80})_\infty}{(q^4,q^4,q^{36},q^{36};q^{40})_\infty}\right)$$

$$
\begin{aligned}
&= \left(\begin{matrix} q^8, q^8, q^{12}, q^{28}, q^{32}, q^{32}, q^{32}, q^{32}, q^{48}, q^{48}, q^{48}, q^{48}, q^{52}, q^{68}, q^{72}, q^{72} \\ q^{16}, q^{16}, q^{20}, q^{20}, q^{20}, q^{20}, q^{24}, q^{24}, q^{56}, q^{56}, q^{60}, q^{60}, q^{60}, q^{60}, q^{64}, q^{64} \end{matrix} ; q^{80} \right) \\
&+ q^6 \left(\begin{matrix} q^8, q^8, q^8, q^8, q^{12}, q^{28}, q^{32}, q^{32}, q^{48}, q^{48}, q^{52}, q^{68}, q^{72}, q^{72}, q^{72}, q^{72} \\ q^4, q^4, q^{20}, q^{20}, q^{20}, q^{20}, q^{36}, q^{36}, q^{44}, q^{44}, q^{60}, q^{60}, q^{60}, q^{60}, q^{76}, q^{76} \end{matrix} ; q^{80} \right)_\infty .
\end{aligned}
\tag{17.3.1}
$$

Similarly, again from (17.1.2),

$$
\begin{aligned}
\sum_{n \geq 0} v_{2n+1} q^{2n+1} &= q \left(\begin{matrix} q^2, q^8, q^{12}, q^{18} \\ q^4, q^{10}, q^{10}, q^{16} \end{matrix} ; q^{20} \right)_\infty \\
&= q \left(\begin{matrix} q^8, q^{12}, q^{12}, q^{12}, q^{28}, q^{28}, q^{28}, q^{32}, q^{48}, q^{52}, q^{52}, q^{52}, q^{68}, q^{68}, q^{68}, q^{72} \\ q^4, q^{16}, q^{20}, q^{20}, q^{20}, q^{20}, q^{24}, q^{36}, q^{44}, q^{56}, q^{60}, q^{60}, q^{60}, q^{60}, q^{64}, q^{76} \end{matrix} ; q^{80} \right)_\infty \\
&- q^3 \left(\begin{matrix} q^8, q^8, q^8, q^{12}, q^{28}, q^{32}, q^{32}, q^{32}, q^{48}, q^{48}, q^{48}, q^{52}, q^{68}, q^{72}, q^{72}, q^{72} \\ q^4, q^{16}, q^{20}, q^{20}, q^{20}, q^{20}, q^{24}, q^{36}, q^{44}, q^{56}, q^{60}, q^{60}, q^{60}, q^{60}, q^{64}, q^{76} \end{matrix} ; q^{80} \right)_\infty .
\end{aligned}
\tag{17.3.2}
$$

(17.3.1) and (17.3.2) together constitute (17.1.4). The proof of (17.1.3) is similar, so omitted.

17.4 Forty Identities of Ramanujan

This seems the appropriate point at which to give a brief mention of a set of 40 identities which Ramanujan compiled and which appear on pages 236–7 of Ramanujan's Lost Notebook [112].

Let

$$
G(q) = \sum_{n \geq 0} \frac{q^{n^2}}{(q;q)_n}, \quad H(q) = \sum_{n \geq 0} \frac{q^{n^2+n}}{(q;q)_n}. \tag{17.4.1}
$$

The Rogers–Ramanujan identities (15.1.1) and (15.1.2) can be written

$$
G(q) = \frac{1}{(q, q^4; q^5)_\infty} \tag{17.4.2}
$$

and

$$
H(q) = \frac{1}{(q^2, q^3; q^5)_\infty}, \tag{17.4.3}
$$

so $G(q)$ and $H(q)$ are sometimes referred to as "the Rogers–Ramanujan functions".

Note that (8.1.2) becomes

$$
R(q) = \frac{H(q)}{G(q)}. \tag{17.4.4}
$$

The 40 identities all involve the Rogers–Ramanujan functions. We will discuss only five of them here.

The five we discuss are

$$G(q)^{11}H(q)-q^2G(q)H(q)^{11}=1+11q\,(G(q)H(q))^6, \tag{17.4.5}$$

$$G(q)H(-q)+G(-q)H(q)=\frac{2}{(q^2;q^4)^2_\infty}=\frac{2\psi(q^2)}{(q^2;q^2)_\infty}, \tag{17.4.6}$$

$$G(q)H(-q)-G(-q)H(q)=\frac{2q\psi(q^{10})}{(q^2;q^2)_\infty}, \tag{17.4.7}$$

$$G(q)G(q^4)+qH(q)H(q^4)=(-q;q^2)^2_\infty=\frac{\phi(q)}{(q^2;q^2)_\infty} \tag{17.4.8}$$

and

$$G(q)G(q^4)-qH(q)H(q^4)=\frac{\phi(q^5)}{(q^2;q^2)_\infty}. \tag{17.4.9}$$

The 40 identities have received a great deal of attention over the last 100 years, and all have now been proved.

For a detailed exposition of the history of the 40 identities and proofs of them all, see the important eight-author paper by Berndt et al. [16] and the papers of Hamza Yesilyurt [126, 127]. See also the paper by Cao and Hu [22].

We simply observe that (17.4.5) is equivalent to (8.4.3), that (17.4.6) and (17.4.7) are equivalent to (17.1.1) and (17.1.2) and that (17.4.8) and (17.4.9) are equivalent to (34.1.7) and (34.1.8).

Exercise: Verify these observations.

Most of the identities in Chapters 40 and 41 can be written in terms of the Rogers–Ramanujan functions. So, for example,

$$G(q)^2H(q^2)+G(q^2)H(q)^2=\frac{2G(q)G(q^2)^2}{(q^5;q^{10})^2_\infty}, \tag{17.4.10}$$

$$G(q)^2H(q^2)-G(q^2)H(q)^2=\frac{2qH(q)H(q^2)^2}{(q^5;q^{10})^2_\infty}, \tag{17.4.11}$$

$$\frac{G(q)H(q^2)-G(q^2)H(q)^2}{G(q)^2H(q^2)+G(q^2)H(q)^2}=q\frac{H(q)H(q^2)^2}{G(q)G(q^2)^2}, \tag{17.4.12}$$

$$G(q)G(q^2)^2+qH(q)H(q^2)^2=(q^5;q^{10})^2_\infty G(q)^2H(q^2), \tag{17.4.13}$$

$$G(q(G(q^2)^2-qH(q)H(q^2)^2=(q^5;q^{10})^2_\infty G(q^2)H(q)^2, \tag{17.4.14}$$

$$\frac{G(q)G(q^2)^2 - qH(q)H(q^2)^2}{G(q)G(q^2)^2 + qH(q)H(q^2)^2} = \frac{G(q^2)H(q)^2}{G(q)^2H(q^2)} \tag{17.4.15}$$

and

$$\left(G(q)^3H(q^3) - G(q^3)H(q)^3\right)\left(G(q)G(q^3)^3 + q^2H(q)H(q^3)^3\right)$$
$$= 3q\left(G(q)H(q)G(q^3)H(q^3)\right)^2. \tag{17.4.16}$$

Exercise: Express (8.4.1) and (8.4.4) in terms of $G(q)$ and $H(q)$.

References

4. G.E. Andrews, Ramanujan's "Lost" notebook III the Rogers-Ramanujan continued fraction. Adv. Math. **41**, 186–208 (1981)
16. B.C. Berndt, G. Choi, Y.-S. Choi, H. Hahn, B.P. Yeap, A.J. Yee, H. Yesilyurt, J. Yi, *Ramanujan's Forty Identities for the Rogers-Ramanujan Functions*, vol. 188, Memoirs of the American Mathematical Society (AMS, Providence, 2007)
22. Z. Cao, Y. Hu, Exact covering systems, quadratic forms, and product identities for theta functions, preprint
71. M.D. Hirschhorn, On the expansion of Ramanujan's continued fraction. Ramanujan J. **2**, 521–527 (1998)
84. M.D. Hirschhorn, On the 2- and 4-dissections of Ramanujan's contiunued fraction and its reciprocal. Ramanujan J. **24**, 85–92 (2011)
103. R.P. Lewis, Z.-G. Liu, A conjecture of Hirschhorn on the 4-dissection of Ramanujan's continued fraction. Ramanujan J. **4**, 347–352 (2000)
112. S. Ramanujan, *The Lost Notebook and Other Unpublished Papers* (Springer, Berlin, 2012)
126. H. Yesilyurt, A generalization of a modular identity of Rogers. J. Number Theory **6**, 1256–1271 (2009)
127. H. Yesilyurt, Elementary proofs of some identities of Ramanujan for the Rogers-Ramanujan functions. JMAA **388**, 420–434 (2009)

Chapter 18
The Series Expansion of the Ramanujan–Göllnitz–Gordon Continued Fraction and Its Reciprocal

18.1 Introduction

Basil Gordon (Gordon (1965) [51]) showed that

$$1+q+\cfrac{q^2}{1+q^3+\cfrac{q^4}{1+q^5+\ddots}} = \left(\begin{matrix} q^3, q^5 \\ q, q^7 \end{matrix}; q^8\right)_\infty. \tag{18.1.1}$$

Richmond and Szekeres (1978) [113] considered the series expansions of the infinite product on the right of (18.1.1), and its reciprocal. Let us define

$$\left(\begin{matrix} q^3, q^5 \\ q, q^7 \end{matrix}; q^8\right)_\infty = \sum_{n\geq 0} a_n q^n, \quad \left(\begin{matrix} q, q^7 \\ q^3, q^5 \end{matrix}; q^8\right)_\infty = \sum_{n\geq 0} b_n q^n. \tag{18.1.2}$$

Using the circle method, they obtained series representations for a_n and b_n valid from some point on, and proved the striking result that for $n \geq 0$,

$$a_{4n+3} = 0, \quad b_{4n+2} = 0. \tag{18.1.3}$$

In this chapter we will not only prove (18.1.3), but we will show that the sign of each of a_n and b_n is periodic, with period 8. Using only JTP, we will show that, apart from the fact that

$$a_4 = 0,$$

we have

$$a_n \begin{cases} > 0 & \text{if } n \equiv 0,\ 1 \text{ or } 2 \pmod 8, \\ < 0 & \text{if } n \equiv 4,\ 5 \text{ or } 6 \pmod 8, \\ = 0 & \text{if } n \equiv 3 \pmod 4), \end{cases}$$

M.D. Hirschhorn, *The Power of q*, Developments in Mathematics 49,
DOI 10.1007/978-3-319-57762-3_18

$$b_n \begin{cases} >0 & \text{if } n \equiv 0,\ 3 \text{ or } 5 \pmod 8, \\ <0 & \text{if } n \equiv 1,\ 4 \text{ or } 7 \pmod 8, \\ =0 & \text{if } n \equiv 2 \pmod 4. \end{cases} \tag{18.1.4}$$

Indeed, on the way, we will find the 2-, 4- and 8-dissections of each of the products.

18.2 The 2-, 4- and 8-dissections

We have, using the techniques detailed in §§17.2, 17.3,

$$\begin{aligned}
\sum_{n\geq 0} a_n q^n &= \left(\begin{matrix} q^3, q^5 \\ q, q^7 \end{matrix}; q^8\right)_\infty = \frac{(-q, q^3, q^5, -q^7; q^8)_\infty}{(q^2, q^{14}; q^{16})_\infty} \\
&= \frac{(-q, q^3, q^5, -q^7, q^8, q^8; q^8)_\infty}{(q^2, q^8, q^8, q^{14}, q^{16}, q^{16}; q^{16})_\infty} \\
&= \frac{1}{(q^2, q^8, q^8, q^{14}, q^{16}, q^{16}; q^{16})_\infty} \\
&\times ((q^4, q^6, q^{10}, q^{12}, q^{16}, q^{16}; q^{16})_\infty + q(q^2, q^4, q^{12}, q^{14}, q^{16}, q^{16}; q^{16})_\infty) \\
&= \left(\begin{matrix} q^4, q^6, q^{10}, q^{12} \\ q^2, q^8, q^8, q^{14} \end{matrix}; q^{16}\right)_\infty + q\left(\begin{matrix} q^4, q^{12} \\ q^8, q^8 \end{matrix}; q^{16}\right)_\infty \qquad (18.2.1) \\
&= \left(\begin{matrix} q^4, q^{12} \\ q^8, q^8 \end{matrix}; q^{16}\right)_\infty \left(\left(\begin{matrix} q^8, q^{12}, q^{20}, q^{24} \\ q^4, q^{16}, q^{16}, q^{28} \end{matrix}; q^{32}\right)_\infty + q^2\left(\begin{matrix} q^8, q^{24} \\ q^{16}, q^{16} \end{matrix}; q^{32}\right)_\infty\right) \\
&\qquad + q\left(\begin{matrix} q^4, q^{12} \\ q^8, q^8 \end{matrix}; q^{16}\right)_\infty \\
&= \left(\begin{matrix} q^{12}, q^{12}, q^{20}, q^{20} \\ q^8, q^{16}, q^{16}, q^{24} \end{matrix}; q^{32}\right)_\infty + q\left(\begin{matrix} q^4, q^{12} \\ q^8, q^8 \end{matrix}; q^{16}\right)_\infty + q^2\left(\begin{matrix} q^4, q^{12}, q^{20}, q^{28} \\ q^8, q^{16}, q^{16}, q^{24} \end{matrix}; q^{32}\right)_\infty \qquad (18.2.2) \\
&= \frac{(q^{12}, q^{12}, q^{20}, q^{20}, q^{32}, q^{32}; q^{32})_\infty}{(q^8, q^{16}, q^{16}; q^{16})_\infty} + q\frac{(q^4, q^{12}, q^{16}; q^{16})_\infty}{(q^8, q^8, q^{16}; q^{16})_\infty} \\
&\qquad + q^2\frac{(q^4, q^{12}, q^{20}, q^{28}, q^{32}, q^{32}; q^{32})_\infty}{(q^8, q^{16}, q^{16}; q^{16})_\infty}
\end{aligned}$$

$$
\begin{aligned}
&= \frac{1}{(q^8,q^{16},q^{16};q^{16})_\infty} \\
&\times ((-q^{24},-q^{32},-q^{32},-q^{40},q^{64},q^{64};q^{64})_\infty \\
&\qquad - 2q^{12}(-q^8,-q^{56},-q^{64},-q^{64},q^{64},q^{64};q^{64})_\infty) \\
&+ \frac{q}{(q^8,q^8,q^{16};q^{16})_\infty} \\
&\times ((-q^{24},-q^{40},q^{64};q^{64})_\infty - q^4(-q^8,-q^{56},q^{64};q^{64})_\infty) \\
&+ \frac{q^2}{(q^8,q^{16},q^{16};q^{16})_\infty} \\
&\times ((-q^{16},-q^{24},-q^{40},-q^{48},q^{64},q^{64};q^{64})_\infty \\
&\qquad - q^4(-q^8,-q^{16},-q^{48},-q^{56},q^{64},q^{64};q^{64})_\infty).
\end{aligned} \tag{18.2.3}
$$

Similarly,

$$
\sum_{n\geq 0} b_n q^n = \begin{pmatrix} q, q^7 \\ q^3, q^5 \end{pmatrix}_{\infty} ; q^8 = \frac{(q,-q^3,-q^5,q^7;q^8)_\infty}{(q^6,q^{10};q^{16})_\infty}
$$

$$
\begin{aligned}
&= \frac{(q,-q^3,-q^5,q^7,q^8,q^8;q^8)_\infty}{(q^6,q^8,q^8,q^{10},q^{16},q^{16};q^{16})_\infty} \\
&= \frac{1}{(q^6,q^8,q^8,q^{10},q^{16},q^{16};q^{16})_\infty} \\
&\times ((q^4,q^6,q^{10},q^{12},q^{16},q^{16};q^{16})_\infty - q(q^2,q^4,q^{12},q^{14},q^{16},q^{16};q^{16})_\infty) \\
&= \begin{pmatrix} q^4,q^{12} \\ q^8,q^8 \end{pmatrix} ; q^{16} \Bigg)_\infty - q \begin{pmatrix} q^2,q^4,q^{12},q^{14} \\ q^6,q^8,q^8,q^{10} \end{pmatrix}_\infty
\end{aligned} \tag{18.2.4}
$$

$$
\begin{aligned}
&= \left(\begin{matrix} q^4,q^{12} \\ q^8,q^8 \end{matrix} ; q^{16} \right)_\infty \\
&- q \left(\begin{matrix} q^4,q^{12} \\ q^8,q^8 \end{matrix} ; q^{16} \right)_\infty \left(\left(\begin{matrix} q^8,q^{24} \\ q^{16},q^{16} \end{matrix} ; q^{32} \right)_\infty - q^2 \left(\begin{matrix} q^4,q^8,q^{24},q^{28} \\ q^{12},q^{16},q^{16},q^{20} \end{matrix} ; q^{32} \right)_\infty \right) \\
&= \left(\begin{matrix} q^4,q^{12} \\ q^8,q^8 \end{matrix} ; q^{16} \right) - q \left(\begin{matrix} q^4,q^{12},q^{20},q^{28} \\ q^8,q^{16},q^{16},q^{24} \end{matrix} ; q^{32} \right)_\infty + q^3 \left(\begin{matrix} q^4,q^4,q^{28},q^{28} \\ q^8,q^{16},q^{16},q^{24} \end{matrix} ; q^{32} \right)_\infty
\end{aligned} \tag{18.2.5}
$$

$$
\begin{aligned}
&= \frac{1}{(q^8,q^8,q^{16};q^{16})_\infty} \\
&\times ((-q^{24},-q^{40},q^{64};q^{64})_\infty - q^4(-q^8,-q^{56},q^{64};q^{64})_\infty) \\
&- \frac{q}{(q^8,q^{16},q^{16};q^{16})_\infty} \\
&\times ((-q^{16},-q^{24},-q^{40},-q^{48},q^{64},q^{64};q^{64})_\infty \\
&\qquad - q^4(-q^8,-q^{16},-q^{48},-q^{56},q^{64},q^{64};q^{64})_\infty) \\
&+ \frac{q^3}{(q^8,q^{16},q^{16};q^{16})_\infty} \\
&\times ((-q^8,-q^{32},-q^{32},-q^{56},q^{64},q^{64};q^{64})_\infty \\
&\qquad - 2q^4(-q^{24},-q^{40},-q^{64},-q^{64},q^{64},q^{64};q^{64})_\infty).
\end{aligned} \tag{18.2.6}
$$

The 2-dissections are (18.2.1) and (18.2.4), the 4-dissections are (18.2.2) and (18.2.5), and the 8-dissections are (18.2.3) and (18.2.6). From (18.2.2) and (18.2.5) it follows that

$$a_{4n+3} = 0, \ b_{4n+2} = 0,$$

from (18.2.3) and (18.2.6) it follows that

$$a_4 = 0$$

and that all of

$$\sum_{n\geq 0} a_{8n}q^n, \ \sum_{n\geq 0} a_{8n+1}q^n = \sum_{n\geq 0} b_{8n}q^n, \ \sum_{n\geq 0} a_{8n+2}q^n = \sum_{n\geq 0} -b_{8n+1}q^n,$$

$$\sum_{n\geq 0} -a_{8n+12}q^n, \ \sum_{n\geq 0} -a_{8n+5}q^n = \sum_{n\geq 0} -b_{8n+4}q^n, \ \sum_{n\geq 0} -a_{8n+6}q^n = \sum_{n\geq 0} b_{8n+5}q^n,$$

$$\sum_{n\geq 0} b_{8n+3}q^n \text{ and } \sum_{n\geq 0} -b_{8n+7}q^n$$

are greater than

$$\frac{1}{1-q} = 1 + q + q^2 + \cdots .$$

(18.1.4) follows.

Endnotes.

The proofs given here arose out of a collaboration with Helmut Prodinger, as yet unpublished. The earlier proofs in Hirschhorn (2001) [76] were far less systematic.

B. Yuttanan (2012) [128] has also proved (18.1.4).

The phenomenon occurring in (18.1.3), that when certain products are dissected, some of the components are zero, has been generalised by Andrews and Bressoud (1979) [8] and Alladi and Gordon (1994) [2].

The phenomenon occurring in (18.1.4), that the sign of the coefficients is eventually periodic, has been generalised by S.H. Chan and H. Yesilyurt (2006) [29].

References

2. K. Alladi, B. Gordon, Vanishing coefficients in the expansion of products of Rogers-Ramanujan type, *The Rademacher Legacy in Mathematics*, vol. 156, Contemporary Mathematics (AMS, Providence, 1994), pp. 129–139
8. G.E. Andrews, D. Bressoud, Vanishing coefficients in infinite product expansions. J. Aust. Math. Soc. Ser. A **27**, 199–202 (1979)
29. S.H. Chan, H. Yesilyurt, The periodicity of the signs of the coefficients of certain infinite products. Pac. J. Math. **225**(1), 13–32 (2006)
51. B. Gordon, Some continued fractions of the Rogers-Ramanujan type. Duke Math. J. **32**, 741–748 (1965)
76. M.D. Hirschhorn, On the expansion of a continued fraction of Gordon. Ramanujan J. **5**, 369–375 (2001)
113. L.B. Richmond, G. Szekeres, The Taylor coefficients of certain infinite products. Acta Scientiarum Mathematicarum **40**, 347–369 (1978)
128. B. Yuttanan, New properties for the Ramanujan-Göllnitz-Gordon continued fraction. Acta Arith. **151**, 293–310 (2012)

Chapter 19
Jacobi's "*aequatio identica satis abstrusa*"

19.1 Introduction

Jacobi [98], p. 90 found the following identity, which he described as "*aequatio identica satis abstrusa*" ("a rather obscure identity").

$$\prod_{n\geq 1}(1+q^{2n-1})^8 - \prod_{n\geq 1}(1-q^{2n-1})^8 = 16q\prod_{n\geq 1}(1+q^{2n})^8. \qquad (19.1.1)$$

This identity has come to play an important role in string and superstring theory (see Green–Schwarz–Witten [53], p. 224, Zwiebach [130], p. 321).

My goal in this chapter is to give as short and simple a proof of (19.1.1) as I can. Indeed, in §§19.2 and 19.3 we give two simple proofs.

In §19.4, we present remarkable formulations of *identica abstrusa* and other identities.

In §19.5 we see a simple generalisation of (19.1.1).

19.2 Our First Proof of Jacobi's *identica abstrusa*

We start with the 2-dissection (1.9.4),

$$\phi(q) = \phi(q^4) + 2q\psi(q^8). \qquad (19.2.1)$$

If in (19.2.1) we put $-q$ for q, we obtain

$$\phi(-q) = \phi(q^4) - 2q\psi(q^8). \qquad (19.2.2)$$

M.D. Hirschhorn, *The Power of q*, Developments in Mathematics 49,
DOI 10.1007/978-3-319-57762-3_19

If we multiply (19.2.1) by (19.2.2) and use (1.5.14), we obtain

$$\phi(-q^2)^2 = \phi(q^4)^2 - 4q^2\psi(q^8)^2. \tag{19.2.3}$$

If in (19.2.3) we replace q^2 by q, we obtain

$$\phi(-q)^2 = \phi(q^2)^2 - 4q\psi(q^4)^2. \tag{19.2.4}$$

If in (19.2.4) we put $-q$ for q, we obtain

$$\phi(q)^2 = \phi(q^2)^2 + 4q\psi(q^4)^2, \tag{19.2.5}$$

which is (1.10.1).

If we multiply (19.2.4) and (19.2.5) and use (1.5.14), we find

$$\phi(-q^2)^4 = \phi(q^2)^4 - 16q^2\psi(q^4)^4. \tag{19.2.6}$$

If in (19.2.6) we put q for q^2, we obtain

$$\phi(-q)^4 = \phi(q)^4 - 16q\psi(q^2)^4,$$

or,

$$\phi(q)^4 - \phi(-q)^4 = 16q\psi(q^2)^4. \tag{19.2.7}$$

By (1.1.1) and (1.5.7), we can write (19.2.7)

$$(-q,-q,q^2;q^2)_\infty^4 - (q,q,q^2;q^2)_\infty^4 = 16q\frac{(q^4;q^4)_\infty^8}{(q^2;q^2)_\infty^4}. \tag{19.2.8}$$

If we divide (19.2.8) by $(q^2;q^2)_\infty^4$, we obtain

$$(-q;q^2)_\infty^8 - (q;q^2)_\infty^8 = 16q\frac{(q^4;q^4)_\infty^8}{(q^2;q^2)_\infty^8} = 16q(-q^2;q^2)_\infty^8,$$

which is (19.1.1).

Note that we can write (19.2.7)

$$\mathbf{O}\left(\phi(q)^4\right) = 8q\psi(q^2)^4, \tag{19.2.9}$$

and indeed, we can write Jacobi's *identica abstrusa*

$$\mathbf{O}\left((-q;q^2)_\infty^8\right) = 8q(-q^2;q^2)_\infty^8 \tag{19.2.10}$$

where

$$\mathbf{O}(f(q)) = \frac{f(q) - f(-q)}{2}$$

is the **O**dd part of $f(q)$.

19.3 Our Second Proof of *identica abstrusa*

Let

$$\rho(q) = \frac{\phi(q)}{\psi(q^2)}. \tag{19.3.1}$$

We have, by (1.10.1) or (19.2.5) and (1.5.13) with q^2 for q,

$$\rho(q)^2 = \frac{\phi(q)^2}{\psi(q^2)^2} = \frac{\phi(q^2)^2 + 4q\psi(q^4)^2}{\phi(q^2)\psi(q^4)} = \frac{\phi(q^2)}{\psi(q^4)} + 4q\frac{\psi(q^4)}{\phi(q^2)} = \rho(q^2) + \frac{4q}{\rho(q^2)}. \tag{19.3.2}$$

If we square (19.3.2) and take odd parts, we find

$$\mathbf{O}\left(\rho(q)^4\right) = 8q, \tag{19.3.3}$$

which is (19.2.9). Now complete the proof as in §19.2.

19.4 Some Remarkable Formulations

If we put $-q$ for q in (19.3.3), we obtain

$$\mathbf{O}\left(\rho(-q)^4\right) = -8q, \tag{19.4.1}$$

$$\mathbf{O}\left(\left(\frac{\phi(-q)}{\psi(q^2)}\right)^4\right) = -8q, \tag{19.4.2}$$

$$\mathbf{O}\left(\left(\frac{(q;q)_\infty^2}{(q^2;q^2)_\infty} \Big/ \frac{(q^4;q^4)_\infty^2}{(q^2;q^2)_\infty}\right)^4\right) = -8q, \tag{19.4.3}$$

$$\mathbf{O}\left(\left(\frac{(q;q)_\infty}{(q^4;q^4)_\infty}\right)^8\right) = -8q, \tag{19.4.4}$$

or,

$$\mathbf{O}\left(\prod_{4\nmid n}(1-q^n)^8\right) = -8q. \tag{19.4.5}$$

To bring home how stunning this is, observe that

$$\prod_{4\nmid n}(1-q^n)^8 = 1-8q+20q^2-62q^4+216q^6-641q^8+1636q^{10}$$
$$-3778q^{12}+8248q^{14}-17277q^{16}+34664q^{18}-\cdots$$

Furthermore, if we define

$$\eta(k\tau) = q^{\frac{k}{24}}E(q^k), \tag{19.4.6}$$

and if H_m denotes the huffing operator modulo m, that is,

$$H_m\left(\sum_n a_n q^n\right) = \sum_n a_{mn}q^{mn}, \tag{19.4.7}$$

then *identica abstrusa* can be written

$$H_2\left(\left(\frac{\eta(\tau)}{\eta(4\tau)}\right)^8\right) = -8. \tag{19.4.8}$$

This should be compared to (see (6.3.3))

$$H_5\left(\left(\frac{\eta(\tau)}{\eta(25\tau)}\right)^4\right) = -5, \tag{19.4.9}$$

(7.4.18),

$$H_7\left(\left(\frac{\eta(\tau)}{\eta(49\tau)}\right)^6\right) = 49, \tag{19.4.10}$$

(13.3.17),

$$H_{11}\left(\left(\frac{\eta(\tau)}{\eta(121\tau)}\right)^{10}\right) = 11^4 \tag{19.4.11}$$

and (see (14.5.8)),

$$H_3\left(\left(\frac{\eta(\tau)\eta(2\tau)}{\eta(9\tau)\eta(18\tau)}\right)^2\right) = -3. \tag{19.4.12}$$

Also, from (1.7.1), we find

$$H_3\left(\left(\frac{\eta(\tau)}{\eta(9\tau)}\right)^3\right) = -3. \tag{19.4.13}$$

Exercises:
Let $a = q^{-\frac{1}{24}}(-q;q^2)_\infty$, $b = q^{-\frac{1}{24}}(q;q^2)_\infty$.

Show that

$$\begin{aligned}
a &= \frac{\eta(2\tau)^2}{\eta(\tau)\eta(4\tau)}, \\
b &= \frac{\eta(\tau)}{\eta(2\tau)}, \\
a^2 - b^2 &= 4\frac{\eta(16\tau)^2}{\eta(2\tau)\eta(8\tau)}, \\
a^2 + b^2 &= 2\frac{\eta(8\tau)^5}{\eta(2\tau)\eta(4\tau)^2\eta(16\tau)^2}, \\
a^4 - b^4 &= 8\frac{\eta(8\tau)^4}{\eta(2\tau)^2\eta(4\tau)^2}, \\
a^4 + b^4 &= 2\frac{\eta(4\tau)^{10}}{\eta(2\tau)^6\eta(8\tau)^4}
\end{aligned}$$

and

$$a^8 - b^8 = 16\frac{\eta(4\tau)^8}{\eta(2\tau)^8}.$$

Show that

$$\begin{aligned}
a + b &= 2q^{-\frac{1}{24}}/(q^4,q^6,q^8,q^{10},q^{22},q^{24},q^{26},q^{28};q^{32})_\infty, \\
a - b &= 2q^{\frac{23}{24}}/(q^2,q^8,q^{12},q^{14},q^{18},q^{20},q^{24},q^{30};q^{30})_\infty,
\end{aligned}$$

Hint:

$$(q;q^2)_\infty = \frac{(q,q^3,q^4;q^4)_\infty}{(q^4;q^4)_\infty}.$$

19.5 A Simple Generalisation of Jacobi's *identica abstrusa*

Jacobi's identity is the special case $a = b = 1$ of the identity

$$
\begin{aligned}
&(-a^{-1}q, -aq, -b^{-1}q, -bq; q^2)_\infty^2 - (a^{-1}q, aq, b^{-1}q, bq; q^2)_\infty^2 \\
&= 4(a^{-1} + a + b^{-1} + b)q(-a^{-1}b^{-1}q^2, -a^{-1}bq^2, -ab^{-1}q^2, -abq^2; q^2)_\infty(-q^2; q^2)_\infty^4.
\end{aligned}
\tag{19.5.1}
$$

The proof of (19.5.1) is left as an exercise.

References

53. M.B. Green, J.H. Schwarz, E. Witten, *Superstring Theory I, Introduction, 25th Anniversary Edition* (Cambridge University Press, Cambridge, 2012)
98. C.G.J. Jacobi, Fundamenta nova theoriae functionum ellipticarum, Königsberg (1829)
130. B. Zwiebach, *A First Course in String Theory*, 2nd edn. (Cambridge University Press, Cambridge, 2009)

Chapter 20
Two Modular Equations

20.1 Introduction

The object of this chapter is to prove the following two identities.

$$\mathbf{O}(\phi(q)\phi(q^3)) = 2q\psi(q^2)\psi(q^6) \tag{20.1.1}$$

and

$$\mathbf{O}(\psi(q)\psi(q^7)) = q\psi(q^2)\psi(q^{14}). \tag{20.1.2}$$

The first of these is described by Venkatachaliengar (Cooper (2011) [35], p. 136) as "a beautiful identity which looks extremely difficult to prove", while the second, as revealed by Farkas and Kra (2000) [44], [45], has a striking combinatorial interpretation.

20.2 Proof of the First Modular Equation

We have

$$\mathbf{O}(\phi(q)\phi(q^3)) = \mathbf{O}\left(\sum_{k,l=-\infty}^{\infty} q^{k^2+3l^2}\right) = \mathbf{O}\left(\sum_{k,l=-\infty}^{\infty} q^{((k+3l)^2+3(k-l)^2)/4}\right). \tag{20.2.1}$$

Define $r = k+3l$, $s = k-l$. Conversely, $k = \dfrac{r+3s}{4}$, $l = \dfrac{r-s}{4}$. Then $r \equiv s \pmod 4$ and

$$\mathbf{O}(\phi(q)\phi(q^3)) = \mathbf{O}\left(\sum_{r\equiv s \pmod 4} q^{(r^2+3s^2)/4}\right). \tag{20.2.2}$$

M.D. Hirschhorn, *The Power of q*, Developments in Mathematics 49,
DOI 10.1007/978-3-319-57762-3_20

If r and s are even and differ by 0 modulo 4 then $\frac{r^2+3s^2}{4}$ is even, so these terms do not contribute to the right of (20.2.2), so we need only consider the case r and s odd. We split the sum into two, depending on the residue of r (and s) mod 4.

In the case $r \equiv 1 \pmod 4$, write $r = 4t+1$, $s = 4u+1$, and in the case $r \equiv -1 \pmod 4$, write $r = -4t-1$, $s = -4u-1$, and we find

$$\mathbf{O}(\phi(q)\phi(q^3)) = 2\sum_{t,u=-\infty}^{\infty} q^{4t^2+2t+12u^2+6u+1} = 2q\psi(q^2)\psi(q^6),$$

as claimed.

20.3 Proof of the Second Modular Equation

We have

$$\begin{aligned}\mathbf{O}(\psi(q)\psi(q^7)) &= \mathbf{O}\left(\sum_{k,l=-\infty}^{\infty} q^{2k^2+k+14l^2+7l}\right)\\ &= \mathbf{O}\left(\sum_{k,l=-\infty}^{\infty} q^{((k+7l+2)^2+7(k-l)^2-4)/4}\right).\end{aligned} \tag{20.3.1}$$

Now define $r = k+7l+2$, $s = k-l$. (Conversely, $k = \frac{r+7s-2}{8}$, $l = \frac{r-s-2}{8}$.) Then $r-s \equiv 2 \pmod 8$ and

$$\mathbf{O}(\psi(q)\psi(q^7)) = \mathbf{O}\left(\sum_{r-s\equiv 2 \pmod 8} q^{(r^2+7s^2-4)/4}\right) \tag{20.3.2}$$

If r and s are even and differ by 2 modulo 8 then $\frac{r^2+7s^2-4}{4}$ is even, so these terms do not contribute to the right of (20.3.2), so we need only consider the case r and s odd. We split the sum into four, depending on the residue of r (and so s) modulo 8.

In the case $r \equiv 1 \pmod 8$, write $r = 8t+1$, $s = -8u-1$, in the case $r \equiv 3 \pmod 8$ write $r = 8t+3$, $s = 8u+1$, in the case $r \equiv -1 \pmod 8$ write $r = -8t-1$, $s = -8u-3$ and in the case $r \equiv -3 \pmod 8$ write $r = -8t-3$, $s = 8u+3$. Then we find

$$\begin{aligned}\mathbf{O}(\psi(q)\psi(q^7)) &= \sum_{t,u=-\infty}^{\infty} q^{16t^2+4t+112u^2+28u+1} + \sum_{t,u=-\infty}^{\infty} q^{16t^2+4t+112u^2+84u+15} \\ &\quad + \sum_{t,u=-\infty}^{\infty} q^{16t^2+12t+112u^2+28u+3} + \sum_{t,u=-\infty}^{\infty} q^{16t^2+12t+112u^2+84u+17} \\ &= q\left(\sum_{-\infty}^{\infty} q^{16t^2+4t} + q^2 \sum_{-\infty}^{\infty} q^{16t^2+12t}\right)\left(\sum_{-\infty}^{\infty} q^{112u^2+28u} + q^{14}\sum_{-\infty}^{\infty} q^{112u^2+84u}\right) \\ &= q\left(\sum_{-\infty}^{\infty} q^{4k^2+2k}\right)\left(\sum_{-\infty}^{\infty} q^{28l^2+14l}\right) \\ &= q\psi(q^2)\psi(q^{14}),\end{aligned}$$

as claimed.

20.4 Combinatorial Interpretations

If we divide (20.1.1) by $(q^2;q^2)_\infty(q^6;q^6)_\infty$, we obtain

$$\mathbf{O}((-q;q^2)_\infty^2(-q^3;q^6)_\infty^2) = 2q(-q^2;q^2)_\infty^2(-q^6;q^6)_\infty^2 \tag{20.4.1}$$

while if we divide (20.1.2) by $(q^4;q^4)_\infty(q^{28};q^{28})_\infty$, we obtain

$$\mathbf{O}\left((-q;q^2)_\infty(-q^7;q^{14})_\infty\right) = q(-q^2;q^2)_\infty(-q^{14};q^{14})_\infty \tag{20.4.2}$$

The identities (20.4.1) and (20.4.2) can be written, respectively,

$$\begin{aligned}&\mathbf{O}((-q,-q,-q^3,-q^3,-q^3,-q^3,-q^5,-q^5;q^6)_\infty) \\ &\qquad = 2q(-q^2,-q^2,-q^4,-q^4,-q^6,-q^6,-q^6,-q^6;q^6)_\infty\end{aligned} \tag{20.4.3}$$

and

$$\begin{aligned}&\mathbf{O}((-q,-q^3,-q^5,-q^7,-q^7,-q^9,-q^{11},-q^{13};q^{14})_\infty) \\ &\qquad = q(-q^2,-q^4,-q^6,-q^8,-q^{10},-q^{12},-q^{14},-q^{14};q^{14})_\infty.\end{aligned} \tag{20.4.4}$$

The first of these has the combinatorial interpretation

The number of partitions of $2n+1$ *into distinct odd parts where the parts divisible by* 3 *come in four colours, while the parts not divisible by* 3 *come in two colours, is twice the number of partitionsof* $2n$ *into distinct even parts, where the parts divisible by* 3 *come in four colours, while the parts not divisible by* 3 *come in two colours* (20.4.5)

while the second has the combinatorial interpretation

The number of partitions of $2n+1$ *into distinct odd parts where the parts divisible by* 7 *come in two colours is equal to the number of partitions of* $2n$ *into distinct even parts, where the parts divisible by* 7 *come in two colours.* (20.4.6)

I heard the second of these from Hershel Farkas in the year 2003 (see Farkas–Kra (2000) [44] or Farkas–Kra (2001) [45]). Combinatorial proofs of both are given in Sandon–Zanello (2013), (2014) [115], [116].

Endnotes.

§20.2 In Chapters 23, 25, 28 and 30 I give different proofs of (20.1.1).

§20.4 I gave proofs of (20.4) in Hirschhorn (2006) [80] and Hirschhorn (2008) [81]. Warnaar (2005) [121] generalised the result.

References

35. S. Cooper, *Development of Elliptic Functions According to Ramanujan, Originally by K. Venkatachaliengar*, vol. 8, Monographs in Number Theory (World Scientific, Singapore, 2011)
44. H.M. Farkas, I. Kra, Partitions and theta constant identities, *Analysis, Geometry, Number Theory, the Mathematics of Leon Ehrenpreis*, vol. 251, Contemporary Mathematics (AMS, Providence, 2000), pp. 197–203
45. H.M. Farkas, I. Kra, *Theta Constants, Riemann Surfaces and the Modular Group*, vol. 37, Graduate Studies in Mathematics (American Mathematical Society, Providence, 2001)
80. M.D. Hirschhorn, The case of the mysterious sevens. Int. J. Number Theory **2**, 213–216 (2006)
81. M.D. Hirschhorn, The case of the mysterious sevens. Aust. Math. Soc. Gaz. **35**, 113–114 (2008)
115. C. Sandon, F. Zanello, Warnaar's bijection and colored partition identities I. J. Comb. Theory Ser. A **120**, 28–38 (2013)
116. C. Sandon, F. Zanello, Warnaar's bijection and colored partition identities II. Ramanujan J. **33**, 83–120 (2014)
121. S.O. Warnaar, A generalization of the Farkas and Kra partition theorem for modulus 7. J. Comb. Theory Ser. A **110**, 43–52 (2005)

Chapter 21
A Letter from Fitzroy House

21.1 Introduction

On a Saturday in 1917 or 1918, Ramanujan sent Hardy a letter from the nursing home Fitzroy House, in which he stated the identity

$$\sum_{m,n=-\infty}^{\infty} q^{m^2+mn+n^2} = 1+6\sum_{n\geq 0}\left(\frac{q^{3n+1}}{1-q^{3n+1}} - \frac{q^{3n+2}}{1-q^{3n+2}}\right). \tag{21.1.1}$$

(see Berndt and Rankin [17], p. 196).

This identity had been stated earlier by L. Lorenz (1871) [104], and proved by the editor of his collected works, H. Valtiner (1904) [120].

This chapter is devoted to a proof of (21.1.1) as well as further identities involving the sum on the left.

21.2 Proof of the Lorenz–Ramanujan Identity (21.1.1)

Let

$$a(q) = \sum_{m,n=-\infty}^{\infty} q^{m^2+mn+n^2}. \tag{21.2.1}$$

Then

$$\begin{aligned} a(q^2) &= \sum_{m,n=-\infty}^{\infty} q^{2m^2+2mn+2n^2} \\ &= \sum_{m,n=-\infty}^{\infty} q^{m^2+n^2+(-m-n)^2} \end{aligned}$$

M.D. Hirschhorn, *The Power of q*, Developments in Mathematics 49,
DOI 10.1007/978-3-319-57762-3_21

$$= \sum_{m+n+p=0} q^{m^2+n^2+p^2}$$

$$= \mathbf{CT}_a \left(\left(\sum_{-\infty}^{\infty} a^n q^{n^2} \right)^3 \right), \tag{21.2.2}$$

where

$$\mathbf{CT}_a \left(\sum_{-\infty}^{\infty} c_n(q) a^n \right) = c_0$$

is the **C**onstant **T**erm of the series.

So we consider

$$\left(\sum_{-\infty}^{\infty} a^n q^{n^2} \right)^3 = (-a^{-1}q, -aq, q^2; q^2)_\infty^3. \tag{21.2.3}$$

We will prove that

$$\begin{aligned}
&(-a^{-1}q, -aq, q^2; q^2)_\infty^3 \\
&= \left(1 + 6 \sum_{n \geq 0} \left(\frac{q^{6n+2}}{1-q^{6n+2}} - \frac{q^{6n+4}}{1-q^{6n+4}} \right) \right) (-a^{-3}q^3, -a^3q^3, q^6; q^6)_\infty \\
&+ 3q \frac{(q^6;q^6)_\infty^3}{(q^2;q^2)_\infty} (a(-a^{-3}q, -a^3q^5, q^6; q^6)_\infty + a^{-1}(-a^{-3}q^5, -a^3q, q^6; q^6)_\infty).
\end{aligned} \tag{21.2.4}$$

By virtue of (21.2.2), it is immediate that

$$a(q^2) = 1 + 6 \sum_{n \geq 0} \left(\frac{q^{6n+2}}{1-q^{6n+2}} - \frac{q^{6n+4}}{1-q^{6n+4}} \right), \tag{21.2.5}$$

and (21.1.1) follows.

We have

$$(-a^{-1}q, -aq, q^2; q^2)_\infty^3 = \sum_{m,n,p=-\infty}^{\infty} a^{m+n+p} q^{m^2+n^2+p^2}. \tag{21.2.6}$$

We split the sum into three, according to the residue modulo 3 of $m+n+p$. Thus if $m+n+p \equiv 0 \pmod 3$, write $m = k+r,\ n = k+s,\ p = k+t$ with $r+s+t = 0$. The corresponding sum is

$$c_0 \sum_{-\infty}^{\infty} a^{3k} q^{3k^2} \text{ where } c_0 = \sum_{r+s+t=0} q^{r^2+s^2+t^2}. \tag{21.2.7}$$

If $m+n+p\equiv 1 \pmod 3$, write $m=k+r$, $n=k+s$, $p=k+t$ with $r+s+t=1$. The corresponding sum is

$$c_1\sum_{-\infty}^{\infty}a^{3k+1}q^{3k^2+2k} \text{ where } c_1=\sum_{r+s+t=1}q^{r^2+s^2+t^2}. \tag{21.2.8}$$

If $m+n+p\equiv -1 \pmod 3$, write $m=-k-r$, $n=-k-s$, $p=-k-t$ with $r+s+t=1$. The corresponding sum is

$$c_1\sum_{-\infty}^{\infty}a^{-3k-1}q^{3k^2+2k}. \tag{21.2.9}$$

Thus we have the identity

$$(-a^{-1}q,-aq,q^2;q^2)_\infty^3$$
$$=c_0\sum_{-\infty}^{\infty}a^{3k}q^{3k^2}+c_1\left(\sum_{-\infty}^{\infty}a^{3k+1}q^{3k^2+2k}+\sum_{-\infty}^{\infty}a^{-3k-1}q^{3k^2-2k}\right). \tag{21.2.10}$$

(21.2.10) may be written

$$\begin{aligned}(-a^{-1}q,-aq,q^2;q^2)_\infty^3&=c_0(-a^{-3}q^3,-a^3q^3,q^6;q^6)_\infty\\&+c_1(a(-a^{-3}q,-a^3q^5,q^6;q^6)_\infty+a^{-1}(-a^{-3}q^5,-a^3q,q^6;q^6)_\infty)\end{aligned} \tag{21.2.11}$$

We now set out to determine c_0 and c_1.

If in (21.2.11) we replace a by a^2q and multiply by a^3, we obtain

$$\begin{aligned}(a+a^{-1})^3(-a^{-2}q^2,-a^2q^2,q^2;q^2)_\infty^3&=c_0(a^3+a^{-3})(-a^{-6}q^6,-a^6q^6,q^6;q^6)_\infty\\&+c_1q^{-1}(a(-a^{-6}q^2,-a^6q^4,q^6;q^6)_\infty+a^{-1}(-a^{-6}q^4,-a^6q^2,q^6;q^6)_\infty).\end{aligned} \tag{21.2.12}$$

If in (21.2.12) we set $a=e^{\frac{\pi i}{6}}$, we find

$$3\sqrt{3}(q^6;q^6)_\infty^3=\sqrt{3}c_1q^{-1}(q^2;q^2)_\infty,$$

or,

$$c_1=3q\frac{(q^6;q^6)_\infty^3}{(q^2;q^2)_\infty}. \tag{21.2.13}$$

If we substitute (21.2.13) into (21.2.12), we obtain

$$
\begin{aligned}
(a+a^{-1})^3(-a^{-2}q^2,-a^2q^2,q^2;q^2)_\infty^3 &= c_0(a^3+a^{-3})(-a^{-6}q^6,-a^6q^6,q^6;q^6)_\infty \\
&+3\frac{(q^6;q^6)_\infty^3}{(q^2;q^2)_\infty}(a(-a^{-6}q^2,-a^6q^4,q^6;q^6)_\infty+a^{-1}(-a^{-6}q^4,-a^6q^2,q^6;q^6)_\infty).
\end{aligned}
\tag{21.2.14}
$$

If we differentiate (21.2.14) with respect to a then multiply by a, we find

$$
\begin{aligned}
&\left(3(a+a^{-1})^2(a-a^{-1})-6(a+a^{-1})^3\sum_{n\geq 1}\left(\frac{a^{-2}q^{2n}}{1+a^{-2}q^{2n}}-\frac{a^2q^{2n}}{1+a^2q^{2n}}\right)\right) \\
&\qquad\qquad\times(-a^{-2}q^2,-a^2q^2,q^2;q^2)_\infty^3 \\
&=c_0\left(3(a^3-a^{-3})-6(a^3+a^{-3})\sum_{n\geq 1}\left(\frac{a^{-6}q^{6n}}{1+a^{-6}q^{6n}}-\frac{a^6q^{6n}}{1+a^6q^{6n}}\right)\right) \\
&\qquad\qquad\times(-a^{-6}q^6,-a^6q^6,q^6;q^6)_\infty \\
&+3\frac{(q^6;q^6)_\infty^3}{(q^2;q^2)_\infty} \\
&\times\left(a\left(1-6\sum_{n\geq 0}\left(\frac{a^{-6}q^{6n+2}}{1+a^{-6}q^{6n+2}}-\frac{a^6q^{6n+4}}{1+a^6q^{6n+4}}\right)\right)(-a^{-6}q^2,-a^6q^4,q^6;q^6)_\infty\right. \\
&\left.-a^{-1}\left(1+6\sum_{n\geq 0}\left(\frac{a^{-6}q^{6n+4}}{1+a^{-6}q^{6n+4}}-\frac{a^6q^{6n+2}}{1+a^6q^{6n+2}}\right)\right)(-a^{-6}q^4,-a^6q^2,q^6;q^6)_\infty\right).
\end{aligned}
\tag{21.2.15}
$$

If in (21.2.15) we set $a=i$, we find

$$
0=-6ic_0(q^6;q^6)_\infty^3+6i\frac{(q^6;q^6)_\infty^3}{(q^2;q^2)_\infty}\left(1+6\sum_{n\geq 0}\left(\frac{q^{6n+2}}{1-q^{6n+2}}-\frac{q^{6n+4}}{1-q^{6n+4}}\right)\right)(q^2;q^2)_\infty,
$$

or,

$$
c_0=1+6\sum_{n\geq 0}\left(\frac{q^{6n+2}}{1-q^{6n+2}}-\frac{q^{6n+4}}{1-q^{6n+4}}\right). \tag{21.2.16}
$$

If we substitute both (21.2.16) and (21.2.13) into (21.2.11), we obtain (21.2.4).

21.3 Further Identities Involving $\sum_{m,n=-\infty}^{\infty} q^{m^2+mn+n^2}$

In this section, we will prove two identities involving $a(q) = \sum_{m,n=-\infty}^{\infty} q^{m^2+mn+n^2}$. They are

$$a(q^3) = \frac{(q;q)_\infty^3}{(q^3;q^3)_\infty} + 3q\frac{(q^9;q^9)_\infty^3}{(q^3;q^3)_\infty} \tag{21.3.1}$$

and

$$a(q)^3 = \left(\frac{(q;q)^3}{(q^3;q^3)_\infty}\right)^3 + 27q\left(\frac{(q^3;q^3)_\infty^3}{(q;q)_\infty}\right)^3. \tag{21.3.2}$$

We have (21.1.1)

$$\begin{aligned}
a(q) &= 1+6\sum_{n\geq 0}\left(\frac{q^{3n+1}}{1-q^{3n+1}} - \frac{q^{3n+2}}{1-q^{3n+2}}\right)\\
&= \left[1+6\sum_{n\geq 0}\left(\frac{a^{-6}q^{3n+1}}{1-a^{-6}q^{3n+1}} - \frac{a^6q^{3n+2}}{1-a^6q^{3n+2}}\right)\right]_{a=1}\\
&= \frac{1}{(q;q)_\infty}\left[(a^{-6}q,a^6q^2,q^3;q^3)_\infty\left(1+6\sum_{n\geq 0}\left(\frac{a^{-6}q^{3n+1}}{1-q^{3n+1}} - \frac{a^6q^{3n+2}}{1-a^6q^{3n+2}}\right)\right)\right]_{a=1}\\
&= \frac{1}{(q;q)_\infty}\left[\frac{d}{da}(a(a^{-6}q,a^6q^2,q^3;q^3)_\infty)\right]_{a=1}\\
&= \frac{1}{(q;q)_\infty}\left[\frac{d}{da}\left(\sum_{-\infty}^{\infty}(-1)^n a^{6n+1}q^{(3n^2+n)/2}\right)\right]_{a=1}\\
&= \frac{1}{(q;q)_\infty}\left[\sum_{-\infty}^{\infty}(-1)^n(6n+1)a^{6n}q^{(3n^2+n)/2}\right]_{a=1}\\
&= \frac{1}{(q;q)_\infty}\sum_{-\infty}^{\infty}(-1)^n(6n+1)q^{(3n^2+n)/2}.
\end{aligned} \tag{21.3.3}$$

It follows that

$$\begin{aligned}
a(q^3) &= \frac{1}{(q^3;q^3)_\infty}\sum_{-\infty}^{\infty}(-1)^n(6n+1)q^{3(3n^2+n)/2}\\
&= \frac{1}{(q^3;q^3)_\infty}(1+5q^3-7q^6-11q^{15}+13q^{21}\ \cdots)\\
&= \frac{1}{(q^3;q^3)_\infty}\left((q;q)_\infty^3+3q(q^9;q^9)_\infty^3\right),
\end{aligned} \tag{21.3.4}$$

where we have used (1.7.1).

We can write (21.3.4)

$$a(q^3) = \frac{(q;q)_\infty^3}{(q^3;q^3)_\infty} + 3q\frac{(q^9;q^9)_\infty^3}{(q^3;q^3)_\infty},$$

which is (21.3.1),
or,

$$a(q^3) - 3q\frac{(q^9;q^9)_\infty^3}{(q^3;q^3)_\infty} = \frac{(q;q)_\infty^3}{(q^3;q^3)_\infty}. \tag{21.3.5}$$

If in (21.3.5) we put q, ωq, $\omega^2 q$ for q in turn, where ω is a cube root of unity other than 1, and multiply the results, we find

$$a(q^3)^3 - 27q^3\frac{(q^9;q^9)_\infty^9}{(q^3;q^3)_\infty^3} = \frac{1}{(q^3;q^3)_\infty^3}\left(\frac{(q^3;q^3)_\infty^4}{(q^9;q^9)_\infty}\right)^3 = \frac{(q^3;q^3)_\infty^9}{(q^9;q^9)_\infty^3}. \tag{21.3.6}$$

If we now replace q by $q^{\frac{1}{3}}$, we obtain (21.3.2).

Note that (21.3.1) gives the 3-dissection

$$(q;q)_\infty^3 = (q^3;q^3)_\infty a(q^3) - 3q(q^9;q^9)_\infty^3. \tag{21.3.7}$$

Alternatively, put $q^{\frac{3}{2}}$ for q, then $a = -q^{\frac{1}{2}}$ in (21.2.4).

Endnotes.

(21.1.1) is given in Borwein and Borwein (1991) [20], p. 695.

References

17. B.C. Berndt, R.A. Rankin, *Ramanujan Letters and Commentary*, vol. 9, History of Mathematics (AMS, Providence, 1995)
20. J.M. Borwein, P.B. Borwein, A cubic counterpart of Jacobi's identity and the AGM. Trans. Am. Math. Soc. **323**, 691–701 (1991)
82. M.D. Hirschhorn, A letter from Fitzroy House. Am. Math. Mon. **115**, 563–566 (2008)
104. L. Lorenz, Contributions à la theorie des nombres. Tidsskrift for Mathematik **3**, 97–114 (1871)
120. H. Valentier (ed.), Oeuvres Scientifique de L. Lorenz, II, Librarie Lehmann & Stage, Copenhagen (1904)

Chapter 22
The Cubic Theta-Function Analogues of Borwein, Borwein and Garvan

22.1 Introduction

In Borwein, Borwein and Garvan (1994) [21], the authors introduce three functions. They are

$$a(q) = \sum_{m,n=-\infty}^{\infty} q^{m^2+mn+n^2}, \tag{22.1.1}$$

which we met in the previous chapter, as well as

$$b(q) = \sum_{m,n=-\infty}^{\infty} \omega^{m-n} q^{m^2+mn+n^2} \tag{22.1.2}$$

where ω is a cube root of unity other than 1, and

$$\begin{aligned} c(q) &= \sum_{m,n=-\infty} q^{(m+\frac{1}{3})^2+(m+\frac{1}{3})(n+\frac{1}{3})+(n+\frac{1}{3})^2} \\ &= q^{\frac{1}{3}} \sum_{m,n=-\infty}^{\infty} q^{m^2+mn+n^2+m+n}. \end{aligned} \tag{22.1.3}$$

In §22.2 we will obtain the 3-dissections of $a(q)$ and $b(q)$, namely

$$a(q) = a(q^3) + 2c(q^3) \tag{22.1.4}$$

and

$$b(q) = a(q^3) - c(q^3). \tag{22.1.5}$$

M.D. Hirschhorn, *The Power of q*, Developments in Mathematics 49,
DOI 10.1007/978-3-319-57762-3_22

In §22.3, we will show that each of $b(q)$ and $c(q)$ is a product, indeed

$$b(q) = \frac{(q;q)_\infty^3}{(q^3;q^3)_\infty} \tag{22.1.6}$$

and

$$c(q) = 3q^{\frac{1}{3}} \frac{(q^3;q^3)_\infty^3}{(q;q)_\infty}. \tag{22.1.7}$$

In §22.4, we will derive the following expressions for $b(q)$ and $c(q)$, analogous to (21.1.1),

$$b(q) = 1 - 3\sum_{n\geq 0}\left(\frac{q^{3n+1}}{1-q^{9n+3}} - \frac{q^{6n+4}}{1-q^{9n+6}}\right) + 6\sum_{n\geq 0}\left(\frac{q^{9n+3}}{1-q^{9n+3}} - \frac{q^{9n+6}}{1-q^{9n+6}}\right) \tag{22.1.8}$$

and

$$c(q) = 3q^{\frac{1}{3}} \sum_{n\geq 0}\left(\frac{q^n}{1-q^{3n+1}} - \frac{q^{2n+1}}{1-q^{3n+2}}\right). \tag{22.1.9}$$

In §22.5, we will derive a result due to Nathan Fine (Fine (1988) [46], 32.35),

$$\sum_{n\geq 0} p(n)q^n = \frac{1}{(q;q)_\infty} = \frac{1}{(q^3;q^3)_\infty^3} \sum_{n\geq 0} (d_{1,3}(3n+1) - d_{2,3}(3n+1))\, q^n \tag{22.1.10}$$

and hence another recurrence (our fifth) for $p(n)$,

$$p(n) - 3p(n-3) + 5p(n-9) - 7p(n-18) + - \cdots = d_{1,3}(3n+1) - d_{2,3}(3n+1), \tag{22.1.11}$$

where 3, 9, 18, and so on are 3 times the triangular numbers and the coefficients 1, 3, 5, 7, and so on are, of course, the odd numbers.

In §22.6, we will obtain the 2-dissections of $a(q)$, $b(q)$ and $q^{-\frac{1}{3}}c(q)$, namely

$$a(q) = a(q^4) + 6q\psi(q^2)\psi(q^6), \tag{22.1.12}$$

$$b(q) = b(q^4) - q^{-\frac{1}{3}} \frac{(q^2;q^2)_\infty^2}{(q^6;q^6)_\infty^2} c(q^4) \tag{22.1.13}$$

and

$$q^{-\frac{1}{3}}c(q) = 3\frac{(q^6;q^6)_\infty^2}{(q^2;q^2)_\infty^2}b(q^4) + q^{-\frac{1}{3}}c(q^4). \tag{22.1.14}$$

In §22.7, we will establish two further relations between $b(q)$ and $c(q)$, namely

$$4b(q^4) - b(q) = q^{-\frac{1}{3}}\frac{(q^2;q^2)_\infty^2}{(q^6;q^6)_\infty^2}c(q) \tag{22.1.15}$$

and

$$c(q) - 4qc(q^4) = 3q^{\frac{1}{3}}\frac{(q^6;q^6)_\infty^2}{(q^2;q^2)_\infty^2}b(q). \tag{22.1.16}$$

In §22.8, we will derive two relations between $a(q)$, $b(q)$ and $c(q)$, namely

$$a(q)^3 = b(q)^3 + c(q)^3 \tag{22.1.17}$$

and

$$a(q)a(q^2) = b(q)b(q^2) + c(q)c(q^2). \tag{22.1.18}$$

In §22.9, we will apply our formulas of §22.1 to the problem of $p^*(n)$, the number of partitions of n where even parts come in two colours, which we first considered in Chapter 14.

Indeed, we will show that

$$\sum_{n\geq 0} p^*(3n)q^n = \frac{(q^3;q^3)_\infty^3}{(q;q)_\infty^4(q^2;q^2)_\infty^3}a(q^2), \tag{22.1.19}$$

$$\sum_{n\geq 0} p^*(3n+1)q^n = \frac{(q^6;q^6)_\infty^3}{(q;q)_\infty^3(q^2;q^2)_\infty^4}a(q), \tag{22.1.20}$$

$$\text{and} \quad \sum_{n\geq 0} p^*(3n+2)q^n = 3\frac{(q^3;q^3)_\infty^3(q^6;q^6)_\infty^3}{(q;q)_\infty^4(q^2;q^2)_\infty^4}. \tag{22.1.21}$$

In §22.10, we will prove the following two identities, the first of which was stated by Ramanujan, namely

$$1 + 27q\left(\frac{(q^3;q^3)_\infty}{(q;q)_\infty}\right)^{12} = \left(1 + 9q\left(\frac{(q^9;q^9)_\infty}{(q;q)_\infty}\right)^3\right)^3 \tag{22.1.22}$$

and

$$\left(1+9q\left(\frac{(q^9;q^9)_\infty}{(q;q)_\infty}\right)^3\right)\left(1+9q^2\left(\frac{(q^{18};q^{18})_\infty}{(q^2;q^2)_\infty}\right)^3\right) \\ =1+9q\left(\frac{(q^3;q^3)_\infty(q^6;q^6)_\infty}{(q;q)_\infty(q^2;q^2)_\infty}\right)^4. \quad (22.1.23)$$

Finally, in §22.11, we compile some more interesting and neat results.

22.2 The 3-Dissections of $a(q)$ and $b(q)$

We have

$$\begin{aligned} a(q) &= \sum_{m,n=-\infty}^{\infty} q^{m^2+mn+n^2} \\ &= \sum_{m\equiv n \pmod 3} q^{m^2+mn+n^2} + \sum_{m\not\equiv n \pmod 3} q^{m^2+mn+n^2} \\ &= \sum_{m\equiv n \pmod 3} q^{m^2+mn+n^2} + \sum_{m-n\equiv 1 \pmod 3} q^{m^2+mn+n^2} + \sum_{m-n\equiv -1 \pmod 3} q^{m^2+mn+n^2} \\ &= \sum_{m\equiv n \pmod 3} q^{m^2+mn+n^2} + 2\sum_{m-n\equiv 1 \pmod 3} q^{m^2+mn+n^2}. \end{aligned} \quad (22.2.1)$$

In the first sum, set $m-n=3u$, $m+2n=3v$, $m=2u+v$, $n=-u+v$, and in the second $m-n=3u+1$, $m+2n=3v+1$, $m=2u+v+1$, $n=-u+v$. Then we have

$$\begin{aligned} \sum_{m\equiv n \pmod 3} q^{m^2+mn+n^2} &= \sum_{u,v=-\infty}^{\infty} q^{(2u+v)^2+(2u+v)(-u+v)+(-u+v)^2} \\ &= \sum_{u,v=-\infty}^{\infty} q^{3u^2+3uv+3v^2} \\ &= a(q^3), \end{aligned} \quad (22.2.2)$$

$$\begin{aligned} \sum_{m-n\equiv 1 \pmod 3} q^{m^2+mn+n^2} &= \sum_{u,v=-\infty}^{\infty} q^{(2u+v+1)^2+(2u+v+1)(-u+v)+(-u+v)^2} \\ &= \sum_{u,v=-\infty}^{\infty} q^{3u^2+3uv+3v^2+3u+3v+1} \\ &= c(q^3), \end{aligned} \quad (22.2.3)$$

and (22.2.1) becomes

$$a(q) = a(q^3) + 2c(q^3),$$

which is (22.1.4).

Similarly,

$$\begin{aligned}
&b(q)\\
&= \sum_{m,n=-\infty}^{\infty} \omega^{m-n} q^{m^2+mn+n^2}\\
&= \sum_{m\equiv n \pmod 3} q^{m^2+mn+n^2} + \omega \sum_{m-n\equiv 1 \pmod 3} q^{m^2+mn+n^2}\\
&\quad + \omega^2 \sum_{m-n\equiv -1 \pmod 3} q^{m^2+mn+n^2}\\
&= \sum_{m\equiv n \pmod 3} q^{m^2+mn+n^2} + \omega \sum_{m-n\equiv 1 \pmod 3} q^{m^2+mn+n^2}\\
&\quad + \omega^2 \sum_{m-n\equiv 1 \pmod 3} q^{m^2+mn+n^2}\\
&= \sum_{m\equiv n \pmod 3} q^{m^2+mn+n^2} - \sum_{m-n\equiv 1 \pmod 3} q^{m^2+mn+n^2}\\
&= a(q^3) - c(q^3)
\end{aligned}$$

where we have used (22.2.2) and (22.2.3), and we have obtained (22.1.5).

22.3 Expression of $b(q)$ and $c(q)$ as Products

We have

$$q^{-\frac{1}{3}} c(q) = \sum_{m,n=-\infty}^{\infty} q^{m^2+mn+n^2+m+n}, \tag{22.3.1}$$

so

$$\begin{aligned}
q^{-\frac{2}{3}} c(q^2) &= \sum_{m,n=-\infty}^{\infty} q^{2m^2+2mn+2n^2+2m+2n}\\
&= \sum_{m,n=-\infty}^{\infty} q^{2m^2+2mn+2n^2-2m-2n}
\end{aligned}$$

and

$$\begin{aligned}
q^{\frac{1}{3}}c(q^2) &= \sum_{m,n=-\infty}^{\infty} q^{m^2+n^2+(-m-n+1)^2} \\
&= \sum_{m+n+p=1} q^{m^2+n^2+p^2} \\
&= c_1 \quad (\text{see } (21.2.8)) \\
&= 3q\frac{(q^6;q^6)_\infty^3}{(q^2;q^2)_\infty} \quad (\text{see } (21.1.13)),
\end{aligned}$$

from which (22.1.7) follows.

By (22.1.5), (22.1.7) and (21.3.1),

$$\begin{aligned}
b(q) &= a(q^3) - c(q^3) \\
&= a(q^3) - 3q\frac{(q^9;q^9)_\infty^3}{(q^3;q^3)_\infty} \\
&= \frac{(q;q)_\infty^3}{(q^3;q^3)_\infty},
\end{aligned}$$

which is (22.1.6).

22.4 Expressions for $b(q)$ and $c(q)$ Analogous to (21.1.1)

It follows from (22.1.4) and (21.1.1) that

$$\begin{aligned}
2c(q^3) &= a(q) - a(q^3) \\
&= 6\sum_{n\geq 0}\left(\frac{q^{3n+1}}{1-q^{3n+1}} - \frac{q^{3n+2}}{1-q^{3n+2}}\right) - 6\sum_{n\geq 0}\left(\frac{q^{9n+3}}{1-q^{9n+3}} - \frac{q^{9n+6}}{1-q^{9n+6}}\right) \\
&= 6\sum_{n\geq 0}\left(\frac{q^{3n+1}+q^{6n+2}}{1-q^{9n+3}} - \frac{q^{3n+2}+q^{6n+4}}{1-q^{9n+6}}\right) \\
&= 6\sum_{n\geq 0}\left(\frac{q^{3n+1}}{1-q^{9n+3}} - \frac{q^{6n+4}}{1-q^{9n+6}}\right). \qquad (22.4.1)
\end{aligned}$$

If we divide (22.4.1) by 2 and replace q by $q^{\frac{1}{3}}$, we obtain (22.1.9).

Now, by (22.1.5) and (22.1.9),

$$b(q) = a(q^3) - c(q^3)$$
$$= 1 + 6\sum_{n\geq 0}\left(\frac{q^{9n+3}}{1-q^{9n+3}} - \frac{q^{9n+6}}{1-q^{9n+6}}\right) - 3q\sum_{n\geq 0}\left(\frac{q^{3n}}{1-q^{9n+3}} - \frac{q^{6n+3}}{1-q^{9n+6}}\right),$$

which is (22.1.8).

22.5 A Fine Recurrence for $p(n)$

We have, by (22.1.7) and (22.1.9),

$$q\frac{(q^9;q^9)_\infty^3}{(q^3;q^3)_\infty} = \sum_{n\geq 0}\left(\frac{q^{3n+1}}{1-q^{9n+3}} - \frac{q^{6n+4}}{1-q^{9n+6}}\right)$$
$$= \sum_{n,k\geq 0}(q^{3n+1+k(9n+3)} - q^{6n+4)+k(9n+6)})$$
$$= \sum_{k,n\geq 0}(q^{(3k+1)(3n+1)} - q^{(3k+2)(3n+2)})$$
$$= \sum_{N\equiv 1 \pmod 3}(d_{1,3}(N) - d_{2,3}(N))\,q^N$$
$$= \sum_{n\geq 0}(d_{1,3}(3n+1) - d_{2,3}(3n+1))\,q^{3n+1}. \tag{22.5.1}$$

If we now divide by $q(q^9;q^9)_\infty^3$ and replace q by $q^{\frac{1}{3}}$, we obtain (22.1.10).

We can write (22.1.10)

$$(q^3;q^3)_\infty^3 \sum_{n\geq 0} p(n)q^n = \sum_{n\geq 0}(d_{1,3}(3n+1) - d_{2,3}(3n+1))\,q^n.$$

That is, by (1.7.1),

$$\sum_{n\geq 0}(-1)^n(2n+1)q^{3(n^2+n)/2}\sum_{n\geq 0} p(n)q^n = \sum_{n\geq 0}(d_{1,3}(3n+1) - d_{2,3}(3n+1))\,q^n.$$

The recurrence (22.1.11) follows.

22.6 The 2-Dissections of $a(q)$, $b(q)$ and $q^{-\frac{1}{3}}c(q)$

In this section we prove

$$a(q) = a(q^4) + 6q\psi(q^2)\psi(q^6), \tag{22.6.1}$$

$$b(q) = b(q^4) - q^{-\frac{1}{3}} \frac{(q^2;q^2)_\infty^2}{(q^6;q^6)_\infty^2} c(q^4) \tag{22.6.2}$$

and

$$q^{-\frac{1}{3}} c(q) = 3\frac{(q^6;q^6)_\infty^2}{(q^2;q^2)_\infty^2} b(q^4) + q^{-\frac{1}{3}} c(q^4). \tag{22.6.3}$$

We have

$$\begin{aligned} a(q) = &\sum_{m,n=-\infty}^{\infty} q^{m^2+mn+n^2} \\ = &\sum_{m,n\equiv 0 \pmod 2} q^{m^2+mn+n^2} + 2 \sum_{m\equiv 1,\ n\equiv 0 \pmod 2} q^{m^2+mn+n^2} \\ &+ \sum_{m,n\equiv 1 \pmod 2} q^{m^2+mn+n^2}. \end{aligned} \tag{22.6.4}$$

The first sum is

$$\sum_{m,n\equiv 0 \pmod 2} q^{m^2+mn+n^2} = \sum_{m,n=-\infty}^{\infty} q^{4m^2+4mn+4n^2} = a(q^4), \tag{22.6.5}$$

the second sum is

$$\begin{aligned} \sum_{m\equiv 1,\ n\equiv 0 \pmod 2} q^{m^2+mn+n^2} &= \sum_{m,n=-\infty}^{\infty} q^{(2m+1)^2+(2m+1)2n+(2n)^2} \\ &= \sum_{m,n=-\infty}^{\infty} q^{4m^2+4mn+4n^2+4m+2n} \\ &= q \sum_{m,n=-\infty}^{\infty} q^{(m-2)^2+3(m+n)^2+(m-n)+3(m+n)} \\ &= q \sum_{u\equiv v \pmod 2} q^{u^2+u+3v^2+3v} \end{aligned}$$

$$
\begin{aligned}
&= q\left(\sum_{u,v\equiv 0 \pmod 2} q^{u^2+u+3v^2+3v} + \sum_{u,v\equiv 1 \pmod 2} q^{u^2+u+3v^2+3v}\right)\\
&= q\left(\sum_{u,v=-\infty}^{\infty} q^{(2u)^2+2u+3(2v)^2+3(2v)} + \sum_{u,v=-\infty}^{\infty} q^{(2u-1)^2+(2u-1)+3(2v-1)^2+3(2v-1)}\right)\\
&= 2q\sum_{u,v=-\infty}^{\infty} q^{4u^2+2u+12v^2+6v}\\
&= 2q\psi(q^2)\psi(q^6) \qquad (22.6.6)
\end{aligned}
$$

and the third sum is

$$
\begin{aligned}
\sum_{m,n\equiv 1 \pmod 2} q^{m^2+mn+n^2} &= \sum_{m,n=-\infty}^{\infty} q^{(2m+1)^2+(2m+1)(2n-1)+(2n-1)^2}\\
&= q\sum_{m,n=-\infty^{\infty}} q^{4m^2+4mn+4n^2+2m-2n}\\
&= \sum_{m,n=-\infty}^{\infty} q^{(m-n+1)^2+3(m+n)^2}\\
&= \sum_{u\not\equiv v \pmod 2} q^{u^2+3v^2}. \qquad (22.6.7)
\end{aligned}
$$

Now, if $u \not\equiv v \pmod 2$ then $u+v \equiv u-3v \equiv \pm 1 \pmod 4$. In the first case, let $u+v=4s+1$, $u-3v=4r+1$, $u=r+3s+1$, $v=s-r$, and in the second let $u+v=-4s-1$, $u-3v=-4r-1$, $u=-r-3s-1$, $v=r-s$, and

$$
\begin{aligned}
\sum_{u\not\equiv v \pmod 2} q^{u^2+3v^2} &= 2\sum_{r,s=-\infty}^{\infty} q^{(r+3s+1)^2+3(s-r)^2}\\
&= 2q\sum_{-\infty}^{\infty} q^{4r^2+12s^2+2r+6s}\\
&= 2q\psi(q^2)\psi(q^6). \qquad (22.6.8)
\end{aligned}
$$

If we substitute (22.6.5), (22.6.6), (22.6.7) and (22.6.8) into (22.6.4), we find

$$
a(q) = a(q^4) + 6q\psi(q^2)\psi(q^6), \qquad (22.6.9)
$$

which is (22.6.1), or (22.1.12).

From (22.1.5), (22.1.4), (22.6.9) and (22.1.5) again,

$$\begin{aligned}
b(q) &= a(q^3) - c(q^3) = a(q^3) + \frac{1}{2}\left(a(q^3) - a(q)\right) = \frac{3}{2}a(q^3) - \frac{1}{2}a(q) \\
&= \frac{3}{2}(a(q^{12}) + 6q^3\psi(q^6)\psi(q^{18})) - \frac{1}{2}(a(q^4) + 6q\psi(q^2)\psi(q^6)) \\
&= b(q^4) - 3q\psi(q^6)\left(\psi(q^2) - 3q^2\psi(q^{18})\right). \qquad (22.6.10)
\end{aligned}$$

Now,

$$\begin{aligned}
\psi(q) - 3q\psi(q^9) &= 1 + q + q^3 + q^6 + q^{10} + q^{15} + q^{21} + q^{28} + \cdots \\
&\qquad - 3q \qquad\qquad - 3q^{10} \qquad\qquad - 3q^{28} - \cdots \\
&= 1 - 2q + q^3 + q^6 - 2q^{10} + q^{15} + q^{21} - 2q^{28} + \cdots \\
&= \sum_{n\geq 0} 2\cos\left((2n+1)\frac{\pi}{3}\right)q^{(n^2+n)/2} \\
&= \sum_{n\geq 0} (\alpha^{2n+1} + \alpha^{-(2n+1)})q^{(n^2+n)/2} \text{ where } \alpha = e^{\frac{\pi i}{3}}, \\
&= \left(\alpha + \frac{1}{\alpha}\right)(-\alpha^{-2}q, -\alpha^2 q, q; q)_\infty \text{ by (1.7.5) with } a = -\alpha^2 \\
&= (-\omega^{-1}q, -\omega q, q; q)_\infty \\
&= \frac{(q;q)_\infty^2 (q^6;q^6)_\infty}{(q^2;q^2)_\infty (q^3;q^3)_\infty}. \qquad (22.6.11)
\end{aligned}$$

From (22.6.10) and (22.6.11) we find

$$b(q) = b(q^4) - 3q\frac{(q^2;q^2)_\infty^2 (q^{12};q^{12})_\infty^3}{(q^4;q^4)_\infty (q^6;q^6)_\infty^2} = b(q^4) - q^{-\frac{1}{3}}\frac{(q^2;q^2)_\infty^2}{(q^6;q^6)_\infty^2}c(q^4),$$

which is (22.6.2), or (22.1.13).

Finally, from (22.1.4), (22.6.9) and (22.1.4) again,

$$\begin{aligned}
c(q^3) &= \frac{1}{2}a(q) - \frac{1}{2}a(q^3) \\
&= \frac{1}{2}(a(q^4) + 6q\psi(q^2)\psi(q^6)) - \frac{1}{2}(a(q^{12}) + 6q^3\psi(q^6)\psi(q^{18})) \\
&= c(q^{12}) + 3q\psi(q^6)\left(\psi(q^2) - q^2\psi(q^{18})\right). \qquad (22.6.12)
\end{aligned}$$

Now,

$$
\begin{aligned}
\psi(q) - q\psi(q^9) &= 1 + q + q^3 + q^6 + q^{10} + q^{15} + q^{21} + q^{28} + \cdots \\
&\quad\;\; - q \qquad\qquad\quad - q^{10} \qquad\qquad\quad - q^{28} - \cdots \\
&= 1 + q^3 + q^6 + q^{15} + q^{21} + \cdots \\
&= \sum_{-\infty}^{\infty} q^{3(3n^2+n)/2} \\
&= \frac{(q^6;q^6)_\infty (q^9;q^9)_\infty^2}{(q^3;q^3)_\infty (q^{18};q^{18})_\infty}.
\end{aligned}
\tag{22.6.13}
$$

From (22.6.12) and (22.6.13) we find

$$
c(q^3) = c(q^{12}) + 3q \frac{(q^{12};q^{12})_\infty^3 (q^{18};q^{18})_\infty^2}{(q^6;q^6)_\infty^2 (q^{36};q^{36})_\infty},
$$

which yields (22.6.3), or (22.1.14).

22.7 Two Further Relations Between $b(q)$ and $c(q)$

In this section, we prove that

$$
4b(q^4) - b(q) = q^{-\frac{1}{3}} \frac{(q^2;q^2)_\infty^2}{(q^6;q^6)_\infty^2} c(q) \tag{22.7.1}
$$

and

$$
c(q) - 4qc(q^4) = 3q^{\frac{1}{3}} \frac{(q^6;q^6)_\infty^2}{(q^2;q^2)_\infty^2} b(q). \tag{22.7.2}
$$

From (22.1.13) and (22.1.14) we have

$$
\begin{aligned}
4b(q^4) - b(q) &= 3b(q^4) + \left(b(q^4) - b(q)\right) \\
&= 3b(q^4) + q^{-\frac{1}{3}} \frac{(q^2;q^2)_\infty^2}{(q^6;q^6)_\infty^2} c(q^4) \\
&= q^{-\frac{1}{3}} \frac{(q^2;q^2)_\infty^2}{(q^6;q^6)_\infty^2} \left(3q^{\frac{1}{3}} \frac{(q^6;q^6)_\infty^2}{(q^2;q^2)_\infty^2} b(q^4) + c(q^4) \right) \\
&= q^{-\frac{1}{3}} \frac{(q^2;q^2)_\infty^2}{(q^6;q^6)_\infty^2} c(q),
\end{aligned}
$$

which is (22.7.1) or (22.1.15), and

$$\begin{aligned} c(q)-4c(q^4) &= \big(c(q)-c(q^4)\big)-3c(q^4) \\ &= 3q^{\frac{1}{3}}\frac{(q^6;q^6)^2_\infty}{(q^2;q^2)^2_\infty}b(q^4)-3c(q^4) \\ &= 3q^{\frac{1}{3}}\frac{(q^6;q^6)^2_\infty}{(q^2;q^2)^2_\infty}\left(b(q^4)-q^{-\frac{1}{3}}\frac{q^2;q^2)^2_\infty}{(q^6;q^6)^2_\infty}c(q^4)\right) \\ &= 3q^{\frac{1}{3}}\frac{(q^6;q^6)^2_\infty}{(q^2;q^2)^2_\infty}b(q), \end{aligned}$$

which is (22.7.2) or (22.1.16).

Note that by (22.1.6) and (22.1.7), (22.1.13) can be written

$$\frac{(q^4;q^4)^3_\infty}{(q^{12};q^{12})_\infty}-\frac{(q;q)^3_\infty}{(q^3;q^3)_\infty}=3q\frac{(q^2;q^2)^2_\infty(q^{12};q^{12})^3_\infty}{(q^4;q^4)_\infty(q^6;q^6)^2_\infty}, \tag{22.7.3}$$

(22.1.15) can be written

$$4\frac{(q^4;q^4)^3_\infty}{(q^{12};q^{12})_\infty}-\frac{(q;q)^3_\infty}{(q^3;q^3)_\infty}=3\frac{(q^2;q^2)^2_\infty(q^3;q^3)^3_\infty}{(q;q)_\infty(q^6;q^6)^2_\infty}, \tag{22.7.4}$$

(22.1.14) can be written

$$\frac{(q^3;q^3)^3_\infty}{(q;q)_\infty}-q\frac{(q^{12};q^{12})^3_\infty}{(q^4;q^4)_\infty}=\frac{(q^4;q^4)^3_\infty(q^6;q^6)^2_\infty}{(q^2;q^2)^2_\infty(q^{12};q^{12})_\infty} \tag{22.7.5}$$

and (22.1.16) can be written

$$\frac{(q^3;q^3)^3_\infty}{(q;q)_\infty}-4q\frac{(q^{12};q^{12})^3_\infty}{(q^4;q^4)_\infty}=\frac{(q;q)^3_\infty(q^6;q^6)^2_\infty}{(q^2;q^2)^2_\infty(q^3;q^3)_\infty}. \tag{22.7.6}$$

22.8 Two Relations Between $a(q)$, $b(q)$ and $c(q)$

In this section, we prove

$$a(q)^3=b(q)^3+c(q)^3 \tag{22.8.1}$$

and

$$a(q)a(q^2)=b(q)b(q^2)+c(q)c(q^2). \tag{22.8.2}$$

The first follows directly from (21.3.2), (22.1.6) and (22.1.7).

The second is harder. We start by setting

$$L(q) = a(q)a(q^2) - b(q)b(q^2) - c(q)c(q^2). \tag{22.8.3}$$

Our aim is to show that

$$L(q) = L(q^2), \tag{22.8.4}$$

for then

$$L(q) = L(q^2) = \cdots = L(q^{2^n}) = \lim_{n\to\infty} L(q^{2^n}) = L(0) = a(0)^2 - b(0)^2 = 1 - 1 = 0 \tag{22.8.5}$$

and (22.1.18) is proved!

We require the following result,

$$\begin{aligned} a(q) &= \frac{3}{2}\frac{\phi(-q^3)^3}{\phi(-q)} - \frac{1}{2}\frac{\phi(-q)^3}{\phi(-q^3)} \\ &= \frac{3}{2}\frac{(q^2;q^2)_\infty (q^3;q^3)_\infty^6}{(q;q)_\infty^2 (q^6;q^6)_\infty^3} - \frac{1}{2}\frac{(q;q)_\infty^6 (q^6;q^6)_\infty}{(q^2;q^2)_\infty^3 (q^3;q^3)_\infty^2}. \end{aligned} \tag{22.8.6}$$

In order to prove (22.8.6), we note that (21.2.12) can be written, after replacing q by $q^{\frac{1}{2}}$,

$$\begin{aligned} (a+a^{-1})^3(-a^{-2}q, -a^2q, q; q)_\infty^3 &= (a^3+a^{-3})a(q)(-a^{-6}q^3, -a^6q^3, q^3; q^3)_\infty \\ &+ 3\frac{(q^3;q^3)_\infty^3}{(q;q)_\infty}(a(-a^{-6}q, -a^6q^2, q^3; q^3)_\infty + a^{-1}(-a^{-6}q^2, -a^6q, q^3; q^3)_\infty). \end{aligned} \tag{22.8.7}$$

If in (22.8.7) we set $a = 1$ and divide by 2, we obtain

$$4(-q, -q, q; q)_\infty^3 = a(q)(-q^3, -q^3, q^3; q^3)_\infty + 3\frac{(q^3;q^3)_\infty^3}{(q;q)_\infty}(-q, -q^2, q^3; q^3)_\infty. \tag{22.8.8}$$

On the other hand, if in (22.8.7) we set $a = e^{\frac{\pi i}{3}}$, we obtain

$$\begin{aligned} (-\omega^{-1}q, -\omega q, q; q)_\infty^3 &= -2a(q)(-q^3, -q^3, q^3; q^3)_\infty \\ &+ 3\frac{(q^3;q^3)_\infty^3}{(q;q)_\infty}(-q, -q^2, q^3; q^3)_\infty. \end{aligned} \tag{22.8.9}$$

We can write (22.8.8) and (22.8.9) as follows:

$$4\frac{(q^2;q^2)_\infty^6}{(q;q)_\infty^3} = \frac{(q^6;q^6)_\infty^2}{(q^3;q^3)_\infty}a(q) + 3\frac{(q^2;q^2)_\infty(q^3;q^3)_\infty^5}{(q;q)_\infty^2(q^6;q^6)_\infty} \tag{22.8.10}$$

and

$$\frac{(q;q)_\infty^6(q^6;q^6)_\infty^3}{(q^2;q^2)_\infty^3(q^3;q^3)_\infty^3} = -2\frac{(q^6;q^6)_\infty^2}{(q^3;q^3)_\infty}a(q) + 3\frac{(q^2;q^2)_\infty(q^3;q^3)_\infty^5}{(q;q)_\infty^2(q^6;q^6)_\infty}. \tag{22.8.11}$$

If we eliminate the term involving $a(q)$ from (22.8.10) and (22.8.11), we obtain

$$8\frac{(q^2;q^2)_\infty^6}{(q;q)_\infty^3} + \frac{(q;q)_\infty^6(q^6;q^6)_\infty^3}{(q^2;q^2)_\infty^3(q^3;q^3)_\infty^3} = 9\frac{(q^2;q^2)_\infty(q^3;q^3)_\infty^5}{(q;q)_\infty^2(q^6;q^6)_\infty}$$

or,

$$\frac{(q^2;q^2)_\infty^6}{(q;q)_\infty^3} = \frac{9}{8}\frac{(q^2;q^2)_\infty(q^3;q^3)_\infty^5}{(q;q)_\infty^2(q^6;q^6)_\infty} - \frac{1}{8}\frac{(q;q)_\infty^6(q^6;q^6)_\infty^3}{(q^2;q^2)_\infty^3(q^3;q^3)_\infty^3}. \tag{22.8.12}$$

On the other hand, if we subtract (22.8.11) from (22.8.10) and divide by 3, we obtain

$$\frac{(q^6;q^6)_\infty^2}{(q^3;q^3)_\infty}a(q) = \frac{4}{3}\frac{(q^2;q^2)_\infty^6}{(q;q)_\infty^3} - \frac{1}{3}\frac{(q;q)_\infty^6(q^6;q^6)_\infty^3}{(q^2;q^2)_\infty^3(q^3;q^3)_\infty^3}. \tag{22.8.13}$$

If we substitute (22.8.12) into (22.8.13) we obtain

$$\frac{(q^6;q^6)_\infty^2}{(q^3;q^3)_\infty}a(q) = \frac{3}{2}\frac{(q^2;q^2)_\infty(q^3;q^3)_\infty^5}{(q;q)_\infty^2(q^6;q^6)_\infty} - \frac{1}{2}\frac{(q;q)_\infty^6(q^6;q^6)_\infty^3}{(q^2;q^2)_\infty^3(q^3;q^3)_\infty^3}. \tag{22.8.14}$$

If we now divide by $\dfrac{(q^6;q^6)_\infty^2}{(q^3;q^3)_\infty}$, we obtain (22.8.6).

It follows from (22.8.6), (22.1.12), (22.1.6), (22.1.13), (22.1.7) and (22.1.14) that

$$\begin{aligned}
&L(q) - L(q^2)\\
&= \left(a(q)a(q^2) - b(q)b(q^2) - c(q)c(q^2)\right) - \left(a(q^2)a(q^4) - b(q^2)b(q^4) - c(q^2)c(q^4)\right)\\
&= a(q^2)\left(a(q) - a(q^4)\right) - b(q^2)\left(b(q) - b(q^4)\right) - c(q^2)\left(c(q) - c(q^4)\right)\\
&= \left(\frac{3}{2}\frac{(q^4;q^4)_\infty(q^6;q^6)_\infty^6}{(q^2;q^2)_\infty^2(q^{12};q^{12})_\infty^3} - \frac{1}{2}\frac{(q^2;q^2)_\infty^6(q^{12};q^{12})_\infty}{(q^4;q^4)_\infty^3(q^6;q^6)_\infty^2}\right)\cdot 6q\frac{(q^4;q^4)_\infty^2(q^{12};q^{12})_\infty^2}{(q^2;q^2)_\infty(q^6;q^6)_\infty}\\
&\quad + \frac{(q^2;q^2)_\infty^3}{(q^6;q^6)_\infty}\cdot 3q\frac{(q^2;q^2)_\infty^2(q^{12};q^{12})_\infty^3}{(q^4;q^4)_\infty(q^6;q^6)_\infty^2} - 3q^{\frac{2}{3}}\frac{(q^6;q^6)_\infty^3}{(q^2;q^2)_\infty}\cdot 3q^{\frac{1}{3}}\frac{(q^4;q^4)_\infty^3(q^6;q^6)_\infty^2}{(q^2;q^2)_\infty^2(q^{12};q^{12})_\infty}\\
&= 0,
\end{aligned}$$

and we are done.

22.9 Partitions Where Even Parts Come in Two Colours

Let $p^*(n)$ be the number of partitions of n where the even parts come in two colours. Chapter 14 was devoted to a study of $p^*(n)$.

We can deduce from (14.4.5) that

$$\sum_{n\geq 0} p^*(3n)q^n = \frac{(q^3;q^3)_\infty^3(q^6;q^6)_\infty^3}{(q;q)_\infty^4(q^2;q^2)_\infty^4}\left(X(q)^{-2} - 2qX(q)\right),$$

$$\sum_{n\geq 0} p^*(3n+1)q^n = \frac{(q^3;q^3)_\infty^3(q^6;q^6)_\infty^3}{(q;q)_\infty^4(q^2;q^2)_\infty^4}\left(X(q)^{-1} + 4qX(q)^2\right),$$

$$\text{and} \quad \sum_{n\geq 0} p^*(3n+2)q^n = 3\frac{(q^3;q^3)_\infty^3(q^6;q^6)_\infty^3}{(q;q)_\infty^4(q^2;q^2)_\infty^4},$$

where

$$X(q) = \left(\begin{matrix} q, q^5 \\ q^3, q^3 \end{matrix}; q^6\right)_\infty.$$

In this section, we will show that

$$\sum_{n\geq 0} p^*(3n)q^n = \frac{(q^3;q^3)_\infty^3}{(q;q)_\infty^4(q^2;q^2)_\infty^3}\, a(q^2) \tag{22.9.1}$$

$$\sum_{n\geq 0} p^*(3n+1)q^n = \frac{(q^6;q^6)_\infty^3}{(q;q)_\infty^3(q^2;q^2)_\infty^4}\, a(q) \tag{22.9.2}$$

$$\text{and} \quad \sum_{n\geq 0} p^*(3n+2)q^n = 3\frac{(q^3;q^3)_\infty^3(q^6;q^6)_\infty^3}{(q;q)_\infty^4(q^2;q^2)_\infty^4}. \tag{22.9.3}$$

We have, using (22.1.7), (22.1.18), (22.1.4), (22.1.5) and (22.1.7) again,

$$\begin{aligned}
\sum_{n\geq 0} p^*(n)q^n &= \frac{1}{(q;q)_\infty(q^2;q^2)_\infty} \\
&= \frac{c(q)c(q^2)}{9q(q^3;q^3)_\infty^3(q^6;q^6)_\infty^3} \\
&= \frac{a(q)a(q^2) - b(q)b(q^2)}{9q(q^3;q^3)_\infty^3(q^6;q^6;)_\infty^2} \\
&= \frac{\left(a(q^3)+2c(q^3)\right)\left(a(q^6)+2c(q^6)\right) - \left(a(q^3)-c(q^3)\right)\left(a(q^6)-c(q^6)\right)}{9q(q^3;q^3)_\infty^3(q^6;q^6)_\infty^3} \\
&= \frac{3a(q^6)c(q^3) + 3a(q^3)c(q^6) + 3c(q^3)c(q^6)}{9q(q^3;q^3)_\infty^3(q^6;q^6)_\infty^3}
\end{aligned}$$

$$= \frac{3\cdot 3q\dfrac{(q^9;q^9)^3_\infty}{(q^3;q^3)_\infty}a(q^6)+3\cdot 3q^2\dfrac{(q^{18};q^{18})^3_\infty}{(q^6;q^6)_\infty}a(q^3)+3\cdot 3q\dfrac{(q^9;q^9)^3_\infty}{(q^3;q^3)_\infty}\cdot 3q^2\dfrac{(q^{18};q^{18})^3_\infty}{(q^6;q^6)_\infty}}{9q(q^3;q^3)^3_\infty(q^6;q^6)^3_\infty}$$

$$= \frac{(q^9;q^9)^3_\infty}{(q^3;q^3)^4_\infty(q^6;q^6)^3_\infty}a(q^6)+q\frac{(q^{18};q^{18})^3_\infty}{(q^3;q^3)^3_\infty(q^6;q^6)^4_\infty}a(q^3)+3q^2\frac{(q^9;q^9)^3_\infty(q^{18};q^{18})^3_\infty}{(q^3;q^3)^4_\infty(q^6;q^6)^4_\infty}, \tag{22.9.4}$$

and the results follow.

22.10 Two Identities

In this section, we make use of the identities involving $a(q)$, $b(q)$ and $c(q)$ to give startlingly simple proofs of (22.1.22), given by Ramanujan [111], p. 241, Berndt [14], p. 346 and (22.1.23).

We have

$$\begin{aligned}
1+27q\left(\frac{(q^3;q^3)_\infty}{(q;q)_\infty}\right)^{12} &= 1+\left(\frac{c(q)}{b(q)}\right)^3 \\
&= \frac{b(q)^3+c(q)^3}{b(q)^3} \\
&= \frac{a(q)^3}{b(q)^3} \\
&= \left(\frac{a(q)}{b(q)}\right)^3 \\
&= \left(1+\frac{a(q)-b(q)}{b(q)}\right)^3 \\
&= \left(1+\frac{3c(q^3)}{b(q)}\right)^3 \\
&= \left(1+9q\left(\frac{(q^9;q^9)_\infty}{(q;q)_\infty}\right)^3\right)^3,
\end{aligned}$$

which is (22.1.22), and

$$\begin{aligned}
1+9q\left(\frac{(q^3;q^3)_\infty(q^6;q^6)_\infty}{(q;q)_\infty(q^2;q^2)_\infty}\right)^4 &= 1+\frac{c(q)c(q^2)}{b(q)b(q^2)} \\
&= \frac{b(q)b(q^2)+c(q)c(q^2)}{b(q)b(q^2)}
\end{aligned}$$

$$
\begin{aligned}
&= \frac{a(q)a(q^2)}{b(q)b(q^2)} \\
&= \left(1+\frac{a(q)-b(q)}{b(q)}\right)\left(1+\frac{a(q^2)-b(q^2)}{b(q^2)}\right) \\
&= \left(1+\frac{3c(q^3)}{b(q)}\right)\left(1+\frac{3c(q^6)}{b(q^2)}\right) \\
&= \left(1+9q\left(\frac{(q^9;q^9)_\infty}{(q;q)_\infty}\right)^3\right)\left(1+9q^2\left(\frac{(q^{18};q^{18})_\infty}{(q^2;q^2)_\infty}\right)^3\right),
\end{aligned}
$$

which is (22.1.23).

22.11 Some More Interesting Results

We saw that (22.8.6)

$$a(q) = \frac{3}{2}\frac{\phi(-q^3)^3}{\phi(-q)} - \frac{1}{2}\frac{\phi(-q)^3}{\phi(-q^3)}. \tag{22.11.1}$$

If we recall (21.1.1),

$$a(q) = 1+6\sum_{n\geq 0}\left(\frac{q^{3n+1}}{1-q^{3n+1}} - \frac{q^{3n+2}}{1-q^{3n+2}}\right) \tag{22.11.2}$$

and (10.8.1),

$$\frac{\phi(-q)^3}{\phi(-q^3)} = 1-6\sum_{n\geq 0}\left(\frac{q^{6n+1}}{1-q^{6n+1}} - \frac{q^{6n+5}}{1-q^{6n+5}}\right) + 18\sum_{n\geq 0}\left(\frac{q^{6n+2}}{1-q^{6n+2}} - \frac{q^{6n+4}}{1-q^{6n+4}}\right). \tag{22.11.3}$$

we can deduce (I leave this as an exercise) the beautiful identity

$$\frac{\phi(-q^3)^3}{\phi(-q)} = 1+2\sum_{n\geq 0}\left(\frac{q^{6n+1}}{1-q^{6n+1}} + \frac{q^{6n+2}}{1-q^{6n+2}} - \frac{q^{6n+4}}{1-q^{6n+4}} - \frac{q^{6n+5}}{1-q^{6n+5}}\right). \tag{22.11.4}$$

This is a companion to (10.8.1) and (10.8.2). When we get to Chapter 28, we will be able to prove the further companion,

$$q\frac{\psi(q^3)^3}{\psi(q)} = \sum_{n\geq 0}\left(\frac{q^{6n+1}}{1-q^{6n+1}} - 2\frac{q^{6n+2}}{1-q^{6n+2}} + 2\frac{q^{6n+4}}{1-q^{6n+4}} - \frac{q^{6n+5}}{1-q^{6n+5}}\right). \tag{22.11.5}$$

Note that (10.8.2), (22.11.5) and (21.1.1) yield the neat identity

$$a(q) = \frac{\psi(q)^3}{\psi(q^3)} + 3q\frac{\psi(q^3)^3}{\psi(q)}. \tag{22.11.6}$$

Exercise: Deduce from (10.9.3), (10.9.4), (22.11.1) and (22.11.6) that

$$\frac{\phi(-q)^3}{\phi(-q^3)} = 2a(q^2) - a(q), \tag{22.11.7}$$

$$\frac{\phi(-q^3)^3}{\phi(-q)} = \frac{1}{3}\left(a(q) + 2a(q^2)\right), \tag{22.11.8}$$

$$\frac{\psi(q)^3}{\psi(q^3)} = \frac{1}{2}\left(a(q) + a(q^2)\right) \tag{22.11.9}$$

and

$$q\frac{\psi(q^3)^3}{\psi(q)} = \frac{1}{6}\left(a(q) - a(q^2)\right). \tag{22.11.10}$$

Endnotes.

§22.1 As stated in the introduction, the definitions of $a(q)$, $b(q)$ and $c(q)$ are due to Borwein, Borwein and Garvan [21]. The important results, (22.1.4)–(22.1.7), (22.1.17) and (22.1.18), together with the applications to (22.1.22) and (22.1.23), are all due to them. The proofs presented here are mine, mostly unpublished.

Most of the results can be generalised. For these generalisations, once again due to Borwein, Borwein and Garvan, and for proofs, see Hirschhorn, Garvan and Borwein [92].

§22.1 I have recently found the following polynomial version of (22.1.6).

$$\sum_{\substack{r+s+t=0\\ r,s,t\geq -n}} \frac{\omega^{r-s} q^{(r^2+s^2+t^2)/2} (q;q)_{3n}}{(q;q)_{n+r}(q;q)_{n+s}(q;q)_{n+t}} = \frac{(q;q)_{3n}}{(q^3;q^3)_n}. \tag{22.11.11}$$

References

14. B.C. Berndt, *Ramanujan's Notebooks, Part III* (Springer, Berlin, 1991)
21. J.M. Borwein, P.B. Borwein, F. Garvan, Some cubic modular identities of Ramanujan. Trans. Am. Math. Soc. **343**, 35–47 (1994)
46. N.J. Fine, *Basic Hypergeometric Series and Applications*, vol. 27, Mathematical Surveys and Monographs (AMS, Providence, 1988)

92. M.D. Hirschhorn, F. Garvan, J.M. Borwein, Cubic analogues of the Jacobian theta function $\theta(z,q)$. Can. J. Math. **45**, 673–694 (1993)
111. S. Ramanujan, Notebooks, II, Tata Institute of Fundamental Research, Mumbai (2012)

Chapter 23
Some Classical Results on Representations

23.1 Introduction

In Chapter 2, we showed (2.1.1) that the number of representations of $n \geq 1$ as a sum of two squares is given by

$$r\{\Box + \Box\}(n) = 4\,(d_{1,4}(n) - d_{3,4}(n)) \tag{23.1.1}$$

where $d_{r,m}(n)$ denotes the number of divisors d of n with $d \equiv r \pmod{m}$.

In this chapter we will prove three further representation theorems. We will show that the number of representations of $n \geq 0$ as a sum of two triangular numbers is given by

$$r\{\triangle + \triangle\}(n) = d_{1,4}(4n+1) - d_{3,4}(4n+1), \tag{23.1.2}$$

as well as

$$r\{\Box + 3\Box\}(n) = 2\,(d_{1,3}(n) - d_{2,3}(n)) + 4\,(d_{4,12}(n) - d_{8,12}(n)) \tag{23.1.3}$$

for $n \geq 1$,
and

$$r\{\triangle + 3\triangle\}(n) = d_{1,3}(2n+1) - d_{2,3}(2n+1) \tag{23.1.4}$$

for $n \geq 0$.

The first result and, I presume, the second, are due to Jacobi (1829) [98]. The third result is explicit in L. Lorenz (1871) [104] and the third and fourth results are implicit in Ramanujan [111], p. 229.

M.D. Hirschhorn, *The Power of q*, Developments in Mathematics 49,
DOI 10.1007/978-3-319-57762-3_23

23.2 Proof of the Two Triangles Result (23.1.2)

We have (2.2.7)

$$\phi(q)^2 = 1 + 4\sum_{k\geq 0}\left(\frac{q^{4k+1}}{1-q^{4k+1}} - \frac{q^{4k+3}}{1-q^{4k+3}}\right). \tag{23.2.1}$$

Also, by (1.10.1),

$$\phi(q)^2 = \phi(q^2)^2 + 4q\psi(q^4)^2. \tag{23.2.2}$$

Thus

$$\phi(q^2)^2 + 4q\psi(q^4)^2 = 1 + 4\sum_{k\geq 0}\left(\frac{q^{4k+1}}{1-q^{4k+1}} - \frac{q^{4k+3}}{1-q^{4k+3}}\right). \tag{23.2.3}$$

If we extract the odd part of both sides of (23.2.3), then divide by 4, we obtain

$$\begin{aligned} q\psi(q^4)^2 &= \sum_{k\geq 0}\left(\frac{q^{4k+1}}{1-q^{8k+2}} - \frac{q^{4k+3}}{1-q^{8k+6}}\right) \\ &= \sum_{k\geq 0}\left(\frac{q^{4k+1}}{1-q^{16k+4}} + \frac{q^{12k+3}}{1-q^{16k+4}} - \frac{q^{4k+3}}{1-q^{16k+12}} - \frac{q^{12k+9}}{1-q^{16k+12}}\right) \\ &= \sum_{k,l\geq 0}\left(q^{16kl+4k+4l+1} + q^{16kl+12k+4l+3} - q^{16kl+4k+12l+3} - q^{16kl+12k+12l+9}\right) \\ &= \sum_{k,l\geq 0}\left(q^{(4k+1)(4l+1)} - q^{(4k+3)(4l+3)}\right). \end{aligned} \tag{23.2.4}$$

The coefficient of q^{4n+1} on the left side of (23.2.4) is $r\{\triangle+\triangle\}(n)$, while the coefficient of q^{4n+1} on the right side of (23.2.4) is $d_{1,4}(4n+1) - d_{3,4}(4n+1)$. This proves (23.1.2).

23.3 Square+3×Square

As in Chapter 21, let

$$a(q) = \sum_{m,n=-\infty}^{\infty} q^{m^2+mn+n^2}. \tag{23.3.1}$$

We start by showing that

$$a(q) + 2a(q^4) = 3\phi(q)\phi(q^3). \tag{23.3.2}$$

We have

$$a(q) = \sum_{\substack{m \text{ odd} \\ n \text{ even}}} q^{m^2+mn+n^2} + \sum_{\substack{m \text{ even} \\ n \text{ odd}}} q^{m^2+mn+n^2} + \sum_{\substack{m \text{ odd} \\ n \text{ odd}}} q^{m^2+mn+n^2} + \sum_{\substack{m \text{ even} \\ n \text{ even}}} q^{m^2+mn+n^2}. \tag{23.3.3}$$

In the first sum, let $m = k - l$, $n = 2l$ (conversely, $k = m + \frac{n}{2}$, $l = \frac{n}{2}$), in the second let $m = 2l$, $n = k - l$ (conversely, $k = \frac{m}{2} + n$, $l = \frac{m}{2}$), in the third and fourth let $m = k + l$, $n = l - k$ (conversely $k = \frac{m-n}{2}$, $l = \frac{m+n}{2}$), and we find

$$a(q) = 3 \sum_{k \not\equiv l \pmod 2} q^{k^2+3l^2} + \sum_{k \equiv l \pmod 2} q^{k^2+3l^2}. \tag{23.3.4}$$

Also,

$$a(q^4) = \sum_{m,n=-\infty}^{\infty} q^{4m^2+4mn+4n^2} = \sum_{\substack{m \text{ even} \\ n \text{ even}}} q^{m^2+mn+n^2} = \sum_{k \equiv l \pmod 2} q^{k^2+3l^2}, \tag{23.3.5}$$

as in the fourth sum above.

From (23.3.4) and (23.3.5) it follows that

$$\begin{aligned} a(q) + 2a(q^4) &= 3 \sum_{k \not\equiv l \pmod 2} q^{k^2+3l^2} + 3 \sum_{k \equiv l \pmod 2} q^{k^2+3l^2} \\ &= 3 \sum_{k,l=-\infty}^{\infty} q^{k^2+3l^2} \\ &= 3\phi(q)\phi(q^3), \end{aligned}$$

which is (23.3.2).

We have (23.3.2)

$$3\phi(q)\phi(q^3) = a(q) + 2a(q^4) \tag{23.3.6}$$

and (21.1.1)

$$a(q) = 1 + 6 \sum_{n \geq 0} \left(\frac{q^{3n+1}}{1 - q^{3n+1}} - \frac{q^{3n+2}}{1 - q^{3n+2}} \right). \tag{23.3.7}$$

It follows from (23.3.6) and (23.3.7) that

$$\begin{aligned}
&\phi(q)\phi(q^3)\\
&=1+2\sum_{n\geq 0}\left(\frac{q^{3n+1}}{1-q^{3n+1}}-\frac{q^{3n+2}}{1-q^{3n+2}}\right)+4\sum_{n\geq 0}\left(\frac{q^{12n+4}}{1-q^{12n+4}}-\frac{q^{12n+8}}{1-q^{12n+8}}\right)\\
&=1+2\sum_{k,l\geq 0}\left(q^{(3k+1)(l+1)}-q^{(3k+2)(l+1)}\right)+4\sum_{k,l\geq 0}\left(q^{(12k+4)(l+1)}-q^{(12k+8)(l+1)}\right).
\end{aligned}\tag{23.3.8}$$

The coefficient of q^n on the left side of (23.3.8) is $r\{\square+3\square\}(n)$, while the coefficient of q^n on the right side of (23.3.8) is

$$2\left(d_{1,3}(n)-d_{2,3}(n)\right)+4\left(d_{4,12}(n)-d_{8,12}(n)\right).$$

The result (23.1.3) follows.

23.4 Triangle+3×Triangle

We have already seen in §22.6 that the 2-dissection of $a(q)$ is

$$a(q)=a(q^4)+6q\psi(q^2)\psi(q^6). \tag{23.4.1}$$

An alternative proof goes as follows. If we subtract (23.3.5) from (23.3.4) we find that

$$a(q)-a(q^4)=3\sum_{k\not\equiv l \pmod 2} q^{k^2+3l^2}. \tag{23.4.2}$$

If we now use the fact (see (22.6.8)) that

$$\sum_{k\not\equiv l \pmod 2} q^{k^2+3l^2}=2q\psi(q^2)\psi(q^6) \tag{23.4.3}$$

we obtain (23.4.1).

We have

$$6q\psi(q^2)\psi(q^6)=a(q)-a(q^4) \tag{23.4.4}$$

and (21.1.1)

$$a(q)=1+6\sum_{n\geq 0}\left(\frac{q^{3n+1}}{1-q^{3n+1}}-\frac{q^{3n+2}}{1-q^{3n+2}}\right). \tag{23.4.5}$$

It follows from (23.4.4) and (23.4.5) that

$$
\begin{aligned}
q\psi(q^2)\psi(q^6) &= \sum_{n\geq 0}\left(\frac{q^{3n+1}}{1-q^{3n+1}} - \frac{q^{3n+2}}{1-q^{3n+2}}\right) - \sum_{n\geq 0}\left(\frac{q^{12n+4}}{1-q^{12n+4}} - \frac{q^{12n+8}}{1-q^{12n+8}}\right) \\
&= \sum_{n\geq 0}\left(\frac{q^{6n+1}}{1-q^{6n+1}} + \frac{q^{6n+4}}{1-q^{6n+4}} - \frac{q^{6n+2}}{1-q^{6n+2}} - \frac{q^{6n+5}}{1-q^{6n+5}}\right) \\
&\qquad - \sum_{n\geq 0}\left(\frac{q^{12n+4}}{1-q^{12n+4}} - \frac{q^{12n+8}}{1-q^{12n+8}}\right) \\
&= \sum_{n\geq 0}\left(\frac{q^{6n+1}}{1-q^{12n+2}} + \frac{q^{12n+2}}{1-q^{12n+2}} + \frac{q^{12n+4}}{1-q^{12n+4}} + \frac{q^{12n+10}}{1-q^{12n+10}}\right. \\
&\qquad \left. - \frac{q^{12n+2}}{1-q^{12n+2}} - \frac{q^{12n+8}}{1-q^{12n+8}} - \frac{q^{6n+5}}{1-q^{12n+10}} - \frac{q^{12n+10}}{1-q^{12n+10}}\right) \\
&\qquad - \sum_{n\geq 0}\left(\frac{q^{12n+4}}{1-q^{12n+4}} - \frac{q^{12n+8}}{1-q^{12n+8}}\right) \\
&= \sum_{n\geq 0}\left(\frac{q^{6n+1}}{1-q^{12n+2}} - \frac{q^{6n+5}}{1-q^{12n+10}}\right) \\
&= \sum_{k,l\geq 0}\left(q^{(6k+1)(2l+1)} - q^{(6l+5)(2l+1)}\right). \qquad (23.4.6)
\end{aligned}
$$

The coefficient of q^{2n+1} on the left side of (23.4.6) is $r\{\triangle + 3\triangle\}(n)$. The coefficient of q^{2n+1} on the right side of (23.4.6) is

$$d_{1,6}(2n+1) - d_{5,6}(2n+1) = d_{1,3}(2n+1) - d_{2,3}(2n+1). \qquad (23.4.7)$$

The result (23.1.4) follows.

23.5 Our Second Proof of the Modular Equation (20.1.1)

We have, by (23.3.2) and (23.4.1),

$$3\phi(q)\phi(q^3) = a(q) + 2a(q^4) \qquad (23.5.1)$$

and

$$6q\psi(q^2)\psi(q^6) = a(q) - a(q^4). \qquad (23.5.2)$$

It follows that

$$\phi(q)\phi(q^3) + 4q\psi(q^2)\psi(q^6) = a(q) \qquad (23.5.3)$$

and

$$\phi(q)\phi(q^3) - 2q\psi(q^2)\psi(q^6) = a(q^4). \tag{23.5.4}$$

Note that (23.5.4) gives the 2-dissection of $\phi(q)\phi(q^3)$,

$$\phi(q)\phi(q^3) = a(q^4) + 2q\psi(q^2)\psi(q^6). \tag{23.5.5}$$

It follows that

$$\mathbf{O}\left(\phi(q)\phi(q^3)\right) = 2q\psi(q^2)\psi(q^6),$$

which is (20.1.1).

References

98. C.G.J. Jacobi, Fundamenta nova theoriae functionum ellipticarum, Königsberg (1829)
104. L. Lorenz, Contributions à la theorie des nombres. Tidsskrift for Mathematik **3**, 97–114 (1871)
111. S. Ramanujan, Notebooks, II, Tata Institute of Fundamental Research, Mumbai (2012)
120. H. Valentier (ed.), Oeuvres Scientifique de L. Lorenz, II, Librarie Lehmann & Stage, Copenhagen (1904)

Chapter 24
Further Classical Results on Representations

24.1 Introduction

In this chapter, we will establish a partial fractions result due to L. Lorenz, and show that it yields a number of representation results, including one due to Dirichlet.

Lorenz's result, in his only paper on number theory (Lorenz (1871) [104]), is the following. If $|q| < 1$ and $|aq| < 1$ then

$$\frac{1}{2-x}\prod_{n\geq 1}\frac{1-aq^nx+a^2q^{2n}}{1-q^nx+q^{2n}} = \frac{(aq;q)_\infty^2}{(q;q)_\infty^2}\left(\frac{1}{2-x}+\sum_{n\geq 1}\frac{a^nq^n(a^{-1};q)_n(1+q^n)}{(aq;q)_n(1-q^nx+q^{2n})}\right) \tag{24.1.1}$$

provided no term in the denominator is zero.

We will show that (24.1.1) yields the following result due to Dirichlet (1840) [41].

For $n \geq 1$,

$$r\{\square+2\square\}(n) = 2\Big(d_{1,8}(n)+d_{3,8}(n)-d_{5,8}(n)-d_{7,8}(n)\Big). \tag{24.1.2}$$

Note that Dirichlet's result can be written as follows.

Suppose the prime factorisation of n is

$$n = 2^\alpha \prod_i p_i^{\alpha_i} \prod_j q_j^{\beta_j},$$

where the $p_i \equiv 1$ or $3 \pmod 8$ and $q_j \equiv 5$ or $7 \pmod 8$ are distinct primes.

Then

$$r\{\square+2\square\}(n) = 2\prod_i(\alpha_i+1)\prod_j\frac{1+(-1)^{\beta_j}}{2}. \tag{24.1.3}$$

M.D. Hirschhorn, *The Power of q*, Developments in Mathematics 49,
DOI 10.1007/978-3-319-57762-3_24

We will show that (24.1.1) also yields Jacobi's two-squares theorem (2.1.1), Jacobi's four-squares theorem (2.1.2) and the Lorenz–Ramanujan result for square+3×square (23.1.3).

24.2 Dirichlet's Result for Square+2×Square

If in (24.1.1) we replace q by q^2, set $x=0$ and $a=-q^{-1}$ and multiply by 2, we find

$$\frac{(q^4;q^4)_\infty^3}{(q^2;q^2)_\infty(q^8;q^8)_\infty^3} = \frac{(q^2;q^2)_\infty^2}{(q;q)_\infty^2(q^4;q^4)_\infty^2}\left(1+2\sum_{n\geq 1}(-1)^n\frac{q^n(1+q^{2n})}{1+q^{4n}}\right). \tag{24.2.1}$$

If we multiply (24.2.1) by $\dfrac{(q;q)_\infty^2(q^4;q^4)_\infty^2}{(q^2;q^2)_\infty^2}$ and make use of (1.5.8) and (1.5.7), we obtain

$$\phi(-q)\phi(q^2) = 1+2\sum_{n\geq 1}(-1)^n\frac{q^n(1+q^{2n})}{1+q^{4n}}. \tag{24.2.2}$$

If in (24.2.2) we set $-q$ for q, we find

$$\begin{aligned}\phi(q)\phi(q^2) &= 1+2\sum_{n\geq 1}\frac{q^n(1+q^{2n})}{1+q^{4n}}\\ &= 1+2\sum_{n\geq 1}\left(\frac{q^n}{1-q^{8n}}+\frac{q^{3n}}{1-q^{8n}}-\frac{q^{5n}}{1-q^{8n}}-\frac{q^{7n}}{1-q^{8n}}\right)\\ &= 1+2\sum_{k,l\geq 0}\left(q^{(8k+1)(l+1)}+q^{(8k+3)(l+1)}-q^{(8k+5)(l+1)}-q^{(8k+7)(l+1)}\right)\end{aligned} \tag{24.2.3}$$

from which Dirichlet's result (24.1.2) follows.

24.3 Our Third Proof of Jacobi's Two-Squares Theorem

If in (24.1.1) we set $a=-1$, $x=0$ and multiply by 2, we find

$$1 = \frac{(-q;q)_\infty^2}{(q;q)_\infty^2}\left(1+4\sum_{n\geq 1}(-1)^n\frac{q^n}{1+q^{2n}}\right), \tag{24.3.1}$$

or,

$$\phi(-q)^2 = 1 + 4\sum_{n\geq 1}(-1)^n \frac{q^n}{1+q^{2n}}. \tag{24.3.2}$$

If in (24.3.2) we set $-q$ for q, we find

$$\begin{aligned}
\phi(q)^2 &= 1 + 4\sum_{n\geq 1}\frac{q^n}{1+q^{2n}} \\
&= 1 + 4\sum_{n\geq 1}\left(\frac{q^n}{1-q^{4n}} - \frac{q^{3n}}{1-q^{4n}}\right) \\
&= 1 + 4\sum_{k,l\geq 0}\left(q^{(4k+1)(l+1)} - q^{(4k+3)(l+1)}\right)
\end{aligned} \tag{24.3.3}$$

which is (2.2.8), and yields Jacobi's two-squares theorem (2.1.1).

24.4 Our Second Proof of Jacobi's Four-Squares Theorem

If in (24.1.1) we set $a = -1$, $x = -2$ and multiply by $4\frac{(q;q)_\infty^2}{(-q;q)_\infty^2}$, we find

$$\phi(-q)^4 = 1 + 8\sum_{n\geq 1}(-1)^n \frac{q^n}{(1+q^n)^2}. \tag{24.4.1}$$

If in (24.4.1) we set $-q$ for q, we find

$$\begin{aligned}
\phi(q)^4 &= 1 + 8\sum_{n \text{ even}}\frac{q^n}{(1+q^n)^2} - 8\sum_{n \text{ odd}}\frac{q^n}{(1-q^n)^2} \\
&= 1 + 8\sum_{n\geq 1}\frac{q^n}{(1-q^n)^2} - 8\sum_{n \text{ even}}\left(\frac{q^n}{(1-q^n)^2} - \frac{q^n}{(1+q^n)^2}\right) \\
&= 1 + 8\sum_{n\geq 1}\frac{q^n}{(1-q^n)^2} - 32\sum_{n\geq 1}\frac{q^{4n}}{(1-q^{4n})^2} \\
&= 1 + 8\sum_{n,k\geq 1}kq^{nk} - 32\sum_{n,k\geq 1}kq^{4nk} \\
&= 1 + 8\sum_{k\geq 1}\frac{kq^k}{1-q^k} - 8\sum_{k\geq 1}\frac{4kq^{4k}}{1-q^{4k}}
\end{aligned} \tag{24.4.2}$$

which is (2.4.6), and yields Jacobi's four-squares theorem (2.1.2).

24.5 Our Second Proof of the Square+3×Square Result

If in (24.1.1) we set $a=-1$, $x=1$ and multiply by $\dfrac{(q;q)_\infty^2}{(-q;q)_\infty^2}$, we obtain

$$\phi(-q)\phi(-q^3) = 1+2\sum_{n\geq 1}(-1)^n\frac{q^n}{1-q^n+q^{2n}}. \tag{24.5.1}$$

If in (24.5.1) we set $-q$ for q, we find

$$\begin{aligned}
\phi(q)\phi(q^3) &= 1+2\sum_{n \text{ even}, \geq 2}\frac{q^n(1+q^n)}{1+q^{3n}}+2\sum_{n \text{ odd}, \geq 1}\frac{q^n(1-q^n)}{1-q^{3n}}\\
&= 1+2\sum_{n\geq 1}\frac{q^n(1-q^n)}{1-q^{3n}}+2\sum_{n \text{ even}, \geq 2}\left(\frac{q^n(1+q^n)}{1+q^{3n}}-\frac{q^n(1-q^n)}{1-q^{3n}}\right)\\
&= 1+2\sum_{n\geq 1}\frac{q^n(1-q^n)}{1-q^{3n}}+4\sum_{n \text{ even}, \geq 2}\frac{q^{2n}(1-q^{2n})}{1-q^{6n}}\\
&= 1+2\sum_{n\geq 1}\frac{q^n(1-q^n)}{1-q^{3n}}+4\sum_{n\geq 1}\frac{q^{4n}(1-q^{4n})}{1-q^{12n}}\\
&= 1+2\sum_{k,l\geq 0}\left(q^{(3k+1)(l+1)}-q^{(3k+2)(l+1)}\right)+4\sum_{k,l\geq 0}\left(q^{(12k+4)(l+1)}-q^{(12k+8)(l+1)}\right)
\end{aligned} \tag{24.5.2}$$

which is (23.3.8), and yields the Lorenz–Ramanujan result (23.1.3).

24.6 Proof of Lorenz's Partial Fractions Result

We start by supposing that

$$\frac{1}{2-x}\prod_{n=1}^{N}\frac{1-aq^nx+a^2q^{2n}}{1-q^nx+q^{2n}} = \frac{A_0}{2-x}+\sum_{n=1}^{N}\frac{A_n}{1-q^nx+q^{2n}}. \tag{24.6.1}$$

It is then an easy exercise to show that

$$A_0 = \frac{(aq;q)_N^2}{(q;q)_N^2} \tag{24.6.2}$$

and that for $1 \leq n \leq N$,

$$A_n = \frac{a^n q^n (a^{-1};q)_n (1+q^n)}{(aq;q)_n} \frac{(aq;q)_{N+n}(aq;q)_{N-n}}{(q;q)_{N+n}(q;q)_{N-n}}. \tag{24.6.3}$$

If we suppose that $|q| < 1$, $|aq| < 1$, and we let $N \to \infty$, we obtain (24.1.1).

References

41. P.G.L. Dirichlet, J. Math. **21** (1840) 3–6, Werke, pp. 463, 466
74. M.D. Hirschhorn, Partial fractions and four classical theorems of number theory. Am. Math. Mon. **107**, 260–264 (2000)
77. M.D. Hirschhorn, Three classical results on representations of a number, in *The Andrews Festschrift, Seventeen Papers on Classical Number Theory and Combinatorics*, ed. by D. Foata, G.-N. Han (Springer, Berlin, 2001), pp. 159–165
98. C.G.J. Jacobi, Fundamenta nova theoriae functionum ellipticarum, Königsberg (1829)
104. L. Lorenz, Contributions à la theorie des nombres. Tidsskrift for Mathematik **3**, 97–114 (1871)
111. S. Ramanujan, Notebooks, II, Tata Institute of Fundamental Research, Mumbai (2012)
120. H. Valentier (ed.), Oeuvres Scientifique de L. Lorenz, II, Librarie Lehmann & Stage, Copenhagen (1904)

Chapter 25
Further Results on Representations

25.1 Introduction

In earlier chapters we have obtained the following representation theorems, among others.

For $n \geq 1$,

$$r\{\Box+\Box\}(n) = 4(d_{1,4}(n) - d_{3,4}(n)) \qquad (25.1.1)$$

(see (2.1.1), §2.2, §2.3, §24.3),

$$r\{\Box+2\Box\}(n) = 2(d_{1,8}(n) + d_{3,8}(n) - d_{5,8}(n) - d_{7,8}(n)) \qquad (25.1.2)$$

(see (24.1.2), §24.2),

$$r\{\Box+3\Box\}(n) = 2(d_{1,3}(n) - d_{2,3}(n)) + 4(d_{4,12}(n) - d_{8,12}(n)) \qquad (25.1.3)$$

(see (23.1.3), §23.3, §24.5) and

$$r\{\Box+\Box+\Box+\Box\}(n) = 8 \sum_{\substack{d|n \\ d\equiv 1,\ 2 \text{ or } 3 \pmod 4}} d \qquad (25.1.4)$$

(see (2.1.2), §23.3, §24.5).

In this chapter, we will make use of three simple identities to deduce from (25.1.1)–(25.1.4) a number of other representation theorems. The tools we require are

$$\phi(q)\psi(q^2) = \psi(q)^2 \qquad (25.1.5)$$

and the 2-dissections,

$$\phi(q) = \phi(q^4) + 2q\psi(q^8) \qquad (25.1.6)$$

M.D. Hirschhorn, *The Power of q*, Developments in Mathematics 49,
DOI 10.1007/978-3-319-57762-3_25

and

$$\psi(q)\psi(q^3) = \phi(q^6)\psi(q^4) + q\phi(q^2)\psi(q^{12}). \tag{25.1.7}$$

The first of these is (1.5.13), the second is (1.9.4), and the third will be proved in §25.2.

Before stating our results, we make the following definition.

For $n \geq 1$, $m \geq 2$, $k \geq 1$, $0 < r_1 < r_2 < \cdots < r_k < m$, $R = \{r_1, \ldots, r_k\}$, let

$$e\{R,m\}(n) = \sum_{r\in R} d_{r,m}(n) - \sum_{r\in R} d_{m-r,m}(n) \tag{25.1.8}$$

be the excess of the number of divisors d of n that are congruent to r_1, $r_2 \ldots$ or r_k modulo m over the number of divisors d of n that are congruent to $-r_1$, $-r_2$, $\ldots$ or $-r_k$ modulo m.

Note that if n is not an integer, $e\{R,m\}(n) = 0$.

Then, for example, Jacobi's 2–squares theorem (25.1.1) can be written

$$r\{\square+\square\}(n) = 4e\{\{1\},4\}(n) \tag{25.1.9}$$

Dirichlet's result (24.1.2) can be written

$$r\{\square+2\square\}(n) = 2e\{\{1,3\},8\}(n) \tag{25.1.10}$$

and Lorenz's result (25.1.3) can be written

$$r\{\square+3\square\}(n) = 2e\{\{1\},3\}(n) + 4e\{\{4\},12\}(n). \tag{25.1.11}$$

The results we will prove are the following. For $n \geq 0$,

$$r\{\triangle+\triangle\}(n) = e\{\{1\},4\}(4n+1), \tag{25.1.12}$$
$$r\{\triangle+2\triangle\}(n) = e\{\{1\},8\}(8n+3), \tag{25.1.13}$$
$$r\{\triangle+3\triangle\}(n) = e\{\{1\},3\}(2n+1), \tag{25.1.14}$$
$$r\{\triangle+4\triangle\}(n) = e\{\{1\},8\}(8n+5), \tag{25.1.15}$$
$$r\{\triangle+12\triangle\}(n) = \frac{1}{2}e\{\{1\},3\}(8n+13), \tag{25.1.16}$$
$$r\{3\triangle+4\triangle\}(n) = \frac{1}{2}e\{\{1\},3\}(8n+7), \tag{25.1.17}$$
$$r\{\triangle+\square\}(n) = e\{\{1,3\},8\}(8n+1), \tag{25.1.18}$$
$$r\{\triangle+2\square\}(n) = e\{\{1\},4\}(8n+1), \tag{25.1.19}$$
$$r\{\triangle+6\square\}(n) = e\{\{1\},3\}(8n+1), \tag{25.1.20}$$
$$r\{2\triangle+\square\}(n) = e\{\{1\},4\}(4n+1), \tag{25.1.21}$$
$$r\{2\triangle+3\square\}(n) = e\{\{1\},3\}(4n+1), \tag{25.1.22}$$
$$r\{3\triangle+2\square\}(n) = e\{\{1\},3\}(8n+3), \tag{25.1.23}$$

$$r\{4\triangle+\square\}(n)=e\{\{1,3\},8\}(2n+1), \tag{25.1.24}$$
$$r\{6\triangle+\square\}(n)=e\{\{1\},3\}(4n+3), \tag{25.1.25}$$
$$r\{\triangle+\triangle+\triangle+\triangle\}(n)=\sigma(2n+1), \tag{25.1.26}$$
$$r\{\triangle+\triangle+2\triangle+2\triangle\}(n)=\frac{1}{4}\sigma(4n+3). \tag{25.1.27}$$

and

$$r\{\triangle+\triangle+\square+\square\}(n)=\sigma(4n+1), \tag{25.1.28}$$

Note that we have already had proofs of (25.1.12) (see (23.1.2), §23.2) and (25.1.14) (see (23.1.4), §23.4).

25.2 Preliminary Result

In this section we prove (25.1.7),

$$\psi(q)\psi(q^3)=\phi(q^6)\psi(q^4)+q\phi(q^2)\psi(q^{12}). \tag{25.2.1}$$

We start by noting that, by (25.1.6),

$$\begin{aligned}\phi(q)\phi(q^3)&=(\phi(q^4)+2q\psi(q^8))(\phi(q^{12})+2q^3\psi(q^{24}))\\&=(\phi(q^4)\phi(q^{12})+4q^4\psi(q^8)\psi(q^{24}))+2q(\phi(q^{12})\psi(q^8)+q^2\phi(q^4)\psi(q^{24})).\end{aligned} \tag{25.2.2}$$

It follows from (25.2.2) that

$$\mathbf{O}(\phi(q)\phi(q^3))=2q\left(\phi(q^{12})\psi(q^8)+q^2\phi(q^4)\psi(q^{24})\right). \tag{25.2.3}$$

But we have already had two proofs (see (20.1.1), §20.2 and §23.5) of the fact that

$$\mathbf{O}(\phi(q)\phi(q^3))=2q\psi(q^2)\psi(q^6). \tag{25.2.4}$$

So (25.2.1) is proved.

25.3 Our Third Proof of the Modular Equation (20.1.1)

Not only does (20.1.1) imply (25.2.1), the converse is, by virtue of (25.2.3), also true. In this section we give a direct proof of (25.2.1), thus yielding a third proof of (20.1.1).

We have

$$\psi(q)\psi(q^3) = \sum_{k,l=-\infty}^{\infty} q^{2k^2+k+6l^2+3l}. \tag{25.3.1}$$

It follows that

$$\begin{aligned} q^4\psi(q^8)\psi(q^{24}) &= \sum_{k,l=-\infty}^{\infty} q^{(4k+1)^2+3(4l+1)^2} \\ &= \sum_{k,l=-\infty}^{\infty} q^{4(k+3l+1)^2+12(k-l)^2} \\ &= \sum_{u-v\equiv 1 \pmod 4} q^{4u^2+12v^2} \end{aligned} \tag{25.3.2}$$

Now consider the two cases, v even, u even. If v is even, $v=2k$, $u=4l+1$ or $-4l-1$, according as k is even or odd, while if u is even, $u=2k$, $v=4l+1$ or $-4l-1$, according as k is odd or even. Thus

$$\begin{aligned} \sum_{u-v\equiv 1 \pmod 4} q^{4u^2+12v^2} &= \sum_{k,l=-\infty}^{\infty} q^{4(4l+1)^2+12(2k)^2} + \sum_{k,l=-\infty}^{\infty} q^{4(2k)^2+12(4l+1)^2} \\ &= q^4\psi(q^{32})\phi(q^{48}) + q^{12}\phi(q^{16})\psi(q^{96}). \end{aligned} \tag{25.3.3}$$

If we substitute (25.3.3) into (25.3.2), we find

$$q^4\psi(q^8)\psi(q^{24}) = q^4\psi(q^{32})\phi(q^{48}) + q^{12}\phi(q^{16})\psi(q^{96}), \tag{25.3.4}$$

which yields (25.2.1), and so (20.1.1).

25.4 Proofs of the Representation Results (25.1.12)–(25.1.28)

We can write (25.1.1)

$$\phi(q)^2 = 1 + 4\sum_{n\geq 1} e\{\{1\},4\}(n)q^n. \tag{25.4.1}$$

We have, by (1.10.1),

$$\phi(q^2)^2 + 4q\psi(q^4)^2 = 1 + 4\sum_{n\geq 1} e\{\{1\},4\}(n)q^n. \tag{25.4.2}$$

If from (25.4.2) we extract those terms in which the power of q is 1 modulo 4, divide by $4q$ and replace q^4 by q, we find

$$\psi(q)^2 = \sum_{n\geq 0} e\{\{1\},4\}(4n+1)q^n, \tag{25.4.3}$$

from which we obtain (25.1.12).

By (25.1.5), (25.4.3) can be written

$$\phi(q)\psi(q^2) = \sum_{n\geq 0} e\{\{1\},4\}(4n+1)q^n. \tag{25.4.4}$$

This yields (25.1.21).

If we substitute (25.1.6) in (25.4.4), it becomes

$$\psi(q^2)(\phi(q^4) + 2q\psi(q^8)) = \sum_{n\geq 0} e\{\{1\},4\}(4n+1)q^n. \tag{25.4.5}$$

If from (25.4.5) we extract those terms in which the power of q is even and replace q^2 by q, we obtain

$$\psi(q)\phi(q^2) = \sum_{n\geq 0} e\{\{1\},4\}(8n+1)q^n \tag{25.4.6}$$

which yields (25.1.19), while if from (25.4.5) we extract those terms in which the power of q is odd, divide by q then replace q^2 by q, we obtain

$$2\psi(q)\psi(q^4) = \sum_{n\geq 0} e\{\{1\},4\}(8n+5)q^n \tag{25.4.7}$$

which yields (25.1.15).

Next, (25.1.2) is equivalent to

$$\phi(q)\phi(q^2) = 1 + 2\sum_{n\geq 1} e\{\{1,3\},8\}(n)q^n. \tag{25.4.8}$$

If we substitute (25.1.6) into (25.4.8), we obtain

$$(\phi(q^4) + 2q\psi(q^8))(\phi(q^8) + 2q^2\psi(q^{16})) = 1 + 2\sum_{n\geq 1} e\{\{1,3\},8\}(n)q^n. \tag{25.4.9}$$

From (25.4.8) we can extract (25.1.18), (25.1.24) and (25.1.13) if we make use of the facts that

$$d_{3,8}(8n+3) = d_{1,8}(8n+3), \;\; d_{5,8}(8n+3) = d_{7,8}(8n+3). \tag{25.4.10}$$

The result (25.1.3) is equivalent to

$$\phi(q)\phi(q^3) = 1 + 2\sum_{n\geq 1} e\{\{1\},3\}(n)q^n + 4\sum_{n\geq 1} e\{\{4\},12\}(n)q^n. \tag{25.4.11}$$

If we substitute (25.1.6) into (25.4.11), we obtain

$$\begin{aligned}(\phi(q^4)+2q\psi(q^8))(\phi(q^{12})+2q^3\psi(q^{24})) =& 1+2\sum_{n\geq 1} e\{\{1\},3\}(n)q^n \\ &+4\sum_{n\geq 1} e\{\{4\},12\}(n)q^n.\end{aligned} \tag{25.4.12}$$

If from (25.4.12) we extract those terms in which the power of q is 0 modulo 4 and replace q^4 by q, we obtain

$$\begin{aligned}\phi(q)\phi(q^3)+4q\psi(q^2)\psi(q^6) &= 1+2\sum_{n\geq 1} e\{\{1\},3\}(4n)q^n + 4\sum_{n\geq 1} e\{\{1\},3\}(n)q^n \\ &= 1+6\sum_{n\geq 1} e\{\{1\},3\}(n)q^n,\end{aligned} \tag{25.4.13}$$

where we have made use of the fact that

$$d_{1,3}(4n) - d_{2,3}(4n) = d_{1,3}(n) - d_{2,3}(n). \tag{25.4.14}$$

If we subtract (25.4.11) from (25.4.13), we obtain

$$\begin{aligned}4q\psi(q^2)\psi(q^6) &= 4\sum_{n\geq 1} e\{\{1\},3\}(n)q^n - 4\sum_{n\geq 1} e\{\{1\},3\}(n)q^{4n}. \\ &= 4\sum_{n\geq 0} e\{\{1\},3\}(2n+1)q^{2n+1}\end{aligned} \tag{25.4.15}$$

where we have made use of (25.4.14) and the fact that

$$d_{1,3}(4n+2) - d_{2,3}(4n+2) = 0. \tag{25.4.16}$$

If we divide (25.4.15) by $4q$, then replace q^2 by q, we obtain

$$\psi(q)\psi(q^3) = \sum_{n\geq 0} e\{\{1\},3\}(2n+1)q^n, \tag{25.4.17}$$

which yields (25.1.14).

If we substitute (25.1.7) into (25.4.17), we obtain

$$\phi(q^6)\psi(q^4)+q\phi(q^2)\psi(q^{12})=\sum_{n\geq 0} e\{\{1\},3\}(2n+1)q^n. \tag{25.4.18}$$

It follows that

$$\phi(q^3)\psi(q^2)=\sum_{n\geq 0} e\{\{1\},3\}(4n+1)q^n \tag{25.4.19}$$

and

$$\phi(q)\psi(q^6)=\sum_{n\geq 0} e\{\{1\},3\}(4n+3)q^n, \tag{25.4.20}$$

which yield (25.1.22) and (25.1.25).

If we substitute (25.1.6) into (25.4.19) and (25.4.20), we obtain

$$\psi(q^2)(\phi(q^{12})+2q^3\psi(q^{24}))=\sum_{n\geq 0} e\{\{1\},3\}(4n+1)q^n \tag{25.4.21}$$

and

$$\psi(q^6)(\phi(q^4)+2q\psi(q^8))=\sum_{n\geq 0} e\{\{1\},3\}(4n+3)q^n. \tag{25.4.22}$$

From (25.4.21) and (25.4.22), we deduce (25.1.16), (25.1.17), (25.1.20) and (25.1.23).

The result (25.1.4) is equivalent to

$$\phi(q)^4=1+8\sum_{n\geq 1}\left(\sum_{\substack{d\mid n\\ d\equiv 1,\ 2 \text{ or } 3 \pmod 4}} d\right)q^n. \tag{25.4.23}$$

If we make use of (25.1.6), this becomes

$$(\phi(q^4)+2q\psi(q^8))^4=1+8\sum_{n\geq 1}\left(\sum_{\substack{d\mid n\\ d\equiv 1,\ 2 \text{ or } 3 \pmod 4}} d\right)q^n. \tag{25.4.24}$$

From (25.4.24), we readily deduce

$$8\phi(q)^2\psi(q)^2=8\sum_{n\geq 0}\left(\sum_{d\mid 4n+1} d\right)q^n=8\sum_{n\geq 0}\sigma(4n+1)q^n, \tag{25.4.25}$$

$$24\psi(q)^4 = 8\sum_{n\geq 0}\left(\sum_{d|4n+2} d\right)q^n = 24\sum_{n\geq 0}\sigma(2n+1)q^n \tag{25.4.26}$$

and

$$32\psi(q)^2\psi(q^2)^2 = 8\sum_{n\geq 0}\left(\sum_{d|4n+3} d\right)q^n = 8\sum_{n\geq 0}\sigma(4n+3)q^n, \tag{25.4.27}$$

which yield (25.1.28), (25.1.26) and (25.1.27) respectively.

Reference

79. M.D. Hirschhorn, The number of representations of a number by various forms. Discret. Math. **298**, 205–211 (2005)

Chapter 26
Even More Representation Results

26.1 Introduction

In this chapter, we indicate how we can use two simple 3-dissection identities to deduce from the 21 representation results of Chapter 25 a further 29 representation theorems.

Let Π represent a pentagonal number (a number of the form $(3n^2+n)/2$), and Ω an octagonal number (a number of the form $3n^2+2n$). Let

$$\Pi(q)=\sum_{-\infty}^{\infty}q^{(3n^2+n)/2},\quad \Omega(q)=\sum_{-\infty}^{\infty}q^{3n^2+2n}.$$

Then it is easy to obtain the 3-dissections

$$\phi(q)=\phi(q^9)+2q\Omega(q^3), \tag{26.1.1}$$
$$\psi(q)=\Pi(q^3)+q\psi(q^9). \tag{26.1.2}$$

We can start with the results (25.1.1)–(25.1.4) and (25.1.12)–(25.1.28) and show that for $n\geq 0$,

$$r\{\triangle+\Pi\}(n)=e\{\{1\},6\}(6n+1), \tag{26.1.3}$$
$$r\{\triangle+4\Pi\}(n)=e\{\{1,19\},24\}(24n+7), \tag{26.1.4}$$
$$r\{3\triangle+\Pi\}(n)=e\{\{1\},12\}(12n+5), \tag{26.1.5}$$
$$r\{3\triangle+2\Pi\}(n)=e\{\{1\},8\}(24n+11), \tag{26.1.6}$$
$$r\{4\triangle+\Pi\}(n)=e\{\{1,19\},24\}(24n+13), \tag{26.1.7}$$
$$r\{6\triangle+\Pi\}(n)=e\{\{1\},8\}(24n+9), \tag{26.1.8}$$
$$r\{\triangle+2\Omega\}(n)=e\{\{1,7\},24\}(24n+19), \tag{26.1.9}$$
$$r\{2\triangle+\Omega\}(n)=e\{\{1\},12\}(12n+7), \tag{26.1.10}$$
$$r\{3\triangle+\Omega\}(n)=e\{\{1,11\},24\}(24n+17), \tag{26.1.11}$$

M.D. Hirschhorn, *The Power of q*, Developments in Mathematics 49,
DOI 10.1007/978-3-319-57762-3_26

$$r\{12\triangle+\Omega\}(n)=\frac{1}{2}e\{\{1,3\},8\}(6n+11) \tag{26.1.12}$$

$$r\{\square+2\Pi\}(n)=e\{\{1\},3\}(12n+1), \tag{26.1.13}$$

$$r\{2\square+\Pi\}(n)=e\{\{1\},3\}(24n+1), \tag{26.1.14}$$

$$r\{3\square+\Pi\}(n)=e\{\{1,3\},8\}(24n+1), \tag{26.1.15}$$

$$r\{3\square+4\Pi\}(n)=e\{\{1,3\},8\}(6n+1), \tag{26.1.16}$$

$$r\{\square+\Omega\}(n)=e\{\{1\},3\}(3n+1)+2e\{\{1\},3\}((3n+1)/4), \tag{26.1.17}$$

$$r\{3\square+\Omega\}(n)=e\{\{1\},4\}(3n+1), \tag{26.1.18}$$

$$r\{3\square+2\Omega\}(n)=e\{\{1,3\},8\}(3n+2), \tag{26.1.19}$$

$$r\{6\square+\Omega\}(n)=e\{\{1,3\},8\}(3n+1), \tag{26.1.20}$$

$$r\{\Pi+\Pi\}(n)=e\{\{1\},4\}(12n+1), \tag{26.1.21}$$

$$r\{\Pi+2\Pi\}(n)=e\{\{1\},8\}(24n+3)-e\{\{1\},8\}((8n+1)/3), \tag{26.1.22}$$

$$r\{\Pi+4\Pi\}(n)=e\{\{1\},8\}(24n+5), \tag{26.1.23}$$

$$r\{\Pi+\Omega\}(n)=\frac{1}{2}\left(e\{\{1,3\},8\}(24n+9)-e\{\{1,3\},8\}((8n+3)/3)\right), \tag{26.1.24}$$

$$r\{\Pi+2\Omega\}(n)=e\{\{1,5\},24\}(24n+17), \tag{26.1.25}$$

$$r\{2\Pi+\Omega\}(n)=e\{\{1\},12\}(12n+5), \tag{26.1.26}$$

$$r\{4\Pi+\Omega\}(n)=\frac{1}{2}\left(e\{\{1,3\},8\}(6n+3)-e\{\{1,3\},8\}((2n+1)/3)\right), \tag{26.1.27}$$

$$r\{\Omega+\Omega\}(n)=e\{\{1\},4\}(3n+2), \tag{26.1.28}$$

$$r\{\Omega+2\Omega\}(n)=\frac{1}{2}\left(e\{\{1,3\},8\}(3n+3)-e\{\{1,3\},8\}((n+1)/3)\right), \tag{26.1.29}$$

$$r\{3\triangle+3\triangle+\Pi+\Pi\}(n)=\frac{1}{6}\sigma(6n+5) \tag{26.1.30}$$

and

$$r\{3\square+3\square+\Omega+\Omega\}(n)=\frac{1}{3}(\sigma(3n+2)-4\sigma((3n+2)/4)). \tag{26.1.31}$$

26.2 Proofs

We will prove (26.1.3), and because proofs of (26.1.4)–(26.1.31) are fairly straightforward we leave them as exercises.

We can write (25.1.14)

$$\psi(q)\psi(q^3) = \sum_{n\geq 0} e\{\{1\},3\}(2n+1)q^n. \tag{26.2.1}$$

If we make use of (26.1.2), we find

$$\left(\Pi(q^3)+q\psi(q^9)\right)\left(\Pi(q^9)+q^3\psi(q^{27})\right) = \sum_{n\geq 0} e\{\{1\},3\}(2n+1)q^n. \tag{26.2.2}$$

If we extract those terms in which the power of q is 1 modulo 9, divide by q and replace q by $q^{\frac{1}{9}}$, we find

$$\begin{aligned}\psi(q)\Pi(q) &= \sum_{n\geq 0} e\{\{1\},3\}(18n+3)q^n \\ &= \sum_{n\geq 0} e\{\{1\},3\}(6n+1)q^n,\end{aligned} \tag{26.2.3}$$

as required.

Endnotes.

The results of this chapter appeared in Hirschhorn (2010) [83].

Reference

83. M.D. Hirschhorn, The number of representations of a number by various forms involving triangles, squares, pentagons and octagons, in *Ramanujan Rediscovered, Proceedings of a Conference on Elliptic Functions, Partitions, and q-series in Memory of K. Venkatachaliengar, Bangalore, 1–5 June, 2009*. RMS Lecture Notes Series, vol. 14 (2010), pp. 113–124

Chapter 27
Representation Results and Lambert Series

27.1 Introduction

Jacobi's two-squares theorem (2.1.1) is equivalent to the identity (2.2.7),

$$\phi(q)^2 = 1 + 4\sum_{n\geq 0}\left(\frac{q^{4n+1}}{1-q^{4n+1}} - \frac{q^{4n+3}}{1-q^{4n+3}}\right). \tag{27.1.1}$$

We have, by (1.9.4),

$$\begin{aligned}\phi(q^2)^2 + 4q\psi(q^4)^2 &= 1 + 4\sum_{n\geq 0}\left(\frac{q^{4n+1}}{1-q^{4n+1}} - \frac{q^{4n+3}}{1-q^{4n+3}}\right)\\ &= 1 + 4\sum_{n\geq 0}\left(\frac{q^{4n+1}(1+q^{4n+1}+q^{8n+2}+q^{12n+3})}{1-q^{16n+4}}\right.\\ &\qquad\left. - \frac{q^{4n+3}(1+q^{4n+3}+q^{8n+6}+q^{12n+9})}{1-q^{16n+12}}\right).\end{aligned} \tag{27.1.2}$$

If from (27.1.2) we extract those terms in which the power of q is 1 modulo 4, divide by $4q$ and replace q^4 by q, we obtain

$$\psi(q)^2 = \sum_{n\geq 0}\left(\frac{q^n}{1-q^{4n+1}} - \frac{q^{3n+2}}{1-q^{4n+3}}\right), \tag{27.1.3}$$

which is equivalent to the two-triangles result (23.1.2).

We shall describe the series appearing on the right of (27.1.1) and (27.1.3) as Lambert series, without giving a rigid definition of the term.

M.D. Hirschhorn, *The Power of q*, Developments in Mathematics 49,
DOI 10.1007/978-3-319-57762-3_27

It happens that we can go further with both (27.1.1) and (27.1.3), and write

$$\phi(q)^2 = 1 + 4\sum_{-\infty}^{\infty} \frac{q^{4n+1}}{1-q^{4n+1}} \tag{27.1.4}$$

and

$$\psi(q)^2 = \sum_{-\infty}^{\infty} \frac{q^n}{1-q^{4n+1}}, \tag{27.1.5}$$

since

$$-\frac{q^{4n+3}}{1-q^{4n+3}} = \frac{q^{4(-n-1)+1}}{1-q^{4(-n-1)+1}} \tag{27.1.6}$$

and

$$-\frac{q^{3n+2}}{1-q^{4n+3}} = \frac{q^{-n-1}}{1-q^{4(-n-1)+1}}. \tag{27.1.7}$$

We will describe the series on the right of (26.1.4) and (27.1.5) as bilateral Lambert series.

In the same way, we can recast (21.1.1), (22.1.8) and (22.1.9) respectively as

$$a(q) = 1 + 6\sum_{-\infty}^{\infty} \frac{q^{3n+1}}{1-q^{3n+1}}, \tag{27.1.8}$$

$$b(q) = 1 - 3\sum_{-\infty}^{\infty} \frac{q^{3n+1}(1-2q^{6n+2})}{1-q^{9n+3}}, \tag{27.1.9}$$

and

$$q^{-\frac{1}{3}}c(q) = 3\sum_{-\infty}^{\infty} \frac{q^n}{1-q^{3n+1}}. \tag{27.1.10}$$

The square+3×square result (23.3.8) becomes

$$\phi(q)\phi(q^3) = 1 + 2\sum_{-\infty}^{\infty} \frac{q^{3n+1}}{1-q^{3n+1}} + 4\sum_{-\infty}^{\infty} \frac{q^{12n+4}}{1-q^{12n+4}} \tag{27.1.11}$$

and the triangle+3×triangle result (23.4.6) becomes

$$\psi(q)\psi(q^3) = \sum_{-\infty}^{\infty} \frac{q^{3n}}{1-q^{6n+1}}. \tag{27.1.12}$$

The square+2×square result (24.2.2) can be written

$$\phi(q)\phi(q^2) = 1 + 2\sum_{-\infty}^{\infty}{}' \frac{q^n(1+q^{2n})}{1-q^{8n}}, \tag{27.1.13}$$

where the prime ($'$) indicates that the term corresponding to $n = 0$ should be omitted from the sum.

The representation results (25.1.9)–(25.1.25) and (26.1.3)–(26.1.29) can all be cast as identities involving Lambert series or bilateral Lambert series.

27.2 Statements of Lambert Series and Bilateral Lambert Series Results

In this section, I state Lambert series and bilateral Lambert series for the representation results of (25.1.12)–(25.1.25). They are as succinct as I can make them. Their derivations are left as exercises. Alternative, direct derivations for at least some of them will be given in chapter 28.

(25.1.12) is equivalent to

$$\psi(q)^2 = \sum_{n\geq 0}\left(\frac{q^n}{1-q^{4n+1}} - \frac{q^{3n+2}}{1-q^{4n+3}}\right) = \sum_{-\infty}^{\infty}\frac{q^n}{1-q^{4n+1}}, \tag{27.2.1}$$

(25.1.13) is equivalent to

$$\psi(q)\psi(q^2) = \sum_{n\geq 0}\left(\frac{q^{3n}}{1-q^{8n+1}} - \frac{q^{5n+4}}{1-q^{8n+7}}\right) = \sum_{-\infty}^{\infty}\frac{q^{3n}}{1-q^{8n+1}}, \tag{27.2.2}$$

(25.1.14) is equivalent to

$$\psi(q)\psi(q^3) = \sum_{n\geq 0}\left(\frac{q^{3n}}{1-q^{6n+1}} - \frac{q^{3n+2}}{1-q^{6n+5}}\right) = \sum_{-\infty}^{\infty}\frac{q^{3n}}{1-q^{6n+1}}, \tag{27.2.3}$$

(25.1.15) is equivalent to

$$\psi(q)\psi(q^4) = \sum_{n\geq 0}\left(\frac{q^{5n}}{1-q^{8n+1}} - \frac{q^{3n+2}}{1-q^{8n+7}}\right) = \sum_{-\infty}^{\infty}\frac{q^{5n}}{1-q^{8n+1}}, \tag{27.2.4}$$

(25.1.16) is equivalent to

$$
\begin{aligned}
2q\psi(q)\psi(q^{12}) &= \sum_{n\geq 0}\left(\frac{q^{15n}}{1-q^{24n+1}} - \frac{q^{3n}}{1-q^{24n+5}} + \frac{q^{9n+2}}{1-q^{24n+7}} - \frac{q^{21n+9}}{1-q^{24n+11}}\right. \\
&\quad \left. + \frac{q^{3n+1}}{1-q^{24n+13}} - \frac{q^{15n+10}}{1-q^{24n+17}} + \frac{q^{21n+16}}{1-q^{24n+19}} - \frac{q^{9n+8}}{1-q^{24n+23}}\right) \\
&= \sum_{-\infty}^{\infty}\left(\frac{q^{15n}}{1-q^{24n+1}} + \frac{q^{9n+2}}{1-q^{25n+7}} + \frac{q^{3n+1}}{1-q^{24n+13}} + \frac{q^{21n+16}}{1-q^{24n+19}}\right),
\end{aligned}
\tag{27.2.5}
$$

(25.1.17) is equivalent to

$$
\begin{aligned}
2\psi(q^3)\psi(q^4) &= \sum_{n\geq 0}\left(\frac{q^{21n}}{1-q^{24n+1}} - \frac{q^{9n+1}}{1-q^{24n+5}} + \frac{q^{3n}}{1-q^{24n+7}} - \frac{q^{15n+6}}{1-q^{24n+11}}\right. \\
&\quad \left. + \frac{q^{9n+4}}{1-q^{24n+13}} - \frac{q^{21n+14}}{1-q^{24n+17}} + \frac{q^{15n+11}}{1-q^{24n+19}} - \frac{q^{3n+2}}{1-q^{24n+23}}\right) \\
&= \sum_{-\infty}^{\infty}\left(\frac{q^{21n}}{1-q^{24n+1}} + \frac{q^{3n}}{1-q^{24n+7}} + \frac{q^{9n+4}}{1-q^{24n+13}} + \frac{q^{15n+11}}{1-q^{24n+19}}\right),
\end{aligned}
\tag{27.2.6}
$$

(25.1.18) is equivalent to

$$
\begin{aligned}
\psi(q)\phi(q) &= \sum_{n\geq 0}\left(\frac{q^{n}}{1-q^{8n+1}} + \frac{q^{3n+1}}{1-q^{8n+3}} - \frac{q^{5n+3}}{1-q^{8n+5}} - \frac{q^{7n+6}}{1-q^{8n+7}}\right) \\
&= \sum_{-\infty}^{\infty}\left(\frac{q^{n}}{1-q^{8n+1}} + \frac{q^{3n+1}}{1-q^{8n+3}}\right),
\end{aligned}
\tag{27.2.7}
$$

(25.1.19) is equivalent to

$$
\begin{aligned}
\psi(q)\phi(q^2) &= \sum_{n\geq 0}\left(\frac{q^{n}}{1-q^{8n+1}} - \frac{q^{3n+1}}{1-q^{8n+3}} + \frac{q^{5n+3}}{1-q^{8n+5}} - \frac{q^{7n+6}}{1-q^{8n+7}}\right) \\
&= \sum_{-\infty}^{\infty}\left(\frac{q^{n}}{1-q^{8n+1}} - \frac{q^{3n+1}}{1-q^{8n+3}}\right),
\end{aligned}
\tag{27.2.8}
$$

(25.1.20) is equivalent to

$$
\begin{aligned}
\psi(q)\phi(q^6) &= \sum_{n\geq 0}\left(\frac{q^{3n}}{1-q^{24n+1}} - \frac{q^{15n+3}}{1-q^{24n+5}} + \frac{q^{21n+6}}{1-q^{24n+7}} - \frac{q^{9n+4}}{1-q^{24n+11}}\right. \\
&\quad \left. + \frac{q^{15n+8}}{1-q^{24n+13}} - \frac{q^{3n+2}}{1-q^{24n+17}} + \frac{q^{9n+7}}{1-q^{24n+19}} - \frac{q^{21n+20}}{1-q^{24n+23}}\right)
\end{aligned}
$$

$$= \sum_{-\infty}^{\infty} \left(\frac{q^{3n}}{1-q^{24n+1}} + \frac{q^{21n+6}}{1-q^{24n+7}} + \frac{q^{15n+8}}{1-q^{24n+13}} + \frac{q^{9n+7}}{1-q^{24n+19}} \right), \tag{27.2.9}$$

(25.1.21) is equivalent to

$$\psi(q^2)\phi(q) = \sum_{n\geq 0} \left(\frac{q^n}{1-q^{4n+1}} - \frac{q^{3n+2}}{1-q^{4n+3}} \right) = \sum_{-\infty}^{\infty} \frac{q^n}{1-q^{4n+1}}, \tag{27.2.10}$$

(25.1.22) is equivalent to

$$\psi(q^2)\phi(q^3) = \sum_{n\geq 0} \left(\frac{q^{3n}}{1-q^{12n+1}} - \frac{q^{3n+1}}{1-q^{12n+5}} + \frac{q^{9n+5}}{1-q^{12n+7}} - \frac{q^{9n+8}}{1-q^{12n+11}} \right)$$
$$= \sum_{-\infty}^{\infty} \left(\frac{q^{3n}}{1-q^{12n+1}} + \frac{q^{9n+5}}{1-q^{12n+7}} \right), \tag{27.2.11}$$

(25.1.23) is equivalent to

$$\psi(q^3)\phi(q^2) = \sum_{n\geq 0} \left(\frac{q^{9n}}{1-q^{24n+1}} - \frac{q^{21n+4}}{1-q^{24n+5}} + \frac{q^{15n+4}}{1-q^{24n+7}} - \frac{q^{3n+1}}{1-q^{24n+11}} \right.$$
$$\left. + \frac{q^{21n+11}}{1-q^{24n+13}} - \frac{q^{9n+6}}{1-q^{24n+17}} + \frac{q^{3n+2}}{1-q^{24n+19}} - \frac{q^{15n+14}}{1-q^{24n+23}} \right)$$
$$= \sum_{-\infty}^{\infty} \left(\frac{q^{9n}}{1-q^{24n+1}} + \frac{q^{15n+4}}{1-q^{24n+7}} + \frac{q^{21n+11}}{1-q^{24n+13}} + \frac{q^{3n+2}}{1-q^{24n+19}} \right), \tag{27.2.12}$$

(25.1.24) is equivalent to

$$\psi(q^4)\phi(q) = \sum_{n\geq 0} \frac{q^n + q^{3n+1} - q^{5n+2} - q^{7n+3}}{1-q^{8n+4}} = \sum_{-\infty}^{\infty} \frac{q^n(1+q^{2n+1})}{1-q^{8n+4}}, \tag{27.2.13}$$

(25.1.25) is equivalent to

$$\psi(q^6)\phi(q) = \sum_{n\geq 0} \left(\frac{q^{9n}}{1-q^{12n+1}} - \frac{q^{9n+3}}{1-q^{12n+5}} + \frac{q^{3n+1}}{1-q^{12n+7}} - \frac{q^{3n+2}}{1-q^{12n+11}} \right)$$
$$= \sum_{-\infty}^{\infty} \left(\frac{q^{9n}}{1-q^{12n+1}} + \frac{q^{3n+1}}{1-q^{12n+7}} \right). \tag{27.2.14}$$

Reference

98. C.G.J. Jacobi, Fundamenta nova theoriae functionum ellipticarum, Königsberg (1829)

Chapter 28
The Jordan–Kronecker Identity

28.1 Introduction

In this chapter, we establish the Jordan–Kronecker identity (see Cooper (2011) [35], p. 35),

$$\sum_{-\infty}^{\infty} \frac{x^n}{1-aq^n} = \left(\begin{matrix} ax, q/(ax), q, q \\ x, q/x, a, q/a \end{matrix} ; q \right)_\infty \tag{28.1.1}$$

for $|q| < x < 1$.

Clearly, the Jordan–Kronecker identity is useful in expanding certain products as Lambert series, and in summing certain Lambert series to products.

We apply it to give a direct proof of some of the identities (27.2.1)–(27.2.14), and to give a fourth proof of the modular identity (20.1.1).

28.2 The Jordan–Kronecker Identity

The q-binomial theorem (Appendix) gives us

$$\left(\begin{matrix} ax \\ x \end{matrix} ; q \right)_\infty = \sum_{m \geq 0} \frac{(a;q)_m}{(q;q)_m} x^m \tag{28.2.1}$$

for $|x| < 1$, and

$$\left(\begin{matrix} q/(ax) \\ q/x \end{matrix} ; q \right)_\infty = \sum_{n \geq 0} \frac{(1/a;q)_n}{(q;q)_n} \frac{q^n}{x^n} \tag{28.2.2}$$

for $|x| > |q|$.

M.D. Hirschhorn, *The Power of q*, Developments in Mathematics 49,
DOI 10.1007/978-3-319-57762-3_28

So, for $|q| < |x| < 1$, we have

$$\begin{aligned}
\left(\begin{matrix} ax, q/(ax) \\ x, q/x \end{matrix}; q\right)_\infty &= \sum_{m\geq 0} \frac{(a;q)_m}{(q;q)_m} x^m \sum_{n\geq 0} \frac{(1/a;q)_n}{(q;q)_n} \frac{q^n}{x^n} \\
&= \sum_{k\geq 0} \frac{(a;q)_k (1/a;q)_k}{(q;q)_k (q;q)_k} q^k \\
&+ \sum_{n\geq 1} x^n \sum_{k\geq 0} \frac{(a;q)_{n+k}(1/a;q)_k}{(q;q)_{n+k}(q;q)_k} q^k \\
&+ \sum_{n\geq 1} q^n x^{-n} \sum_{k\geq 0} \frac{(a;q)_k (1/a;q)_{n+k}}{(q;q)_k (q;q)_{n+k}} q^k \\
&= \sum_{k\geq 0} \frac{(a;q)_k (1/a;q)_k}{(q;q)_k (q;q)_k} q^k \\
&+ \sum_{n\geq 1} x^n \frac{(a;q)_n}{(q;q)_n} \sum_{k\geq 0} \frac{(aq^n;q)_k (1/a;q)_k}{(q^{n+1};q)_k (q;q)_k} q^k \\
&+ \sum_{n\geq 1} q^n x^{-n} \frac{(1/a;q)_n}{(q;q)_n} \sum_{k\geq 0} \frac{(a;q)_k (q^n/a;q)_k}{(q;q)_k (q^{n+1};q)_k} q^k \\
&= {}_2\phi_1\left(\begin{matrix} a, 1/a \\ q \end{matrix}; q\right) \\
&+ \sum_{n\geq 1} x^n \frac{(a;q)_n}{(q;q)_n}\, {}_2\phi_1\left(\begin{matrix} aq^n, 1/a \\ q^{n+1} \end{matrix}; q\right) \\
&+ \sum_{n\geq 1} q^n x^{-n} \frac{(1/a;q)_n}{(q;q)_n}\, {}_2\phi_1\left(\begin{matrix} a, q^n/a \\ q^{n+1} \end{matrix}; q\right) \\
&= \left(\begin{matrix} aq, q/a \\ q, q \end{matrix}; q\right)_\infty \\
&+ \sum_{n\geq 1} \frac{x^n (a;q)_n}{(q;q)_n} \cdot \left(\begin{matrix} aq^{n+1}, q/a \\ q^{n+1}, q \end{matrix}; q\right)_\infty \\
&+ \sum_{n\geq 1} \frac{q^n x^{-n} (1/a;q)_n}{(q;q)_n} \cdot \left(\begin{matrix} q^{n+1}/a, aq \\ q^{n+1}, q \end{matrix}; q\right)_\infty \\
&= \left(\begin{matrix} aq, q/a \\ q, q \end{matrix}; q\right)_\infty \left(1 + \sum_{n\geq 1} \frac{x^n (1-a)}{1-aq^n} + \sum_{n\geq 1} \frac{q^n x^{-n}(1-1/a)}{1-q^n/a}\right)
\end{aligned}$$

$$= \left(\begin{matrix} aq, q/a \\ q, q \end{matrix} ; q \right)_\infty \times \sum_{-\infty}^{\infty} \frac{x^n(1-a)}{1-aq^n}$$

$$= \left(\begin{matrix} a, q/a \\ q, q \end{matrix} ; q \right)_\infty \times \sum_{-\infty}^{\infty} \frac{x^n}{1-aq^n}, \tag{28.2.3}$$

where we have made use of the q-Gauss identity (Appendix)

$$ {}_2\phi_1 \left(\begin{matrix} a, b \\ c \end{matrix} ; q; \frac{c}{ab} \right) = \left(\begin{matrix} c/a, c/b \\ c, c/(ab) \end{matrix} ; q \right)_\infty .$$

It follows from (28.2.3) that

$$\left(\begin{matrix} ax, q/(ax), q, q \\ x, q/x, a, q/a \end{matrix} ; q \right)_\infty = \sum_{-\infty}^{\infty} \frac{x^n}{1-aq^n},$$

which is (28.1.1).

28.3 The Two Triangles Result

If in (28.1.1) we replace q by q^4 and set $a = q$, $x = q$, we obtain

$$\begin{aligned} \sum_{-\infty}^{\infty} \frac{q^n}{1-q^{4n+1}} &= \left(\begin{matrix} q^2, q^2, q^4, q^4 \\ q, q^3, q, q^3 \end{matrix} ; q^4 \right)_\infty \\ &= \left(\begin{matrix} q^2, q^4 \\ q, q^3 \end{matrix} ; q^4 \right)_\infty^2 \\ &= \frac{(q^2;q^2)_\infty^2}{(q;q^2)_\infty^2} \\ &= \frac{(q^2;q^2)_\infty^4}{(q;q)_\infty^2} \\ &= \psi(q)^2, \end{aligned}$$

which is (27.2.1), or (25.1.12).

28.4 Triangle+2×Triangle

In (28.1.1), replace q by q^8 and set $a=q$, $x=q^3$ and we obtain

$$\begin{aligned}\sum_{-\infty}^{\infty}\frac{q^{3n}}{1-q^{8n+1}} &= \left(\begin{matrix}q^4,q^4,q^8,q^8\\ q^3,q^5,q,q^7\end{matrix};q^8\right)_\infty\\ &= \frac{(q^4;q^4)_\infty^2}{(q;q^2)_\infty}\\ &= \frac{(q^2;q^2)_\infty^2}{(q;q)_\infty}\frac{(q^4;q^4)_\infty^2}{(q^2;q^2)_\infty}\\ &= \psi(q)\psi(q^2),\end{aligned}$$

which is (27.2.2), or (25.1.13).

28.5 Triangle+3×Triangle

In (28.1.1), replace q by q^6, set $a=q$, $x=q^3$, and we obtain

$$\begin{aligned}\sum_{-\infty}^{\infty}\frac{q^{3n}}{1-q^{6n+1}} &= \left(\begin{matrix}q^4,q^2,q^6,q^6\\ q^3,q^3,q,q^5\end{matrix};q^6\right)_\infty\\ &= \frac{(q^2;q^2)_\infty(q^6;q^6)_\infty}{(q;q^2)_\infty(q^3;q^6)_\infty}\\ &= \frac{(q^2;q^2)_\infty^2}{(q;q)_\infty}\frac{(q^6;q^6)_\infty^2}{(q^3;q^3)_\infty}\\ &= \psi(q)\psi(q^3),\end{aligned}$$

which is (27.2.3), or (25.1.14).

28.6 Triangle+4×Triangle

In (28.1.1), replace q by q^8, set $a=q$, $x=q^5$, and we obtain

$$\begin{aligned}\sum_{-\infty}^{\infty}\frac{q^{5n}}{1-q^{8n+1}} &= \left(\begin{matrix}q^6,q^2,q^8,q^8\\ q^5,q^3,q,q^7\end{matrix};q^8\right)_\infty\\ &= \frac{(q^2;q^4)_\infty(q^8;q^8)_\infty^2}{(q,q^2)_\infty}\end{aligned}$$

$$= \frac{(q^2;q^2)_\infty}{(q;q^2)_\infty} \frac{(q^8;q^8)_\infty^2}{(q^4;q^4)_\infty}$$
$$= \frac{(q^2;q^2)_\infty^2}{(q;q)_\infty} \frac{(q^8;q^8)_\infty^2}{(q^4;q^4)_\infty}$$
$$= \psi(q)\psi(q^4),$$

which is (27.2.4), or (25.1.15).

28.7 Triangle+Square

We use (28.1.1) to prove (27.2.7).

We have

$$\sum_{-\infty}^{\infty}\left(\frac{q^n}{1-q^{8n+1}}+\frac{q^{3n+1}}{1-q^{8n+3}}\right) = \left(\begin{matrix} q^2,q^6,q^8,q^8 \\ q,q^7,q,q^7 \end{matrix};q^8\right)_\infty + q\left(\begin{matrix} q^6,q^2,q^8,q^8 \\ q^3,q^5,q^3,q^5 \end{matrix};q^8\right)_\infty$$
$$= \frac{(q^2;q^4)_\infty}{(q;q^2)_\infty^2}((q^3,q^5,q^8;q^8)_\infty^2 + q(q,q^7,q^8;q^8)_\infty^2)$$
$$= \frac{(q^2;q^2)_\infty^3}{(q;q)_\infty^2(q^4;q^4)_\infty}((q^3,q^5,q^8;q^8)_\infty^2 + q(q,q^7,q^8;q^8)_\infty^2). \tag{28.7.1}$$

We now show that

$$(q^3,q^5,q^8;q^8)_\infty^2 + q(q,q^7,q^8;q^8)_\infty^2 = \psi(q)\phi(-q^2) = \frac{(q^2;q^2)_\infty^4}{(q;q)_\infty(q^4;q^4)_\infty}. \tag{28.7.2}$$

Let

$$f(q) = (q^3,q^5,q^8;q^8)_\infty^2 + q(q,q^7,q^8;q^8)_\infty^2$$
$$= \sum_{m,n=-\infty}^{\infty}(-1)^{m+n}q^{4m^2+m+4n^2+n} + q\sum_{m,n=-\infty}^{\infty}(-1)^{m+n}q^{4m^2-3m+4n^2-3n}.$$

Then

$$f(-q) = \sum_{m,n=-\infty}^{\infty} q^{4m^2+m+4n^2+n} - q\sum_{m,n=-\infty}^{\infty} q^{4m^2-3m+4n^2-3n}$$

and

$$\begin{aligned}
q^2 f(-q^{16}) &= \sum_{m,n=-\infty}^{\infty} q^{(8m+1)^2+(8n+1)^2} - \sum_{m,n=-\infty}^{\infty} q^{(8m-3)^2+(8n-3)^2} \\
&= \sum_{m,n\,\text{even}} q^{(4m+1)^2+(4n+1)^2} - \sum_{m,n\,\text{odd}} q^{(4m+1)^2+(4n+1)^2} \\
&= \sum_{m\equiv n \pmod 2} (-1)^m q^{(4m+1)^2+(4n+1)^2} \\
&= \sum_{r,s=-\infty}^{\infty} (-1)^{r+s} q^{(4(r+s)+1)^2+(4(r-s)+1)^2} \\
&= \sum_{r,s=-\infty}^{\infty} (-1)^{r+s} q^{32r^2+16r+32s^2+2} \\
&= q^2 \psi(-q^{16})\phi(-q^{32}),
\end{aligned}$$

from which (28.7.2) follows.

If we substitute (28.7.2) into (28.7.1), we find

$$\begin{aligned}
\sum_{-\infty}^{\infty} \left(\frac{q^n}{1-q^{8n+1}} + \frac{q^{3n+1}}{1-q^{8n+3}} \right) &= \frac{(q^2;q^2)_\infty^7}{(q;q)_\infty^3 (q^4;q^4)_\infty^2} \\
&= \frac{(q^2;q^2)_\infty^2}{(q;q)_\infty} \frac{(q^2;q^2)_\infty^5}{(q;q)_\infty^2 (q^4;q^4)_\infty^2} \\
&= \psi(q)\phi(q),
\end{aligned}$$

which is (27.2.7), or (25.1.18).

28.8 Triangle+2×Square

We use (28.1.1) to prove (27.2.8).

We have

$$\begin{aligned}
\sum_{-\infty}^{\infty} \left(\frac{q^n}{1-q^{8n+1}} - \frac{q^{3n+1}}{1-q^{8n+3}} \right) &= \left(\begin{matrix} q^2, q^6, q^8, q^8 \\ q, q^7, q, q^7 \end{matrix} ; q^8 \right)_\infty - q \left(\begin{matrix} q^6, q^2, q^8, q^8 \\ q^3, q^5, q^3, q^5 \end{matrix} ; q^8 \right)_\infty \\
&= \frac{(q^2;q^4)_\infty}{(q;q^2)_\infty^2} ((q^3,q^5,q^8;q^8)_\infty^2 - q(q,q^7,q^8;q^8)_\infty^2) \\
&= \frac{(q^2;q^2)_\infty^3}{(q;q)_\infty^2 (q^4;q^4)_\infty} ((q^3,q^5,q^8;q^8)_\infty^2 - q(q,q^7,q^8;q^8)_\infty^2).
\end{aligned} \tag{28.8.1}$$

It can be shown (this is left as an exercise, see §28.7) that

$$(q^3,q^5,q^8;q^8)^2_\infty - q(q,q^7,q^8;q^8)^2_\infty = \psi(-q)\phi(q^2) = \frac{(q;q)_\infty(q^4;q^4)^6_\infty}{(q^2;q^2)^3_\infty(q^8;q^8)^2_\infty}. \tag{28.8.2}$$

If we substitute (28.8.2) into (28.8.1), we find that

$$\begin{aligned}\sum_{-\infty}^{\infty}\left(\frac{q^n}{1-q^{8n+1}} - \frac{q^{3n+1}}{1-q^{8n+3}}\right) &= \frac{(q^4;q^4)^5_\infty}{(q;q)_\infty(q^8;q^8)^2_\infty}\\ &= \frac{(q^2;q^2)^2_\infty}{(q;q)_\infty}\frac{(q^4;q^4)^5_\infty}{(q^2;q^2)^2_\infty(q^8;q^8)^2_\infty}\\ &= \psi(q)\phi(q^2),\end{aligned}$$

which is (27.2.8), or (25.1.19).

28.9 2×Triangle+3×Square

We have, by (28.1.1),

$$\begin{aligned}\sum_{-\infty}^{\infty}\left(\frac{q^{3n}}{1-q^{12n+1}} + \frac{q^{9n+5}}{1-q^{12n+7}}\right) &= \left(\begin{matrix}q^4,q^8,q^{12},q^{12}\\ q^3,q^9,q,q^{11}\end{matrix}\right)_\infty + q^5\left(\begin{matrix}q^{16},q^{-4},q^{12},q^{12}\\ q^9,q^3,q^7,q^5\end{matrix};q^{12}\right)_\infty\\ &= \left(\begin{matrix}q^4,q^8,q^{12},q^{12}\\ q,q^{11},q^3,q^9\end{matrix};q^{12}\right)_\infty - q\left(\begin{matrix}q^4,q^8,q^{12},q^{12}\\ q^5,q^7,q^3,q^9\end{matrix};q^{12}\right)_\infty\\ &= \left(\begin{matrix}q^4,q^8,q^{12}\\ q,q^3,q^5,q^7,q^9,q^{11}\end{matrix};q^{12}\right)_\infty((q^5,q^7,q^{12};q^{12})_\infty - q(q,q^{11},q^{12};q^{12})_\infty)\\ &= \frac{(q^4;q^4)_\infty}{(q;q^2)_\infty}((q^5,q^7,q^{12};q^{12})_\infty - q(q,q^{11},q^{12};q^{12})_\infty)\\ &= \frac{(q^2;q^2)_\infty(q^4;q^4)_\infty}{(q;q)_\infty}f(q),\end{aligned} \tag{28.9.1}$$

where

$$f(q) = \sum_{-\infty}^{\infty}(-1)^n q^{6n^2+n} - q\sum_{-\infty}^{\infty}(-1)^n q^{6n^2-5n}, \tag{28.9.2}$$

$$f(-q) = \sum_{-\infty}^{\infty} q^{6n^2+n} + q\sum_{-\infty}^{\infty} q^{6n^2-5n},$$

$$qf(-q^{24}) = \sum_{-\infty}^{\infty} q^{(12n+1)^2} + \sum_{-\infty}^{\infty} q^{(12n-5)^2}$$
$$= \sum_{-\infty}^{\infty} q^{(6n+1)^2}$$
$$= q\sum_{-\infty}^{\infty} q^{36n^2+12n},$$

$$f(-q) = \sum_{-\infty}^{\infty} q^{(3n^2+n)/2}$$
$$= (-q, -q^2, q^3; q^3)_\infty$$
$$= (-q, -q^2, q^3, -q^4, -q^5, q^6; q^6)_\infty$$

and

$$f(q) = (q, -q^2, -q^3, -q^4, q^5, q^6; q^6)_\infty$$
$$= \frac{(q;q^2)_\infty}{(q^3;q^6)_\infty}(-q^3;q^6)_\infty \frac{(-q^2;q^2)_\infty}{(-q^6;q^6)_\infty}(q^6;q^6)_\infty$$
$$= \frac{(q;q)_\infty}{(q^2;q^2)_\infty}\frac{(q^6;q^6)_\infty}{(q^3;q^3)_\infty}\frac{(q^6;q^{12})_\infty}{(q^3;q^6)_\infty}\frac{(q^4;q^4)_\infty}{(q^2;q^2)_\infty}\frac{(q^6;q^6)_\infty}{(q^{12};q^{12})_\infty}(q^6;q^6)_\infty$$
$$= \frac{(q;q)_\infty}{(q^2;q^2)_\infty}\frac{(q^6;q^6)_\infty}{(q^3;q^3)_\infty}\frac{(q^6;q^6)_\infty}{(q^{12};q^{12})_\infty}\frac{(q^6;q^6)_\infty}{(q^3;q^3)_\infty}\frac{(q^4;q^4)_\infty}{(q^2;q^2)_\infty}\frac{(q^6;q^6)_\infty}{(q^{12};q^{12})_\infty}(q^6;q^6)_\infty$$
$$= \frac{(q;q)_\infty(q^4;q^4)_\infty(q^6;q^6)_\infty^5}{(q^2;q^2)_\infty^2(q^3;q^3)_\infty^2(q^{12};q^{12})_\infty^2}. \tag{28.9.3}$$

If we substitute (28.9.3) into (28.9.1), we find

$$\sum_{-\infty}^{\infty}\left(\frac{q^{3n}}{1-q^{12n+1}} + \frac{q^{9n+5}}{1-q^{12n+7}}\right) = \frac{(q^2;q^2)_\infty(q^4;q^4)_\infty}{(q;q)_\infty}\frac{(q;q)_\infty(q^4;q^4)_\infty(q^6;q^6)_\infty^5}{(q^2;q^2)_\infty^2(q^3;q^3)_\infty^2(q^{12};q^{12})_\infty^2}$$
$$= \frac{(q^4;q^4)_\infty^2}{(q^2;q^2)_\infty}\frac{(q^6;q^6)_\infty^5}{(q^3;q^3)_\infty^2(q^{12};q^{12})_\infty^2}$$
$$= \psi(q^2)\phi(q^3),$$

which is (27.2.11), or (25.1.22).

28.10 4×Triangle+Square

We have, by (28.1.1),

$$\begin{aligned}\sum_{-\infty}^{\infty}\frac{q^n+q^{3n+1}}{1-q^{8n+4}} &= \left(\begin{matrix}q^5,q^3,q^8,q^8\\ q,q^7,q^4,q^4\end{matrix};q^8\right)_\infty + q\left(\begin{matrix}q^7,q,q^8,q^8\\ q^3,q^5,q^4,q^4\end{matrix};q^8\right)_\infty \\ &= \frac{1}{(q;q^2)_\infty(q^4;q^8)_\infty^2}\left((q^3,q^5,q^8;q^8)_\infty^2+q(q,q^7,q^8;q^8)_\infty^2\right) \\ &= \frac{(q^2;q^2)_\infty}{(q;q)_\infty}\frac{(q^8;q^8)_\infty^2}{(q^4;q^4)_\infty^2}\frac{(q^2;q^2)_\infty^4}{(q;q)_\infty(q^4;q^4)_\infty} \qquad \text{(see §27.7)} \\ &= \frac{(q^2;q^2)_\infty^5(q^8;q^8)_\infty^2}{(q;q)_\infty^2(q^4;q^4)_\infty^3} \\ &= \frac{(q^8;q^8)_\infty^2}{(q^4;q^4)_\infty}\frac{(q^2;q^2)_\infty^5}{(q;q)_\infty^2(q^4;q^4)_\infty^2} \\ &= \psi(q^4)\phi(q),\end{aligned}$$

which is (27.2.13), or (25.1.24).

28.11 6×Triangle+Square

We have, by (28.1.1),

$$\begin{aligned}\sum_{-\infty}^{\infty}\left(\frac{q^{9n}}{1-q^{12n+1}}+\frac{q^{3n+1}}{1-q^{12n+7}}\right) &= \left(\begin{matrix}q^{10},q^2,q^{12},q^{12}\\ q^9,q^3,q,q^{11}\end{matrix};q^{12}\right)_\infty \\ &\qquad + q\left(\begin{matrix}q^{10},q^2,q^{12},q^{12}\\ q^3,q^9,q^7,q^5\end{matrix};q^{12}\right)_\infty \\ &= \frac{(q^2,q^{10},q^{12};q^{12})_\infty}{(q;q^2)_\infty}((q^5,q^7,q^{12};q^{12})_\infty+q(q,q^{11},q^{12};q^{12})_\infty) \\ &= \frac{(q^2;q^4)_\infty}{(q^6;q^{12})_\infty}\frac{(q^{12};q^{12})_\infty}{(q;q^2)_\infty}((q^5,q^7,q^{12};q^{12})_\infty+q(q,q^{11},q^{12};q^{12})_\infty) \\ &= \frac{(q^2;q^2)_\infty^2(q^{12};q^{12})_\infty^2}{(q;q)_\infty(q^4;q^4)_\infty(q^6;q^6)_\infty}f(q), \qquad (28.11.1)\end{aligned}$$

where

$$f(q)=\sum_{-\infty}^{\infty}(-1)^n q^{6n^2+n}+q\sum_{-\infty}^{\infty}(-1)^n q^{6n^2-5n},$$

$$f(-q)=\sum_{-\infty}^{\infty}q^{6n^2+n}-q\sum_{-\infty}^{\infty}q^{6n^2-5n},$$

$$\begin{aligned}qf(-q^{24})&=\sum_{-\infty}^{\infty}q^{(12n+1)^2}-\sum_{-\infty}^{\infty}q^{(12n-5)^2}\\&=\sum_{-\infty}^{\infty}(-1)^nq^{(6n+1)^2}\\&=q\sum_{-\infty}^{\infty}(-1)^nq^{36n^2+12n},\end{aligned}$$

$$\begin{aligned}f(-q)&=\sum_{-\infty}^{\infty}(-1)^nq^{(3n^2+n)/2}\\&=(q;q)_\infty\\&=(q,q^2;q^2)_\infty,\end{aligned}$$

and

$$\begin{aligned}f(q)&=(-q,q^2;q^2)_\infty\\&=(-q;q^2)_\infty(q^2;q^2)_\infty\\&=\frac{(q^2;q^4)_\infty}{(q;q^2)_\infty}(q^2;q^2)_\infty\\&=\frac{(q^2;q^2)^3_\infty}{(q;q)_\infty(q^4;q^4)_\infty}.\end{aligned}\tag{28.11.2}$$

If we substitute (28.11.2) into (28.11.1), we find

$$\begin{aligned}\sum_{-\infty}^{\infty}\left(\frac{q^{9n}}{1-q^{12n+1}}+\frac{q^{3n+1}}{1-q^{12n+7}}\right)&=\frac{(q^2;q^2)^2_\infty(q^{12};q^{12})^2_\infty}{(q;q)_\infty(q^4;q^4)_\infty(q^6;q^6)_\infty}\frac{(q^2;q^2)^3_\infty}{(q;q)_\infty(q^4;q^4)_\infty}\\&=\frac{(q^2;q^2)^5_\infty(q^{12};q^{12})^2_\infty}{(q;q)^2_\infty(q^4;q^4)^2_\infty(q^6;q^6)_\infty}\\&=\frac{(q^2;q^2)^5_\infty}{(q;q)^2_\infty(q^4;q^4)^2_\infty}\frac{(q^{12};q^{12})^2_\infty}{(q^6;q^6)_\infty}\\&=\phi(q)\psi(q^6),\end{aligned}$$

which is (27.2.14), or (25.1.25).

28.12 Our Fourth Proof of the Modular Equation (20.1.1)

We have proved (§§28.5, 28.9, 28.11) that

$$\psi(q)\psi(q^3) = \sum_{-\infty}^{\infty} \left(\frac{q^{6n}}{1-q^{12n+1}} + \frac{q^{6n+3}}{1-q^{12n+7}} \right), \tag{28.12.1}$$

$$\phi(q^3)\psi(q^2) = \sum_{-\infty}^{\infty} \left(\frac{q^{3n}}{1-q^{12n+1}} + \frac{q^{9n+5}}{1-q^{12n+7}} \right) \tag{28.12.2}$$

and

$$\phi(q)\psi(q^6) = \sum_{-\infty}^{\infty} \left(\frac{q^{9n}}{1-q^{12n+1}} + \frac{q^{3n+1}}{1-q^{12n+7}} \right). \tag{28.12.3}$$

From these we find

$$\begin{aligned}
\psi(q)\psi(q^3) &= \sum_{-\infty}^{\infty} \left(\frac{q^{6n}(1+q^{12n+1})}{1-q^{24n+2}} + \frac{q^{6n+3}(1+q^{12n+7})}{1-q^{24n+14}} \right) \\
&= \sum_{-\infty}^{\infty} \left(\frac{q^{6n}}{1-q^{24n+2}} + \frac{q^{18n+10}}{1-q^{24n+14}} \right) + q\sum_{-\infty}^{\infty} \left(\frac{q^{18n}}{1-q^{24n+2}} + \frac{q^{6n+2}}{1-q^{24n+14}} \right) \\
&= \phi(q^6)\psi(q^4) + q\phi(q^2)\psi(q^{12}).
\end{aligned} \tag{28.12.4}$$

This is (25.2.1), and (see §§28.2, 28.3) implies (20.1.1).

28.13 Proof of (22.11.5)

We now turn to the proof of (22.11.5).

We have

$$\begin{aligned}
&\sum_{n\geq 0} \left(\frac{q^{6n+1}}{1-q^{6n+1}} - 2\frac{q^{6n+2}}{1-q^{6n+2}} + 2\frac{q^{6n+4}}{1-q^{6n+4}} - \frac{q^{6n+5}}{1-q^{6n+5}} \right) \\
&\quad = \sum_{n\geq 0} \left(\frac{q^{6n+1}}{1-q^{6n+1}} + \frac{q^{6n+4}}{1-q^{6n+4}} - \frac{q^{6n+2}}{1-q^{6n+2}} - \frac{q^{6n+5}}{1-q^{6n+5}} \right) \\
&\qquad - \sum_{n\geq 0} \left(\frac{q^{6n+2}}{1-q^{6n+2}} - \frac{q^{6n+4}}{1-q^{6n+4}} \right)
\end{aligned}$$

$$
\begin{aligned}
&= \sum_{n\geq 0}\left(\frac{q^{3n+1}}{1-q^{3n+1}}-\frac{q^{3n+2}}{1-q^{3n+2}}\right)-\sum_{n\geq 0}\left(\frac{q^{6n+2}}{1-q^{6n+2}}-\frac{q^{6n+4}}{1-q^{6n+4}}\right)\\
&= \sum_{n\geq 0}\left(\frac{q^{3n+1}(1+q^{3n+1})-q^{6n+2}}{1-q^{6n+2}}-\frac{q^{3n+2}(1+q^{3n+2})-q^{6n+4}}{1-q^{6n+4}}\right)\\
&= \sum_{n\geq 0}\left(\frac{q^{3n+1}}{1-q^{6n+2}}-\frac{q^{3n+2}}{1-q^{6n+4}}\right)\\
&= \sum_{-\infty}^{\infty}\frac{q^{3n+1}}{1-q^{6n+2}}\\
&= q\left(\begin{matrix}q,q^5,q^6,q^6\\ q^2,q^3,q^3,q^4\end{matrix};q^6\right)_\infty\\
&= q\left(\begin{matrix}q,q^2,q^3,q^4,q^5,q^6,q^6,q^6,q^6,q^6,q^6,q^6\\ q^2,q^2,q^3,q^3,q^3,q^4,q^4,q^6,q^6,q^6,q^6,q^6\end{matrix};q^6\right)_\infty\\
&= q\frac{(q;q)_\infty}{(q^2;q^2)_\infty^2}\frac{(q^6;q^6)_\infty^6}{(q^3;q^3)_\infty^3}\\
&= q\frac{\psi(q^3)^3}{\psi(q)},
\end{aligned}
$$

as required.

Reference

35. S. Cooper, *Development of Elliptic Functions According to Ramanujan, Originally by K. Venkatachaliengar*, vol. 8, Monographs in Number Theory (World Scientific, Singapore, 2011)

Chapter 29
Melham's Identities

29.1 Introduction

In 2007, Ray Melham [106] published a list of 298 conjectured identities in which a product is equal to a Lambert series. Each of these can be interpreted as a representation theorem of the sort we have studied in Chapters 25–28.

I managed to prove upwards of 20 of them using the methods of this book. Pee Choon Toh (2013) [119] proves 21 of them and indicates proofs of the rest.

For this chapter, I have chosen to present just three of Melham's list. They can be written as follows:

$$\psi(q)\psi(q^5) = \sum_{n\geq 0}\left(\frac{q^{5n}}{1-q^{20n+3}} + \frac{q^{5n+1}}{1-q^{20n+7}} - \frac{q^{15n+9}}{1-q^{20n+13}} - \frac{q^{15n+12}}{1-q^{20n+17}}\right) \tag{29.1.1}$$

$$= \sum_{n=-\infty}^{\infty}\left(\frac{q^{5n}}{1-q^{20n+3}} + \frac{q^{5n+1}}{1-q^{20n+7}}\right), \tag{29.1.2}$$

$$\psi(q)\psi(q^6) = \sum_{n\geq 0}\left(\frac{q^{7n}}{1-q^{24n+3}} + \frac{q^{5n+1}}{1-q^{24n+9}} - \frac{q^{19n+11}}{1-q^{24n+15}} - \frac{q^{17n+14}}{1-q^{24n+21}}\right) \tag{29.1.3}$$

$$= \sum_{-\infty}^{\infty}\left(\frac{q^{7n}}{1-q^{24n+3}} + \frac{q^{9n+1}}{1-q^{24n+5}}\right), \tag{29.1.4}$$

M.D. Hirschhorn, *The Power of q*, Developments in Mathematics 49,
DOI 10.1007/978-3-319-57762-3_29

and

$$\Pi(q)\Pi(q^5) = \sum_{n\geq 0}\left(\frac{q^{15n}}{1-q^{60n+1}} + \frac{q^{45n+5}}{1-q^{60n+7}} + \frac{q^{45n+17}}{1-q^{60n+23}} + \frac{q^{15n+7}}{1-q^{60n+29}}\right.$$
$$+\frac{q^{15n+10}}{1-q^{60n+41}} + \frac{q^{45n+32}}{1-q^{60n+43}} + \frac{q^{45n+35}}{1-q^{60n+47}} + \frac{q^{15n+12}}{1-q^{60n+49}}$$
$$-\frac{q^{45n+8}}{1-q^{60n+11}} - \frac{q^{15n+3}}{1-q^{60n+13}} - \frac{q^{15n+4}}{1-q^{60n+17}} - \frac{q^{45n+14}}{1-q^{60n+19}}$$
$$\left.-\frac{q^{45n+23}}{1-q^{60n+31}} - \frac{q^{15n+9}}{1-q^{60n+37}} - \frac{q^{15n+13}}{1-q^{60n+53}} - \frac{q^{45n+44}}{1-q^{60n+59}}\right) \tag{29.1.5}$$

$$= \sum_{-\infty}^{\infty}\left(\frac{q^{15n}}{1-q^{60n+1}} + \frac{q^{45n+5}}{1-q^{60n+7}} + \frac{q^{45n+17}}{1-q^{60n+23}} + \frac{q^{15n+7}}{1-q^{60n+29}}\right.$$
$$\left.+\frac{q^{15n+10}}{1-q^{60n+41}} + \frac{q^{45n+32}}{1-q^{60n+43}} + \frac{q^{45n+35}}{1-q^{60n+47}} + \frac{q^{15n+12}}{1-q^{60n+49}}\right). \tag{29.1.6}$$

They give rise to the following representation theorems. The first two are natural companions to (25.1.12)–(25.1.15).

$$r\{\triangle + 5\triangle\}(n) = e\{\{3,7\},20\}(4n+3), \tag{29.1.7}$$

$$r\{\triangle + 6\triangle\}(n) = e\{\{5,7\},24\}(8n+7) \tag{29.1.8}$$

and

$$r\{\Pi + 5\Pi\}(n) = e\{\{1,7,23,29,41,43,47,49\},60\}(4n+1). \tag{29.1.9}$$

In §§29.2–29.4 I will present proofs of these three identities.

29.2 Triangle+5×Triangle

We have

$$\sum_{-\infty}^{\infty}\left(\frac{q^{5n}}{1-q^{20n+3}} + \frac{q^{5n+1}}{1-q^{20n+7}}\right)$$
$$= \left(\begin{matrix} q^8, q^{12}, q^{20}, q^{20} \\ q^5, q^{15}, q^3, q^{17} \end{matrix}; q^{20}\right)_\infty + q\left(\begin{matrix} q^{12}, q^8, q^{20}, q^{20} \\ q^5, q^{15}, q^7, q^{13} \end{matrix}; q^{20}\right)_\infty$$

$$= \left(\begin{matrix} q^8, q^{12}, q^{20} \\ q^3, q^5, q^7, q^{13}, q^{15}, q^{17} \end{matrix}; q^{20} \right)_\infty ((q^7, q^{13}, q^{20}; q^{20})_\infty + q(q^3, q^{17}, q^{20}; q^{20})_\infty)$$

$$= \left(\begin{matrix} q^8, q^{12}, q^{20} \\ q^3, q^5, q^7, q^{13}, q^{15}, q^{17} \end{matrix}; q^{20} \right)_\infty f(q) \tag{29.2.1}$$

where

$$f(q) = \sum_{-\infty}^{\infty} (-1)^n q^{10n^2+3n} + q \sum_{-\infty}^{\infty} (-1)^n q^{10n^2-7n},$$

$$f(-q) = \sum_{-\infty}^{\infty} q^{10n^2+3n} - q \sum_{-\infty}^{\infty} q^{10n^2-7n},$$

$$\begin{aligned} q^9 f(-q^{40}) &= \sum_{-\infty}^{\infty} q^{(20n+3)^2} - \sum_{-\infty}^{\infty} q^{(20n-7)^2} \\ &= \sum_{-\infty}^{\infty} (-1)^n q^{(10n+3)^2} \\ &= q^9 \sum_{-\infty}^{\infty} (-1)^n q^{100n^2+60n}, \end{aligned}$$

$$\begin{aligned} f(-q) &= \sum_{-\infty}^{\infty} (-1)^n q^{(5n^2+3n)/2} \\ &= (q, q^4, q^5; q^5)_\infty \\ &= (q, q^4, q^5, q^6, q^9, q^{10}; q^{10})_\infty \end{aligned}$$

and

$$\begin{aligned} f(q) &= (-q, q^4, -q^5, q^6, -q^9, q^{10}; q^{10})_\infty \\ &= \frac{(q^2; q^{20})_\infty}{(q; q^{10})_\infty} \frac{(q^{10}; q^{20})_\infty}{(q^5; q^{10})_\infty} \frac{(q^{18}; q^{20})_\infty}{(q^9; q^{10})_\infty} (q^4, q^6, q^{10}; q^{10})_\infty \\ &= \left(\begin{matrix} q^2, q^4, q^6, q^{10}, q^{10}, q^{14}, q^{16}, q^{18}, q^{20} \\ q, q^5, q^9, q^{11}, q^{15}, q^{19} \end{matrix}; q^{20} \right)_\infty . \end{aligned} \tag{29.2.2}$$

If we substitute (29.2.2) into (29.2.1), we find

$$\begin{aligned}
&\sum_{-\infty}^{\infty}\left(\frac{q^{5n}}{1-q^{20n+3}}+\frac{q^{5n+1}}{1-q^{20n+7}}\right)\\
&=\left(\begin{matrix}q^{8},q^{12},q^{20}\\ q^{3},q^{5},q^{7},q^{13},q^{15},q^{17}\end{matrix};q^{20}\right)_\infty\left(\begin{matrix}q^{2},q^{4},q^{6},q^{10},q^{10},q^{14},q^{16},q^{18},q^{20}\\ q,q^{5},q^{9},q^{11},q^{15},q^{19}\end{matrix};q^{20}\right)_\infty\\
&=\left(\begin{matrix}q^{2},q^{4},q^{6},q^{8},q^{10},q^{10},q^{12},q^{14},q^{16},q^{18},q^{20},q^{20}\\ q,q^{3},q^{5},q^{5},q^{7},q^{9},q^{11},q^{13},q^{15},q^{15},q^{17},q^{19}\end{matrix};q^{20}\right)_\infty\\
&=\left(\begin{matrix}q^{2},q^{4},q^{6},q^{8},q^{10},q^{10}\\ q,q^{3},q^{5},q^{5},q^{7},q^{9}\end{matrix};q^{10}\right)_\infty\\
&=\frac{(q^2;q^2)_\infty}{(q;q^2)_\infty}\frac{(q^{10};q^{10})_\infty}{(q^5;q^{10})_\infty}\\
&=\frac{(q^2;q^2)_\infty^2}{(q;q)_\infty}\frac{(q^{10};q^{10})_\infty^2}{(q^5;q^5)_\infty}\\
&=\psi(q)\psi(q^5),
\end{aligned}$$

which is (29.1.2).

We have

$$\begin{aligned}
&\sum_{n\geq 0} r\{\triangle+5\triangle\}(n)q^n=\psi(q)\psi(q^5)\\
&=\sum_{n\geq 0}\left(\frac{q^{5n}}{1-q^{20n+3}}+\frac{q^{5n+1}}{1-q^{20n+7}}-\frac{q^{15n+9}}{1-q^{20n+13}}-\frac{q^{15n+12}}{1-q^{20n+17}}\right)\\
&=\sum_{k,l\geq 0} q^{20k+5k+3l}+\sum_{k,l\geq 0} q^{20kl+5k+7l+1}-\sum_{k,l\geq 0} q^{20kl+15k+13l+9}\\
&\qquad -\sum_{k,l\geq 0} q^{20kl+15k+17l+12},
\end{aligned}$$

$$\begin{aligned}
&\sum_{n\geq 0} r\{\triangle+5\triangle\}(n)q^{4n+3}\\
&=\sum_{k,l\geq 0} q^{80kl+20k+12l+3}+\sum_{k,l\geq 0} q^{80kl+20k+28l+7}-\sum_{k,l\geq 0} q^{80kl+60k+52l+39}\\
&\qquad -\sum_{k,l\geq 0} q^{80kl+60k+68l+51}
\end{aligned}$$

$$= \sum_{k,l\ge 0} q^{(20k+3)(4l+1)} + \sum_{k,l\ge 0} q^{(20k+7)(4l+1)} - \sum_{k,l\ge 0} q^{(20k+13)(4l+3)} - \sum_{k,l\ge 0} q^{(20k+17)(4l+3)}$$

$$= \sum_{n\ge 0} e\{\{3,7\},20\}(4n+3)q^{4n+3},$$

from which (29.1.7) follows.

29.3 Triangle+6×Triangle

We have

$$\begin{aligned}
&\sum_{-\infty}^{\infty}\left(\frac{q^{7n}}{1-q^{24n+3}}+\frac{q^{5n+1}}{1-q^{24n+5}}\right)\\
&=\left(\begin{matrix}q^{10},q^{14},q^{24},q^{24}\\ q^{7},q^{17},q^{3},q^{21}\end{matrix};q^{24}\right)_\infty + q\left(\begin{matrix}q^{14},q^{10},q^{24},q^{24}\\ q^{9},q^{15},q^{5},q^{19}\end{matrix};q^{24}\right)_\infty\\
&=\left(\begin{matrix}q^{10},q^{14}\\ q^{3},q^{5},q^{7},q^{9},q^{15},q^{17},q^{19},q^{21}\end{matrix};q^{24}\right)_\infty\\
&\quad\times((q^5,q^9,q^{15},q^{19},q^{24},q^{24};q^{24})_\infty + q(q^3,q^7,q^{17},q^{21},q^{24},q^{24};q^{24})_\infty)\\
&=\left(\begin{matrix}q^{10},q^{14}\\ q^{3},q^{5},q^{7},q^{9},q^{15},q^{17},q^{19},q^{21}\end{matrix};q^{24}\right)_\infty f(q)
\end{aligned}\tag{29.3.1}$$

where

$$f(q) = \sum_{r,s=-\infty}^{\infty} (-1)^{r+s} q^{12r^2-7r+12s^2-3s} + q\sum_{r,s=-\infty}^{\infty} (-1)^{r+s} q^{12r^2+5r+12s^2+9s},$$

$$\begin{aligned}
f(-q) &= \sum_{r,s=-\infty}^{\infty} q^{12r^2-7r+12s^2-3s} - q\sum_{r,s=-\infty}^{\infty} q^{12r^2+5r+12s^2+9s}\\
&= \sum_{r,s=-\infty}^{\infty} q^{6(r+s)^2-5(r+s)+6(r-s)^2-2(r-s)} - \sum_{r,s=-\infty}^{\infty} q^{6(r+s+1)^2-5(r+s+1)+6(r-s)^2-2(r-s)}\\
&= \sum_{m\equiv n \pmod 2} q^{6m^2-5m+6n^2-2n} - \sum_{m\not\equiv n \pmod 2} q^{6m^2-5m+6n^2-2n}\\
&= \sum_{m,n=-\infty}^{\infty} (-1)^{m+n} q^{6m^2-5m+6n^2-2n}\\
&= (q,q^4,q^8,q^{11},q^{12},q^{12};q^{12})_\infty
\end{aligned}$$

and

$$\begin{aligned} f(q) &= (-q,q^4,q^8,-q^{11},q^{12},q^{12};q^{12})_\infty \\ &= \frac{(q^2;q^{24})_\infty}{(q;q^{12})_\infty}\frac{(q^{22};q^{24})_\infty}{(q^{11};q^{12})_\infty}(q^4;q^4)_\infty(q^{12};q^{12})_\infty \\ &= \left(\begin{matrix} q^2,q^4,q^8,q^{12},q^{12},q^{16},q^{20},q^{22},q^{24},q^{24} \\ q,q^{11},q^{13},q^{23} \end{matrix};q^{24}\right)_\infty. \end{aligned} \tag{29.3.2}$$

If we substitute (29.3.2) into (29.3.1), we find

$$\begin{aligned} &\sum_{-\infty}^{\infty}\left(\frac{q^{7n}}{1-q^{24n+3}}+\frac{q^{5n+1}}{1-q^{24n+5}}\right) \\ &= \left(\begin{matrix} q^{10},q^{14} \\ q^3,q^5,q^7,q^9,q^{15},q^{17},q^{19},q^{21} \end{matrix};q^{24}\right)_\infty \\ &\quad\times\left(\begin{matrix} q^2,q^4,q^8,q^{12},q^{12},q^{16},q^{20},q^{22},q^{24},q^{24} \\ q,q^{11},q^{13},q^{23} \end{matrix};q^{24}\right)_\infty \\ &= \left(\begin{matrix} q^2,q^4,q^8,q^{10},q^{12},q^{12},q^{14},q^{16},q^{20},q^{22},q^{24},q^{24} \\ q,q^3,q^5,q^7,q^9,q^{11},q^{13},q^{15},q^{17},q^{19},q^{21},q^{23} \end{matrix};q^{24}\right)_\infty \\ &= \left(\begin{matrix} q^2,q^4,q^8,q^{10},q^{12},q^{12} \\ q,q^3,q^5,q^7,q^9,q^{11} \end{matrix};q^{12}\right)_\infty \\ &= \frac{(q^2;q^2)_\infty}{(q;q^2)_\infty}\frac{(q^{12};q^{12})_\infty}{(q^6;q^{12})_\infty} \\ &= \frac{(q^2;q^2)^2_\infty}{(q;q)_\infty}\frac{(q^{12};q^{12})^2_\infty}{(q^6;q^6)_\infty} \\ &= \psi(q)\psi(q^6), \end{aligned}$$

which is (29.1.4).

We have

$$\begin{aligned} &\sum_{n\geq 0} r\{\triangle+6\triangle\}(n)q^n = \psi(q)\psi(q^6) \\ &= \sum_{n\geq 0}\left(\frac{q^{7n}}{1-q^{24n+3}}+\frac{q^{5n+1}}{1-q^{24n+9}}-\frac{q^{19n+11}}{1-q^{24n+15}}-\frac{q^{17n+14}}{1-q^{24n+21}}\right) \\ &= \sum_{k,l\geq 0} q^{24kl+7k+3l}+\sum_{k,l\geq 0} q^{24kl+5k+9l+1}-\sum_{k,l\geq 0} q^{24kl+19k+15l+11} \\ &\qquad -\sum_{k,l\geq 0} q^{24kl+14k+21l+14}, \end{aligned}$$

$$
\begin{aligned}
&\sum_{n\geq 0} r\{\triangle+6\triangle\}(n)q^{8n+7}\\
&=\sum_{k,l\geq 0} q^{192kl+56k+24l+7}+\sum_{k,l\geq 0} q^{192kl+40k+72l+15}\\
&\quad-\sum_{k,l\geq 0} q^{192kl+152k+120l+95}-\sum_{k,l\geq 0} q^{192kl+136k+168l+119}\\
&=\sum_{k,l\geq 0} q^{(24l+7)(8k+1)}+\sum_{k,l\geq 0} q^{(24l+5)(8k+3)}\\
&\quad-\sum_{k,l\geq 0} q^{(24l+19)(8k+5)}-\sum_{k,l\geq 0} q^{(24l+17)(8k+7)}\\
&=\sum_{n\geq 0} e\{\{5,7\},24\}(8n+7)q^{8n+7},
\end{aligned}
$$

from which (29.1.8) follows.

29.4 Pentagon+5×Pentagon

In this section, we will provide only an outline of the proof of (29.1.6). The details, which are analogous to the calculations in the previous sections, are left as exercises.

We have

$$
\begin{aligned}
&\sum_{-\infty}^{\infty}\left(\frac{q^{15n}}{1-q^{60n+1}}+\frac{q^{45n+5}}{1-q^{60n+7}}+\frac{q^{45n+17}}{1-q^{60n+23}}+\frac{q^{15n+7}}{1-q^{60n+29}}\right.\\
&\qquad\left.+\frac{q^{15n+10}}{1-q^{60n+41}}+\frac{q^{45n+32}}{1-q^{60n+43}}+\frac{q^{45n+35}}{1-q^{60n+47}}+\frac{q^{15n+12}}{1-q^{60n+49}}\right)\\
&=\left(\begin{matrix}q^{16},q^{44},q^{60},q^{60}\\ q,q^{15},q^{45},q^{59}\end{matrix};q^{60}\right)_\infty-q^3\left(\begin{matrix}q^{28},q^{32},q^{60},q^{60}\\ q^{13},q^{15},q^{45},q^{47}\end{matrix};q^{60}\right)_\infty\\
&-q^4\left(\begin{matrix}q^{28},q^{32},q^{60},q^{60}\\ q^{15},q^{17},q^{43},q^{45}\end{matrix};q^{60}\right)_\infty+q^7\left(\begin{matrix}q^{16},q^{44},q^{60},q^{60}\\ q^{15},q^{29},q^{31},q^{45}\end{matrix};q^{60}\right)_\infty\\
&-q^9\left(\begin{matrix}q^{8},q^{52},q^{60},q^{60}\\ q^{15},q^{23},q^{37},q^{45}\end{matrix};q^{60}\right)_\infty+q^{10}\left(\begin{matrix}q^{4},q^{56},q^{60},q^{60}\\ q^{15},q^{19},q^{41},q^{45}\end{matrix};q^{60}\right)_\infty\\
&+q^5\left(\begin{matrix}q^{8},q^{52},q^{60},q^{60}\\ q^{7},q^{15},q^{45},q^{53}\end{matrix};q^{60}\right)_\infty-q^8\left(\begin{matrix}q^{4},q^{56},q^{60},q^{60}\\ q^{11},q^{15},q^{45},q^{49}\end{matrix};q^{60}\right)_\infty\\
&=\left(\begin{matrix}q^{16},q^{44},q^{60}\\ q,q^{15},q^{29},q^{31},q^{45},q^{59}\end{matrix};q^{60}\right)_\infty((q^{29},q^{31},q^{60};q^{60})_\infty+q^7(q,q^{59},q^{60};q^{60})_\infty)
\end{aligned}
$$

$$
\begin{aligned}
&-q^3\left(\begin{matrix} q^{28},q^{32},q^{60} \\ q^{13},q^{15},q^{17},q^{43},q^{45},q^{47} \end{matrix};q^{60}\right)_\infty ((q^{17},q^{43},q^{60};q^{60})_\infty + q(q^{13},q^{47},q^{60};q^{60})_\infty) \\
&+q^5\left(\begin{matrix} q^{8},q^{52},q^{60} \\ q^{7},q^{15},q^{23},q^{37},q^{45},q^{53} \end{matrix};q^{60}\right)_\infty ((q^{23},q^{37},q^{60};q^{60})_\infty - q^4(q^{7},q^{53},q^{60};q^{60})_\infty) \\
&-q^8\left(\begin{matrix} q^{4},q^{56},q^{60} \\ q^{11}.q^{15},q^{19},q^{41},q^{45},q^{49} \end{matrix};q^{60}\right)_\infty ((q^{19},q^{41},q^{60};q^{60})_\infty - q^2(q^{11},q^{49},q^{60};q^{60})_\infty)
\end{aligned}
$$

$$
\begin{aligned}
&=\left(\begin{matrix} q^{16},q^{44},q^{60} \\ q,q^{15},q^{29},q^{31},q^{45},q^{59} \end{matrix};q^{60}\right)_\infty \left(\begin{matrix} q^{8},q^{14},q^{22},q^{30},q^{30},q^{38},q^{46},q^{52},q^{60} \\ q^{7},q^{15},q^{23},q^{37},q^{45},q^{53} \end{matrix};q^{60}\right)_\infty \\
&-q^3\left(\begin{matrix} q^{28},q^{32},q^{60} \\ q^{13},q^{15},q^{17},q^{43},q^{45},q^{47} \end{matrix};q^{60}\right)_\infty \left(\begin{matrix} q^{2},q^{14},q^{16},q^{30},q^{30},q^{44},q^{46},q^{58},q^{60} \\ q,q^{15},q^{29},q^{31},q^{45},q^{59} \end{matrix};q^{60}\right)_\infty \\
&+q^5\left(\begin{matrix} q^{8},q^{52},q^{60} \\ q^{7},q^{15},q^{23},q^{37},q^{45},q^{53} \end{matrix};q^{60}\right)_\infty \left(\begin{matrix} q^{4},q^{22},q^{26},q^{30},q^{30},q^{34},q^{38},q^{56},q^{60} \\ q^{11},q^{15},q^{19},q^{41},q^{45},q^{49} \end{matrix};q^{60}\right)_\infty \\
&-q^8\left(\begin{matrix} q^{4},q^{56},q^{60} \\ q^{11}.q^{15},q^{19},q^{41},q^{45},q^{49} \end{matrix};q^{60}\right)_\infty \left(\begin{matrix} q^{2},q^{26},q^{28},q^{30},q^{30},q^{32},q^{34},q^{58},q^{60} \\ q^{13},q^{15},q^{17},q^{43},q^{45},q^{47} \end{matrix};q^{60}\right)_\infty
\end{aligned}
$$

$$
\begin{aligned}
&=\left(\begin{matrix} q^{8},q^{14},q^{16},q^{22},q^{30},q^{30} \\ q,\ q^{7},q^{15},q^{15},q^{23},q^{29} \end{matrix};q^{30}\right)_\infty - q^3\left(\begin{matrix} q^{2},q^{14},q^{16},q^{28},q^{30},q^{30} \\ q,q^{13},q^{15},q^{15},q^{17},q^{29} \end{matrix};q^{30}\right)_\infty \\
&+q^5\left(\begin{matrix} q^{4},\ q^{8},q^{22},q^{26},q^{30},q^{30} \\ q^{7},q^{11},q^{15},q^{15},q^{19},q^{23} \end{matrix};q^{30}\right)_\infty - q^8\left(\begin{matrix} q^{2},\ q^{4},q^{26},q^{28},q^{30},q^{30} \\ q^{11},q^{13},q^{15},q^{15},q^{17},q^{19} \end{matrix};q^{30}\right)_\infty
\end{aligned}
$$

$$
\begin{aligned}
&=\frac{1}{(q,q^{7},q^{11},q^{13},q^{15},q^{15},q^{17},q^{19},q^{23},q^{29},q^{30},q^{30};q^{30})_\infty} \\
&\times((q^{8},q^{13},q^{17},q^{22},q^{30},q^{30};q^{30})_\infty - q^3(q^{2},q^{7},q^{23},q^{28},q^{30},q^{30};q^{30})_\infty) \\
&\times((q^{11},q^{14},q^{16},q^{19},q^{30},q^{30};q^{30})_\infty + q^5(q,q^{4},q^{26},q^{29},q^{30},q^{30};q^{30})_\infty)
\end{aligned}
$$

$$
\begin{aligned}
&=\frac{1}{(q,q^{7},q^{11},q^{13},q^{15},q^{15},q^{17},q^{19},q^{23},q^{29},q^{30},q^{30};q^{30})_\infty} \\
&\times\left(\begin{matrix} q^{3},q^{12},q^{15},q^{15},q^{18},q^{27},q^{30},q^{30} \\ q^{5},q^{25} \end{matrix};q^{30}\right)_\infty \left(\begin{matrix} q^{6},q^{9},q^{15},q^{15},q^{21},q^{24},q^{30},q^{30} \\ q^{5},q^{25} \end{matrix};q^{30}\right)_\infty
\end{aligned}
$$

$$
=\left(\begin{matrix} q^{3},q^{6},q^{9},q^{12},q^{15},q^{15},q^{18},q^{21},q^{24},q^{27},q^{30},q^{30} \\ q,q^{5},q^{5},\ q^{7},q^{11},q^{13},q^{17},q^{19},q^{23},q^{25},q^{25},q^{29} \end{matrix};q^{30}\right)_\infty
$$

$$
=(q^3;q^3)_\infty(q^{15};q^{15})_\infty \Big/ \frac{(q,q^2)_\infty(q^5;q^{10})_\infty}{(q^3;q^6)_\infty(q^{15};q^{30})_\infty}
$$

$$= \frac{(q^2;q^2)_\infty (q^3;q^3)_\infty^2 (q^{10};q^{10})_\infty (q^{15};q^{15})_\infty^2}{(q;q)_\infty (q^5;q^5)_\infty (q^6;q^6)_\infty (q^{30};q^{30})_\infty}$$

$$= \frac{(q^2;q^2)_\infty (q^3;q^3)_\infty^2}{(q;q)_\infty (q^6;q^6)_\infty} \frac{(q^{10};q^{10})_\infty (q^{15};q^{15})_\infty^2}{(q^5;q^5)_\infty (q^{30};q^{30})_\infty}$$

$$= \Pi(q)\Pi(q^5),$$

which is (29.1.6).

The proof of (29.1.9) is left as an (easy) exercise.

References

106. R.S. Melham, Analogues of Jacobi's two-square theorem, preprint, 16 August 2007, http://maths.science.uts.edu.au/maths/wiki/RayMelham
107. R.S. Melham, Analogues of Jacobi's two-square theorem: an informal account. Integers **10**, 83–100 (2010)
119. P.C. Toh, On representations by figurate numbers: a uniform approach to the conjectures of Melham. Int. J. Number Theory **9**, 1055–1071 (2013)

Chapter 30
Partitions into Four Squares

30.1 Introduction

In Chapter 2, we proved Jacobi's four-squares theorem, for $n \geq 1$,

$$r\{\square+\square+\square+\square\}(n) = 8 \sum_{\substack{d|n \\ d\equiv 1,2 \text{ or } 3 \pmod 4}} d,$$

which shows that every positive integer is the sum of four squares.

In this chapter, we determine the generating function for the number of **partitions** of a number into four squares of non-negative integers, and for certain other, related partition functions.

James Sellers expanded these generating functions, and was led to conjecture certain identities and congruences which we subsequently proved, some of them arithmetically/combinatorially, and later all by q-series [94]. These proofs are presented here.

30.2 The Generating Functions

Let $p_{4\square}(n)$ denote the number of partitions of n into four squares of non-negative integers,
$p^{+}_{4\square}(n)$ the number of partitions of n into four squares of positive integers,
$p^{d}_{4\square}(n)$ the number of partitions of n into four squares of distinct non-negative integers
and $p^{d+}_{4\square}(n)$ the number of partitions of n into four squares of distinct positive integers.

The original version of this chapter has been revised. The erratum to this chapter is available at https://doi.org/10.1007/978-3-319-57762-3_44.

M.D. Hirschhorn, *The Power of q*, Developments in Mathematics 49,
DOI 10.1007/978-3-319-57762-3_30

We start by showing that

$$\sum_{n\geq 0} p_{4\square}(n)q^n = \frac{1}{384}\left(\phi(q)^4+4\phi(q)^3+12\phi(q)^2\phi(q^2)+18\phi(q)^2+24\phi(q)\phi(q^2)\right.$$
$$\left.+32\phi(q)\phi(q^3)+12\phi(q^2)^2+60\phi(q)+36\phi(q^2)+32\phi(q^3)+48\phi(q^4)+105\right), \tag{30.2.1}$$

$$\sum_{n\geq 1} p^{+}_{4\square}(n)q^n = \frac{1}{384}\left(\phi(q)^4-4\phi(q)^3+12\phi(q)^2\phi(q^2)-6\phi(q)^2-24\phi(q)\phi(q^2)\right.$$
$$\left.+32\phi(q)\phi(q^3)+12\phi(q^2)^2-12\phi(q)-12\phi(q^2)-32\phi(q^3)+48\phi(q^4)-15\right), \tag{30.2.2}$$

$$\sum_{n\geq 1} p^{d}_{4\square}(n)q^n = \frac{1}{384}\left(\phi(q)^4+4\phi(q)^3-12\phi(q)^2\phi(q^2)-6\phi(q)^2-24\phi(q)\phi(q^2)\right.$$
$$\left.+32\phi(q)\phi(q^3)+12\phi(q^2)^2+12\phi(q)+12\phi(q^2)+32\phi(q^3)-48\phi(q^4)-15\right), \tag{30.2.3}$$

and

$$\sum_{n\geq 1} p^{d+}_{4\square}(n)q^n = \frac{1}{384}\left(\phi(q)^4-4\phi(q)^3-12\phi(q)^2\phi(q^2)+18\phi(q)^2+24\phi(q)\phi(q^2)\right.$$
$$\left.+32\phi(q)\phi(q^3)+12\phi(q^2)^2-60\phi(q)-36\phi(q^2)-32\phi(q^3)-48\phi(q^4)+105\right). \tag{30.2.4}$$

For $n>0$, let $p\{\square_a+\square_b+\square_c+\square_d\}(n)$ be the number of partitions of n, $n=a^2+b^2+c^2+d^2$ with $a>b>c>d>0$,
let $p\{\square_a+\square_a+\square_b+\square_c\}(n)$ be the number of partitions of n, $n=a^2+a^2+b^2+c^2$ with $a\neq b$, $a\neq c$, $a>0$ and $b>c>0$,
let $p\{\square_a+\square_a+\square_b+\square_b\}(n)$ be the number of partitions of n, $n=a^2+a^2+b^2+b^2$ with $a>b>0$,
let $p\{\square_a+\square_a+\square_a+\square_b\}(n)$ be the number of partitions of n, $n=a^2+a^2+a^2+b^2$ with $a\neq b$, $a>0$ and $b>0$,
and so on.

Then we have the following 11 identities:

$$\left(\frac{1}{2}(\phi(q)-1)\right)^4 = \left(\sum_{n\geq 1} q^{n^2}\right)^4$$
$$= 24\sum_{n\geq 1} p\{\square_a+\square_b+\square_c+\square_d\}(n)q^n + 12\sum_{n\geq 1} p\{\square_a+\square_a+\square_b+\square_c\}(n)q^n$$

$$+6\sum_{n\geq 1} p\{\square_a+\square_a+\square_b+\square_b\}(n)q^n+4\sum_{n\geq 1} p\{\square_a+\square_a+\square_a+\square_b\}(n)q^n$$
$$+\sum_{n\geq 1} p\{\square_a+\square_a+\square_a+\square_a\}(n)q^n, \tag{30.2.5}$$

$$\left(\frac{1}{2}(\phi(q)-1)\right)^2\cdot\frac{1}{2}(\phi(q^2)-1)=\left(\sum_{n\geq 1} q^{n^2}\right)^2\left(\sum_{n\geq 1} q^{2n^2}\right)$$
$$=2\sum_{n\geq 1} p\{\square_a+\square_a+\square_b+\square_c\}(n)q^n+2\sum_{n\geq 1} p\{\square_a+\square_a+\square_b+\square_b\}(n)q^n$$
$$+2\sum_{n\geq 1} p\{\square_a+\square_a+\square_a+\square_b\}(n)q^n+\sum_{n\geq 1} p\{\square_a+\square_a+\square_a+\square_a\}(n)q^n, \tag{30.2.6}$$

$$\left(\frac{1}{2}(\phi(q^2)-1)\right)^2=\left(\sum_{n\geq 1} q^{2n^2}\right)^2$$
$$=2\sum_{n\geq 1} p\{\square_a+\square_a+\square_b+\square_b\}(n)q^n+\sum_{n\geq 1} p\{\square_a+\square_a+\square_a+\square_a\}(n)q^n, \tag{30.2.7}$$

$$\frac{1}{2}(\phi(q)-1)\cdot\frac{1}{2}(\phi(q^3)-1)=\left(\sum_{n\geq 1} q^{n^2}\right)\left(\sum_{n\geq 1} q^{3n^2}\right)$$
$$=\sum_{n\geq 1} p\{\square_a+\square_a+\square_a+\square_b\}(n)q^n+\sum_{n\geq 1} p\{\square_a+\square_a+\square_a+\square_a\}(n)q^n, \tag{30.2.8}$$

$$\frac{1}{2}(\phi(q^4)-1)=\sum_{n\geq 1} q^{4n^2}=\sum_{n\geq 1} p\{\square_a+\square_a+\square_a+\square_a\}(n)q^n, \tag{30.2.9}$$

$$\left(\frac{1}{2}(\phi(q)-1)\right)^3=\left(\sum_{n\geq 1} q^{n^2}\right)^3$$
$$=6\sum_{n\geq 1} p\{\square_a+\square_b+\square_c\}(n)q^n+3\sum_{n\geq 1} p\{\square_a+\square_a+\square_b\}(n)q^n$$
$$+\sum_{n\geq 1} p\{\square_a+\square_a+\square_a\}(n)q^n, \tag{30.2.10}$$

$$\frac{1}{2}(\phi(q)-1)\cdot\frac{1}{2}(\phi(q^2)-1)=\left(\sum_{n\geq 1} q^{n^2}\right)\left(\sum_{n\geq 1} q^{2n^2}\right)$$
$$=\sum_{n\geq 1} p\{\square_a+\square_a+\square_b\}(n)q^n+\sum_{n\geq 1} p\{\square_a+\square_a+\square_a\}(n)q^n, \tag{30.2.11}$$

$$\frac{1}{2}(\phi(q^3)-1)=\sum_{n\geq 1} q^{3n^2}=\sum_{n\geq 1} p\{\square_a+\square_a+\square_a\}(n)q^n, \tag{30.2.12}$$

$$\left(\frac{1}{2}(\phi(q)-1)\right)^2=\left(\sum_{n\geq 1} q^{n^2}\right)^2$$

$$= 2\sum_{n\geq 1} p\{\square_a+\square_b\}(n)q^n + \sum_{n\geq 1} p\{\square_a+\square_a\}(n)q^n, \tag{30.2.13}$$

$$\frac{1}{2}\left(\phi(q^2)-1\right) = \sum_{n\geq 1} q^{2n^2} = \sum_{n\geq 1} p\{\square_a+\square_a\}(n)q^n \tag{30.2.14}$$

and

$$\frac{1}{2}\left(\phi(q)-1\right) = \sum_{n\geq 1} q^{n^2} = \sum_{n\geq 1} p\{\square_a\}(n)q^n. \tag{30.2.15}$$

From these we find

$$\sum_{n\geq 1} p\{\square_a\}(n)q^n = \frac{1}{2}\left(\phi(q)-1\right), \tag{30.2.16}$$

$$\sum_{n\geq 1} p\{\square_a+\square_a\}(n)q^n = \frac{1}{2}\left(\phi(q^2)-1\right), \tag{30.2.17}$$

$$\sum_{n\geq 1} p\{\square_a+\square_b\}(n)q^n = \frac{1}{8}\left(\phi(q)^2-2\phi(q)-2\phi(q^2)+3\right), \tag{30.2.18}$$

$$\sum_{n\geq 1} p\{\square_a+\square_a+\square_a\}(n)q^n = \frac{1}{2}\left(\phi(q^3)-1\right), \tag{30.2.19}$$

$$\sum_{n\geq 1} p\{\square_a+\square_a+\square_b\}(n)q^n = \frac{1}{4}\left(\phi(q)\phi(q^2)-\phi(q)-\phi(q^2)-2\phi(q^3)+3\right), \tag{30.2.20}$$

$$\sum_{n\geq 1} p\{\square_a+\square_b+\square_c\}(n)q^n$$
$$= \frac{1}{48}\left(\phi(q)^3-3\phi(q)^2-6\phi(q)\phi(q^2)+9\phi(q)+6\phi(q^2)+8\phi(q^3)-15\right), \tag{30.2.21}$$

$$\sum_{n\geq 1} p\{\square_a+\square_a+\square_a+\square_a\})n)q^n = \frac{1}{2}\left(\phi(q^4)-1\right), \tag{30.2.22}$$

$$\sum_{n\geq 1} p\{\square_a+\square_a+\square_a+\square_b\}(n)q^n = \frac{1}{4}\left(\phi(q)\phi(q^3)-\phi(q)-\phi(q^3)-2\phi(q^4)+3\right), \tag{30.2.23}$$

$$\sum_{n\geq 1} p\{\square_a+\square_a+\square_b+\square_b\}(n)q^n = \frac{1}{8}\left(\phi(q^2)^2-2\phi(q^2)-2\phi(q^4)+3\right), \tag{30.2.24}$$

$$\sum_{n\geq 1} p\{\square_a+\square_a+\square_b+\square_c\}(n)q^n$$
$$= \frac{1}{16}\left(\phi(q)^2\phi(q^2)-\phi(q)^2-2\phi(q)\phi(q^2)-4\phi(q)\phi(q^3)\right.$$

$$-2\phi(q^2)^2+6\phi(q)+5\phi(q^2)+4\phi(q^3)+8\phi(q^4)-15\Big) \qquad (30.2.25)$$

and

$$\sum_{n\geq 1} p\{\Box_a+\Box_b+\Box_c+\Box_d\}(n)q^n$$
$$=\frac{1}{384}\Big(\phi(q)^4-4\phi(q)^3-12\phi(q)^2\phi(q^2)+8\phi(q)^2+24\phi(q)\phi(q^2)+32\phi(q)\phi(q^3)$$
$$+12\phi(q^2)^2-60\phi(q)-36\phi(q^2)-32\phi(q^3)-48\phi(q^4)+105\Big). \qquad (30.2.26)$$

If we add all of (30.2.16)–(30.2.26) and add 1, we obtain

$$\sum_{n\geq 0} p_{4\Box}(n)q^n=\frac{1}{384}\Big(\phi(q)^4+4\phi(q)^3+12\phi(q)^2\phi(q^2)+18\phi(q)^2+24\phi(q)\phi(q^2)$$
$$+32\phi(q)\phi(q^3)+12\phi(q^2)^2+60\phi(q)+36\phi(q^2)+32\phi(q^3)+48\phi(q^4)+105\Big),$$

which is (30.2.1).

If we add (30.2.22)–(30.2.26), we obtain (30.2.2), if we add (30.2.21) and (30.2.26), we obtain (30.2.3), while (30.2.26) is (30.2.4).

As an interesting aside, note that if $\alpha\geq 1$, $r\{\Box+\Box+\Box+\Box\}(2^\alpha)=24$, and we can deduce that $p_{4\Box}(2^\alpha)=1$ if α is odd, and $=2$ if α is even.

For $\beta\geq 0$,

$$\begin{aligned}2^{2\beta+1}&=(2^\beta)^2+(2^\beta)^2+0^2+0^2 \ \text{(24 representations)},\\ 2^{2\beta+2}&=(2^\beta)^2+(2^\beta)^2+(2^\beta)^2+(2^\beta)^2 \ \text{(16 representations)},\\ &=(2^{\beta+1})^2+0^2+0^2+0^2 \ \text{(8 representations)}.\end{aligned}$$

This should be contrasted with the fact that the average value of $p_{4\Box}(n)$ for $n\leq N$ is

$$\frac{1}{N}\sum_{n=1}^{N} p_{4\Box}(n)\sim\frac{\pi^2}{768}N \text{ as } N\to\infty.$$

30.3 Some Relations

In this section, we employ (30.2.1)–(30.2.4) to prove the following relations. The first four have obvious arithmetic proofs.

$$p_{4\Box}(8n)=p_{4\Box}(2n), \qquad (30.3.1)$$

$$p_{4\square}^{+}(8n) = p_{4\square}^{+}(2n), \tag{30.3.2}$$
$$p_{4\square}^{d}(8n) = p_{4\square}^{d}(2n), \tag{30.3.3}$$
$$p_{4\square}^{d+}(8n) = p_{4\square}^{d+}(2n), \tag{30.3.4}$$

$$p_{4\square}(8n+4) = 2p_{4\square}(2n+1) + p_{4\square}^{+}(2n+1), \tag{30.3.5}$$
$$p_{4\square}^{+}(8n+4) = p_{4\square}(2n+1) + 2p_{4\square}^{+}(2n+1), \tag{30.3.6}$$
$$p_{4\square}^{d}(8n+4) = 2p_{4\square}^{d}(2n+1) + p_{4\square}^{d+}(2n+1), \tag{30.3.7}$$
$$p_{4\square}^{d+}(8n+4) = p_{4\square}^{d}(2n+1) + 2p_{4\square}^{d+}(2n+1). \tag{30.3.8}$$

In order to prove (30.3.1), we 4-dissect (30.2.1) by use of (1.9.4),

$$\phi(q) = \phi(q^4) + 2q\psi(q^8).$$

We have

$$\begin{aligned}\sum_{n\geq 1} p_{4\square}(n)q^n = \frac{1}{384}&((\phi(q^4)+2q\psi(q^8))^4 + 4(\phi(q^4)+2q\psi(q^8))^3\\ &+12(\phi(q^4)+2q\psi(q^8))^2(\phi(q^8)+2q^2\psi(q^{16}))\\ &+18(\phi(q^4)+2q\psi(q^8))^2\\ &+24(\phi(q^4)+2q\psi(q^8))(\phi(q^8)+2q^2\psi(q^{16}))\\ &+32(\phi(q^4)+2q\psi(q^8))(\phi(q^{12})+2q^3\psi(q^{24}))\\ &+12(\phi(q^8)+2q^2\psi(q^{16}))^2+60(\phi(q^4)+2q\psi(q^8))\\ &+36(\phi(q^8)+2q^2\psi(q^{16}))+32(\phi(q^{12})+2q^3\psi(q^{24}))\\ &+48(\phi(q^{16})+2q^4\psi(q^{32}))+105).\end{aligned} \tag{30.3.9}$$

From (30.3.9) we extract

$$\begin{aligned}\sum_{n\geq 1} p_{4\square}(4n)q^n = \frac{1}{384}&\left(\phi(q)^4+16q\psi(q^2)^4+4\phi(q)^3+12\phi(q)^2\phi(q^2)\right.\\ &+96q\psi(q^2)^2\psi(q^4)+18\phi(q)^2+24\phi(q)\phi(q^2)+32\phi(q)\phi(q^3)\\ &+128q\psi(q^2)\psi(q^6)+12\phi(q^2)^2+48q\psi(q^4)^2\\ &\left.+60\phi(q)+36\phi(q^2)+32\phi(q^3)+48\phi(q^4)+96q\psi(q^8)+105\right).\end{aligned} \tag{30.3.10}$$

If we subtract (30.2.1) from (30.3.10), we obtain

$$\sum_{n\geq 1}(p_{4\square}(4n)-p_{4\square}(n))\,q^n=\frac{1}{24}q(\psi(q^2)^4+6\psi(q^2)^2\psi(q^4)+8\psi(q^2)\psi(q^6)+3\psi(q^4)^2+6\psi(q^8)). \tag{30.3.11}$$

The right side of (30.3.11) is odd, and we deduce (30.3.1).
The proofs of (30.3.2)–(30.3.4) are similar, so we omit them.

It follows from (30.3.11) that

$$\sum_{n\geq 1}(p_{4\square}(8n+4)-p_{4\square}(2n+1))q^n=\frac{1}{24}(\psi(q)^4+6\psi(q)^2\psi(q^2)+8\psi(q)\psi(q^3)+3\psi(q^2)^2+6\psi(q^4)). \tag{30.3.12}$$

On the other hand, if we add (30.2.1) and (30.2.2), we find

$$\sum_{n\geq 1}(p_{4\square}(n)+p_{4\square}^{+}(n))q^n=\frac{1}{192}(\phi(q)^4+12\phi(q)^2\phi(q^2)+6\phi(q)^2+32\phi(q)\phi(q^2)+12\phi(q^2)^2+24\phi(q)+12\phi(q^2)+48\phi(q^4)+45). \tag{30.3.13}$$

We now 2-dissect (30.3.13) by use of (1.9.4) and (1.10.1).

We have

$$\begin{aligned}\sum_{n\geq 1}(p_{4\square}(n)+p_{4\square}^{+}(n))q^n=\frac{1}{192}&((\phi(q^2)^2+4q\psi(q^4)^2)^2\\&+12(\phi(q^2)^2+4q\psi(q^4)^2)\phi(q^2)+6(\phi(q^2)^2+4q\psi(q^4)^2)\\&+32(\phi(q^4)+2q\psi(q^8))(\phi(q^{12}+2q\psi(q^{24}))+12\phi(q^2)^2\\&+24(\phi(q^4)+2q\psi(q^8))+12\phi(q^2)+48\phi(q^4)+45).\end{aligned} \tag{30.3.14}$$

From (30.3.14) we extract

$$\sum_{n\geq 1}(p_{4\square}(2n+1)+p_{4\square}^{+}(2n+1))q^n=\frac{1}{24}(\phi(q)^2\psi(q^2)^2+6\phi(q)\psi(q^2)^2+3\psi(q^2)^2+8\phi(q^6)\psi(q^4)+8q\phi(q^2)\psi(q^{12})+6\psi(q^4)). \tag{30.3.15}$$

If we now make use of (1.5.13),

$$\phi(q)\psi(q^2) = \psi(q)^2$$

and (25.2.1),

$$\phi(q^6)\psi(q^4) + q\phi(q^2)\psi(q^{12}) = \psi(q)\psi(q^3),$$

then (30.3.15) becomes

$$\sum_{n\geq 1} \left(p_{4\square}(2n+1) + p_{4\square}^{+}(2n+1)\right) q^n = \frac{1}{24}\left(\psi(q)^4 + 6\psi(q)^2\psi(q^2) + 3\psi(q^2)^2 \right.$$
$$\left. +8\psi(q)\psi(q^3) + 6\psi(q^4)\right). \qquad (30.3.16)$$

The right sides of (30.3.12) and (30.3.16) are identical, and we deduce (30.3.5). The proofs of (30.3.6)–(30.3.8) are similar, so we omit them.

30.4 An Important Corollary

We obtain the following theorem.

A number of the form $4^{\alpha}(8m+7)$ *is not the sum of fewer than four squares.* (30.4.1)

First we observe that if $8m+7$ is the sum of four squares, three are 1 modulo 8 and the other is 4 modulo 8, so all are positive.
So

$$p_{4\square}(8m+7) = p_{4\square}^{+}(8m+7). \qquad (30.4.2)$$

Next,

$$\begin{aligned}
p_{4\square}(4(8m+7)) &= p_{4\square}(8(4m+3)+4) \\
&= 2p_{4\square}(2(4m+3)+1) + p_{4\square}^{+}(2(4m+3)+1) \text{ by (30.3.5)} \\
&= 2p_{4\square}(8m+7) + p_{4\square}^{+}(8m+7) \\
&= p_{4\square}(8m+7) + 2p_{4\square}^{+}(8m+7) \text{ by (30.4.2)} \\
&= p_{4\square}(2(4m+3)+1) + 2p_{4\square}^{+}(2(4m+3)+1) \\
&= p_{4\square}^{+}(8(4m+3)+4) \text{ by (30.3.6)} \\
&= p_{4\square}^{+}(4(8m+7)). \qquad (30.4.3)
\end{aligned}$$

(30.4.1) now follows by induction on α for $\alpha \geq 1$, if we make use of (30.3.1) and (30.3.2).

30.5 Some Congruences

In this and the next few sections, we prove the following congruences:

$$p_{4\square}(72n+69) \equiv 0 \pmod{2} \tag{30.5.1}$$

$$p_{4\square}^{+}(72n+69) \equiv 0 \pmod{2} \tag{30.5.2}$$

$$p_{4\square}^{d}(72n+69) \equiv 0 \pmod{2} \tag{30.5.3}$$

$$p_{4\square}^{d+}(72n+69) \equiv 0 \pmod{2}. \tag{30.5.4}$$

Indeed we go further and prove the following identities, akin to Ramanujan's identity (5.1.1):

$$\sum_{n\geq 0} p_{4\square}(72n+69)q^n = 2\left(\Pi_1+\Pi_2+\Pi_4\right) \tag{30.5.5}$$

$$\sum_{n\geq 0} p_{4\square}^{+}(72n+69)q^n = 2\left(\Pi_1+\Pi_3+\Pi_4\right) \tag{30.5.6}$$

$$\sum_{n\geq 0} p_{4\square}^{d}(72n+69)q^n = 2\left(\Pi_1-\Pi_3+\Pi_4\right) \tag{30.5.7}$$

$$\sum_{n\geq 0} p_{4\square}^{d+}(72n+69)q^n = 2\left(\Pi_1-\Pi_2+\Pi_4\right). \tag{30.5.8}$$

where

$$\begin{aligned}
\Pi_1 &= \left(\begin{matrix} q^6,q^6,q^{10},q^{12},q^{12},q^{14},q^{18},q^{18},q^{24},q^{24},q^{24},q^{24} \\ q,q,q^5,q^7,q^{11},q^{11},q^{13},q^{13},q^{17},q^{19},q^{23},q^{23} \end{matrix};q^{24}\right)_\infty, \\
\Pi_2 &= \left(\begin{matrix} q^4,q^8,q^{12},q^{16},q^{20},q^{24},q^{24},q^{24} \\ q,q^2,q^7,q^{10},q^{14},q^{17},q^{22},q^{23} \end{matrix};q^{24}\right)_\infty, \\
\Pi_3 &= q^2\left(\begin{matrix} q^4,q^8,q^{12},q^{16},q^{20},q^{24},q^{24},q^{24} \\ q^2,q^5,q^{10},q^{11},q^{13},q^{14},q^{19},q^{22} \end{matrix};q^{24}\right)_\infty, \\
\Pi_4 &= q^3\left(\begin{matrix} q^2,q^6,q^6,q^{12},q^{12},q^{18},q^{18},q^{22},q^{24},q^{24},q^{24},q^{24} \\ q,q^5,q^5,q^7,q^7,q^{11},q^{13},q^{17},q^{17},q^{19},q^{19},q^{23} \end{matrix};q^{24}\right)_\infty.
\end{aligned} \tag{30.5.9}$$

We start with the following definitions:

$$\begin{aligned}
&\phi(q)^4 = \sum_{n\geq 0} a(n)q^n, \quad \phi(q)^3 = \sum_{n\geq 0} b(n)q^n, \quad \phi(q)^2\phi(q^2) = \sum_{n\geq 0} c(n)q^n,\\
&\phi(q)^2 = \sum_{n\geq 0} d(n)q^n, \quad \phi(q)\phi(q^2) = \sum_{n\geq 0} e(n)q^n, \quad \phi(q)\phi(q^3) = \sum_{n\geq 0} f(n)q^n,\\
&\phi(q^2)^2 = \sum_{n\geq 0} g(n)q^n, \quad \phi(q) = \sum_{n\geq 0} h(n)q^n, \quad \phi(q^2) = \sum_{n\geq 0} j(n)q^n,\\
&\phi(q^3) = \sum_{n\geq 0} k(n)q^n, \quad \phi(q^4) = \sum_{n\geq 0} l(n)q^n. \qquad (30.5.10)
\end{aligned}$$

We note that by Jacobi's two-squares theorem (2.2.12),

$$d(72n+69) = 0, \qquad (30.5.11)$$

by Dirichlet's square+2×square theorem (24.1.2),

$$e(72n+69) = 0 \qquad (30.5.12)$$

and by Lorenz's square+3×square theorem (23.1.3),

$$f(72n+69) = 0. \qquad (30.5.13)$$

Also, since $72n+69$ is odd and neither a square nor three times a square,

$$g(72n+69) = h(72n+69) = j(72n+69) = k(72n+69) = l(72n+69) = 0. \qquad (30.5.14)$$

It follows from (30.2.1)–(30.2.4), (30.5.10)–(30.5.14) that

$$\begin{aligned}
p_{4\square}(72n+69) &= \frac{1}{384}\left(a(72n+69)+4b(72n+69)+12c(72n+69)\right), \qquad (30.5.15)\\
p_{4\square}^{+}(72n+69) &= \frac{1}{384}\left(a(72n+69)-4b(72n+69)+12c(72n+69)\right), \qquad (30.5.16)\\
p_{4\square}^{d}(72n+69) &= \frac{1}{384}\left(a(72n+69)+4b(72n+69)-12c(72n+69)\right), \qquad (30.5.17)\\
p_{4\square}^{d+}(72n+69) &= \frac{1}{384}\left(a(72n+69)-4b(72n+69)-12c(72n+69)\right). \qquad (30.5.18)
\end{aligned}$$

We now turn to determining

$$\sum_{n\geq 0} a(72n+69)q^n, \quad \sum_{n\geq 0} b(72n+69)q^n \ \text{ and } \ \sum_{n\geq 0} c(72n+69)q^n.$$

30.6 The Generating Function for $a(72n+69)$

We show that

$$\sum_{n\geq 0} a(72n+69)q^n = 768\,(\Pi_1+\Pi_4), \tag{30.6.1}$$

where Π_1 and Π_4 are defined in (30.5.9).

We know that

$$a(n) = 8 \sum_{\substack{d|n \\ d\equiv 1,2 \text{ or } 3 \pmod 4}} d.$$

It follows that

$$a(72n+69) = 8 \sum_{d|72n+69} d = 32 \sum_{d|24n+23} d. \tag{30.6.2}$$

Now observe that if $d|24n+23$ then $d = 24k+r$ for some $k\geq 0$ and $r\in R = \{1,5,7,11,13,17,19,23\}$, and then the codivisor d' of d is $d' = 24l+r'$ for some $l\geq 0$ and $r' = 24-r$.

Conversely, if $r\in R$ and $r' = 24-r$, and $k,l\geq 0$ then $(24k+r)(24l+r')\equiv 23 \pmod{24}$.

It follows that

$$\sum_{n\geq 0} a(72n+69)q^{24n+23} = 32 \sum_{\substack{k,l\geq 0\\ r\in R}} (24k+r)q^{(24k+r)(24l+r')}$$

$$= 32 \sum_{\substack{k,l\geq 0\\ r\in\{1,5,7,11\}}} (24k+r+24l+r')q^{(24k+r)(24l+r')}$$

$$= 768 \sum_{\substack{k,l\geq 0\\ r\in\{1,5,7,11\}}} (k+l+1)q^{576kl+24r'k+24rl+rr'}$$

$$= 768 \sum_{r\in\{1,5,7,11\}} q^{rr'} \sum_{k,l\geq 0} (k+l+1)q^{576kl+24r'k+24rl}$$

$$= 768 \Bigg(q^{23} \sum_{k,l\geq 0} (k+l+1)q^{576kl+552k+24l} + q^{95} \sum_{k,l\geq 0} (k+l+1)q^{576kl+456k+120l}$$

$$+ q^{119} \sum_{k,l\geq 0} (k+l+1)q^{576kl+408k+168l} + q^{143} \sum_{k,l\geq 0} (k+l+1)q^{576kl+312k+264l} \Bigg) \tag{30.6.3}$$

and hence that

$$\sum_{n\geq 0} a(72n+69)q^n$$
$$= 768\left(\sum_{k,l\geq 0}(k+l+1)q^{24kl+23k+l} + q^3\sum_{k,l\geq 0}(k+l+1)q^{24kl+19k+5l}\right.$$
$$\left. + q^4\sum_{k,l\geq 0}(k+l+1)q^{24kl+17k+7l} + q^5\sum_{k,l\geq 0}(k+l+1)q^{kl+13k+11l}\right). \quad (30.6.4)$$

We now require the fact that for $|q| < |x| < 1$,

$$\sum_{k,l\geq 0}(k+l+1)q^{kl}x^k(q/x)^l = \left(\begin{matrix} q, q \\ x, q/x \end{matrix}; q\right)_\infty^2. \quad (30.6.5)$$

We defer the proof of this remarkable identity to §30.7.

It follows from (30.6.4) and (30.6.5) with q replaced by q^{24} that

$$\sum_{n\geq 0} a(72n+69)q^n = 768\left(\left(\begin{matrix} q^{24}, q^{24} \\ q, q^{23} \end{matrix}; q^{24}\right)_\infty^2 + q^3\left(\begin{matrix} q^{24}, q^{24} \\ q^5, q^{19} \end{matrix}; q^{24}\right)_\infty^2\right.$$
$$\left. + q^4\left(\begin{matrix} q^{24}, q^{24} \\ q^7, q^{17} \end{matrix}; q^{24}\right)_\infty^2 + q^5\left(\begin{matrix} q^{24}, q^{24} \\ q^{11}, q^{143} \end{matrix}; q^{24}\right)_\infty^2\right). \quad (30.6.6)$$

We can go further, since

$$\left(\begin{matrix} q^{24}, q^{24} \\ q, q^{23} \end{matrix}; q^{24}\right)_\infty^2 + q^5\left(\begin{matrix} q^{24}, q^{24} \\ q^{11}, q^{13} \end{matrix}; q^{24}\right)_\infty^2$$
$$= \left(\begin{matrix} q^{24} \\ q, q^{11}, q^{13}, q^{23} \end{matrix}; q^{24}\right)_\infty^2 \left((q^{11}, q^{13}, q^{24}; q^{24})_\infty^2 + q^5(q, q^{23}, q^{24}; q^{24})_\infty^2\right)$$
$$= \left(\begin{matrix} q^{24} \\ q, q^{11}, q^{13}, q^{23} \end{matrix}; q^{24}\right)_\infty^2 \left(\begin{matrix} q^6, q^6, q^{10}, q^{12}, q^{12}, q^{14}, q^{18}, q^{18}, q^{24}, q^{24} \\ q^5, q^7, q^{17}, q^{19} \end{matrix}; q^{24}\right)_\infty$$
$$= \left(\begin{matrix} q^6, q^6, q^{10}, q^{12}, q^{12}, q^{14}, q^{18}, q^{18}, q^{24}, q^{24}, q^{24}, q^{24} \\ q, q, q^5, q^7, q^{11}, q^{11}, q^{13}, q^{13}, q^{17}, q^{19}, q^{23}, q^{23} \end{matrix}; q^{24}\right)_\infty \quad (30.6.7)$$

(the penultimate step is left as an exercise) and

$$\begin{aligned}
&\left(\begin{matrix}q^{24},q^{24}\\ q^5,q^{19}\end{matrix};q^{24}\right)_\infty^2+q\left(\begin{matrix}q^{24},q^{24}\\ q^7,q^{17}\end{matrix};q^{24}\right)_\infty^2\\
&=\left(\begin{matrix}q^{24}\\ q^5,q^7,q^{17},q^{19}\end{matrix};q^{24}\right)_\infty^2\left((q^7,q^{17},q^{24};q^{24})_\infty^2+q(q^5,q^{19},q^{24};q^{24})_\infty^2\right)\\
&=\left(\begin{matrix}q^{24}\\ q^5,q^7,q^{17},q^{19}\end{matrix};q^{24}\right)_\infty^2\left(\begin{matrix}q^2,q^6,q^6,q^{12},q^{12},q^{18},q^{18},q^{22},q^{24},q^{24}\\ q,q^{11},q^{13},q^{23}\end{matrix};q^{24}\right)_\infty\\
&=\left(\begin{matrix}q^2,q^6,q^6,q^{12},q^{12},q^{18},q^{18},q^{22},q^{24},q^{24},q^{24},q^{24}\\ q,q^5,q^5,q^7,q^7,q^{11},q^{13},q^{17},q^{17},q^{19},q^{19},q^{23}\end{matrix};q^{24}\right)_\infty. \qquad (30.6.8)
\end{aligned}$$

If we combine (30.6.6), (30.6.7) and (30.6.8), we obtain (30.6.1).

30.7 A Remarkable Identity

We show that for $|q|<|x|<1$,

$$\sum_{k,l\geq 0} q^{kl}x^k(q/x)^l=\left(\begin{matrix}q,q\\ x,q/x\end{matrix};q\right)_\infty^2. \qquad (30.7.1)$$

Our proof is due to Shaun Cooper.

We start with the Jordan–Kronecker identity (28.1.1),

For $|q|<|x|<1,\ \ |q|<|a|<1/|q|$,

$$\begin{aligned}
\left(\begin{matrix}ax,q/(ax),q,q\\ x,q/x,a,q/a\end{matrix};q\right)_\infty&=\sum_{-\infty}^{\infty}\frac{x^n}{1-aq^n}\\
&=\frac{1}{1-a}+\sum_{n\geq 1}\frac{x^n}{1-aq^n}+\sum_{n\geq 1}\frac{x^{-n}}{1-aq^{-n}}. \qquad (30.7.2)
\end{aligned}$$

The left side of (30.7.2) is zero when $ax=1$. We now write the right side in such a way that it becomes obvious that it also is zero when $ax=1$.

The right side is

$$\begin{aligned}
&\frac{1}{1-a}+\sum_{n\geq 1}x^n\left(1+\frac{aq^n}{1-aq^n}\right)+\sum_{n\geq 1}\frac{x^{-n}}{1-aq^{-n}}\\
&=\frac{1}{1-a}+\frac{x}{1-x}+\sum_{k\geq 1}\frac{ax^kq^k}{1-aq^k}-\sum_{l\geq 1}\frac{a^{-1}x^{-l}q^l}{1-a^{-1}q^l}
\end{aligned}$$

$$
\begin{aligned}
&= \frac{1-ax}{(1-a)(1-x)} + \sum_{k,l\geq 1} q^{kl}a^l x^k - \sum_{k,l\geq 1} q^{kl}a^{-k}x^{-l} \\
&= \frac{1-ax}{(1-a)(1-x)} - \sum_{k,l\geq 1} q^{kl}a^{-k}x^{-l}\left(1-(ax)^{k+l}\right). \qquad (30.7.3)
\end{aligned}
$$

So we have

$$
\left(\begin{matrix} ax, q/(ax), q, q \\ x, q/x, a, q/a \end{matrix}; q\right)_\infty = \frac{1-ax}{(1-a)(1-x)} - \sum_{k,l\geq 1} q^{kl}a^{-k}x^{-l}\left(1-(ax)^{k+l}\right). \quad (30.7.4)
$$

If we multiply by $\dfrac{1-a}{1-a^{-1}} = -a$, we obtain

$$
\left(\begin{matrix} ax, q/(ax), q, q \\ x, q/x, aq, 1/a \end{matrix}; q\right)_\infty = \frac{1-ax}{(1-a^{-1})(1-x)} + a \sum_{k,l\geq 1} q^{kl}a^{-k}x^{-l}\left(1-(ax)^{k+l}\right). \quad (30.7.5)
$$

If we now divide by $1-ax$, we obtain

$$
\begin{aligned}
&\left(\begin{matrix} axq, q/(ax), q, q \\ x, q/x, aq, 1/a \end{matrix}; q\right)_\infty \\
&= \frac{1}{(1-a^{-1})(1-x)} + a \sum_{k,l\geq 1} q^{kl}a^{-k}x^{-l}\left(1+ax+\cdots+(ax)^{k+l-1}\right). \quad (30.7.6)
\end{aligned}
$$

Now let $a \to x^{-1}$, and we obtain

$$
\begin{aligned}
\left(\begin{matrix} q, q \\ x, q/x \end{matrix}; q\right)_\infty^2 &= \frac{1}{(1-x)^2} + \sum_{k,l\geq 1} (k+l)q^{kl}x^{k-l-1} \\
&= \sum_{k\geq 0} (k+1)x^k + \sum_{k\geq 0, l\geq 1} (k+l+1)q^{(k+1)l}x^{k-l} \\
&= \sum_{k,l\geq 0} (k+l+1)q^{kl}x^k (q/x)^l,
\end{aligned}
$$

which is (30.7.1).

30.8 The Generating Function for $b(72n+69)$

We show that

$$\sum_{n\geq 0} b(72n+69)q^n = 96\begin{pmatrix} q^4,q^6,q^8,q^{12},q^{12},q^{12} \\ q,q^3,q^5,q^7,q^9,q^{11} \end{pmatrix};q^{12}\Bigg)_\infty. \tag{30.8.1}$$

We have

$$\sum_{n\geq 0} b(n)q^n = \phi(q)^3, \tag{30.8.2}$$

and we want to 72-dissect this.

It is easy to check that, modulo 72,

$$n^2 \equiv 0,\ 1,\ 4,\ 9,\ 16,\ 25,\ 28,\ 36,\ 40,\ 49,\ 52 \text{ or } 64,$$

according as $n \equiv 0 \pmod{12}$, $\pm 1 \pmod{18}$, $\pm 2 \pmod{36}$, $3 \pmod{6}$, $\pm 4 \pmod{36}$, $\pm 5 \pmod{18}, \pm 10 \pmod{36}$, $6 \pmod{12}$, $\pm 16 \pmod{36}$, $\pm 7 \pmod{18}$, $\pm 14 \pmod{36}$, $\pm 8 \pmod{36}$.

It follows that

$$\begin{aligned}\phi(q) &= \sum_{-\infty}^{\infty} q^{(12n)^2} + 2\sum_{-\infty}^{\infty} q^{(18n+1)^2} + 2\sum_{-\infty}^{\infty} q^{(36n+2)^2} + \sum_{-\infty}^{\infty} q^{(6n+3)^2} \\ &\quad + 2\sum_{-\infty}^{\infty} q^{(36n+4)^2} + 2\sum_{-\infty}^{\infty} q^{(18n-5)^2} + 2\sum_{-\infty}^{\infty} q^{(36n-10)^2} + \sum_{-\infty}^{\infty} q^{(12n+6)^2} \\ &\quad + 2\sum_{-\infty}^{\infty} q^{(36n+16)^2} + 2\sum_{-\infty}^{\infty} q^{(18n+7)^2} + 2\sum_{-\infty}^{\infty} q^{(36n+14)^2} + 2\sum_{-\infty}^{\infty} q^{(36n-8)^2}. \end{aligned} \tag{30.8.3}$$

We substitute (30.8.3) into (30.8.2) and extract

$$\begin{aligned}&\sum_{n\geq 0} b(72n+69)q^{72n+69} \\ &= 48\Bigg(\sum_{r,s,t=-\infty}^{\infty} q^{(18r+1)^2+(36s-8)^2+(36t+2)^2} + \sum_{r,s,t=-\infty}^{\infty} q^{(18r+7)^2+(36s+4)^2+(36t+2)^2} \\ &\quad + \sum_{r,s,t=-\infty}^{\infty} q^{(18r-5)^2+(36s+4)^2+(36t-10)^2} + \sum_{r,s,t=-\infty}^{\infty} q^{(18r+7)^2+(36s-8)^2+(36t-10)^2} \\ &\quad + \sum_{r,s,t=-\infty}^{\infty} q^{(18r+1)^2+(36s+4)^2+(36t+14)^2} + \sum_{r,s,t=-\infty}^{\infty} q^{(18r-5)^2+(36s+16)^2+(36t+2)^2} \end{aligned}$$

$$+\sum_{r,s,t=-\infty}^{\infty} q^{(18r-5)^2+(36s-8)^2+(36t+14)^2}+\sum_{r,s,t=-\infty}^{\infty} q^{(18r+1)^2+(36s+16)^2+(36t-10)^2}$$

$$\left.+\sum_{r,s,t=-\infty}^{\infty} q^{(18r+7)^2+(36s+16)^2+(36t+14)^2}\right). \tag{30.8.4}$$

Remarkably, (30.8.4) can be written

$$\sum_{n\geq 0} b(72n+69)q^{72n+69}=48\sum_{a+b+c\equiv 1 \pmod 3} q^{(6a+1)^2+(12b-4)^2+(12b+2)^2}. \tag{30.8.5}$$

We now split this sum, according as a is even or odd.

$$\sum_{n\geq 0} b(72n+69)q^{72n+69}=48\left(\sum_{2a+b+c\equiv 1 \pmod 3} q^{(12a+1)^2+(12b-4)^2+(12c+2)^2}\right.$$
$$\left.+\sum_{2a-1+b+c\equiv 1 \pmod 3} q^{(12a-5)^2+(12b-4)^2+(12c+2)^2}\right). \tag{30.8.6}$$

Now, $2a+b+c\equiv 1 \pmod 3$ is equivalent to $a-b-c\equiv -1 \pmod 3$, while $2a-1+b+c\equiv 1 \pmod 3$ is equivalent to $a-b-c\equiv 1 \pmod 3$. So, if we replace b by $-b$, c by $-c$,

$$\sum_{n\geq 0} b(72n+69)q^{72n+69}=48\left(\sum_{a+b+c\equiv -1 \pmod 3} q^{(12a+1)^2+(12b+4)^2+(12c-2)^2}\right.$$
$$\left.+\sum_{a+b+c\equiv 1 \pmod 3} q^{(12a-5)^2+(12b+4)^2+)12c-2)^2}\right). \tag{30.8.7}$$

In the first sum, set $(a,b,c)=(r+s,r+t-1,r+u)$ and in the second sum set $(a,b,c)=(-r+s+1,-r+u,-r+t)$, where in each case $s+t+u=0$.

We obtain

$$\sum_{n\geq 0} b(72n+69)q^{72n+69}$$
$$=48\left(\sum_{\substack{r=-\infty\\ s+t+u=0}}^{\infty} q^{432r^2-216r+144s^2+144t^2+144u^2+24s-192t-48u+69}\right.$$

$$+ \sum_{\substack{r=-\infty \\ s+t+u=0}}^{\infty} q^{432r^2-216r+144s^2+144t^2+144u^2+168s-48t+96u+69}\Bigg) \tag{30.8.8}$$

$$= 96q^{69}\sum_{-\infty}^{\infty} q^{432r^2-216r} \sum_{s+t+u=0} q^{144s^2+144t^2+144u^2+144s-72t+72u}. \tag{30.8.9}$$

It follows that

$$\begin{aligned}\sum_{n\geq 0} b(72n+69)q^n &= 96\sum_{-\infty}^{\infty} q^{6r^2-3r} \sum_{s+t+u=0} q^{2s^2+2t^2+2u^2+2s-t+u} \\ &= 96\psi(q^3)\sum_{-\infty}^{\infty} q^{2s^2+2s} \sum_{t+u=-s} q^{2t^2+2u^2-t+u}.\end{aligned} \tag{30.8.10}$$

Now, $\sum_{t+u=-s} q^{2t^2+2u^2-t+u}$ is the coefficient of a^{-s} in

$$\begin{aligned}\sum_{-\infty}^{\infty} a^t q^{2t^2-t} \sum_{-\infty}^{\infty} a^u q^{2u^2+u} &= (-aq, -a^{-1}q^3, q^4; q^4)_\infty(-aq^3, -a^{-1}q, q^4; q^4)_\infty \\ &= \frac{(q^4;q^4)_\infty^2}{(q^2;q^2)_\infty}(-aq, -a^{-1}q, q^2; q^2)_\infty \\ &= \psi(q^2)\sum_{-\infty}^{\infty} a^{-s}q^{s^2}.\end{aligned}$$

That is,

$$\sum_{t+u=-s} q^{2t^2+2u^2-t+u} = \psi(q^2)q^{s^2}. \tag{30.8.11}$$

If we substitute (30.8.11) into (30.8.10), we find

$$\begin{aligned}\sum_{n\geq 0} b(72n+69)q^n &= 96\psi(q^3)\psi(q^2)\sum_{-\infty}^{\infty} q^{3s^2+2s} \\ &= 96\psi(q^2)\psi(q^3)\Omega(q) \\ &= 96\frac{(q^4;q^4)_\infty^2}{(q^2;q^2)_\infty}\frac{(q^6;q^6)_\infty^2}{(q^3;q^3)_\infty}\frac{(q^2;q^2)_\infty^2(q^3;q^3)_\infty(q^{12};q^{12})_\infty}{(q;q)_\infty(q^4;q^4)_\infty(q^6;q^6)_\infty} \\ &= 96\frac{(q^2;q^2)_\infty(q^4;q^4)_\infty(q^6;q^6)_\infty(q^{12};q^{12})_\infty}{(q;q)_\infty} \\ &= 96\left(\begin{matrix} q^4, q^6, q^8, q^{12}, q^{12}, q^{12} \\ q, q^3, q^5, q^7, q^9, q^{11} \end{matrix}; q^{12}\right)_\infty,\end{aligned}$$

which is (30.8.1).

30.9 The Generating Function for $c(72n+69)$

We show that

$$\sum_{n\geq 0} c(72n+69)q^n$$
$$= 32\left(\begin{matrix} q^3,q^4,q^4,q^8,q^9,q^{12},q^{12},q^{12},q^{15},q^{16},q^{20},q^{20},q^{24},q^{24},q^{24},q^{24} \\ q,q^2,q^2,q^5,q^7,q^{10},q^{10},q^{11},q^{13},q^{14},q^{14},q^{17},q^{19},q^{22},q^{22},q^{23} \end{matrix} ;q^{24}\right)_\infty . \tag{30.9.1}$$

We have

$$\sum_{n\geq 0} c(n)q^n = \phi(q)^2\phi(q^2). \tag{30.9.2}$$

It is easy to check that, modulo 72,

$$2n^2 \equiv 0,\ 2,\ 8,\ 18,\ 26,\ 32,\ 50 \text{ or } 56$$

according as $n \equiv 0 \pmod 6$, $\equiv \pm 1 \pmod{18}$, $\equiv \pm 2 \pmod{18}$, $\equiv 3 \pmod 6$, $\equiv \pm 7 \pmod{18}$, $\equiv \pm 4 \pmod{18}$, $\equiv \pm 5 \pmod{18}$, $\equiv \pm 8 \pmod{18}$.

It follows that

$$\phi(q^2) = \sum_{-\infty}^{\infty} q^{2(6n)^2} + 2\sum_{-\infty}^{\infty} q^{(18m+1)^2} + 2\sum_{-\infty}^{\infty} q^{2(18n+2)^2} + \sum_{-\infty}^{\infty} q^{2(6n+3)^2}$$
$$+ 2\sum_{-\infty}^{\infty} q^{2(18n+7)^2} + 2\sum_{-\infty}^{\infty} q^{2(18n-4)^2} + 2\sum_{-\infty}^{\infty} q^{2(18n-5)^2}. + 2\sum_{-\infty}^{\infty} q^{2(18n+8)^2} \tag{30.9.3}$$

We substitute (30.8.3) and (30.9.3) into (30.9.2) and extract

$$\sum_{n\geq 0} c(72n+69)q^{72n+69}$$
$$= 8\left(\sum_{r,s,t=-\infty}^{\infty} q^{(6r+3)^2+(36s+14)^2+2(18t+2)^2} + \sum_{r,s,t=-\infty}^{\infty} q^{(12r+6)^2+(18s-5)^2+2(18t+2)^2}\right.$$
$$+ \sum_{r,s,t=-\infty}^{\infty} q^{(6r+3)^2+(36s-10)^2+2(18t-4)^2} + \sum_{r,s,t=-\infty}^{\infty} q^{(12r+6)^2+(18s+1)^2+2(18t-4)^2}$$
$$\left.+ \sum_{r,s,t=-\infty}^{\infty} q^{(6r+3)^2+(36s+2)^2+2(18t+8)^2} + \sum_{r,s,t=-\infty}^{\infty} q^{(12r+6)^2+(18s+7)^2+2(18t+8)^2}\right)$$

$$
\begin{aligned}
&= 8\sum_{-\infty}^{\infty} q^{(6r+3)^2}\Bigg(\sum_{s,t=-\infty}^{\infty} q^{(36s+14)^2+2(18t+2)^2} + \sum_{s,t=-\infty}^{\infty} q^{(36s-10)^2+2(18t-4)^2} \\
&\qquad\qquad + \sum_{s,t=-\infty}^{\infty} q^{(36s+2)^2+2(18t+8)^2}\Bigg) \\
&+ 8\sum_{-\infty}^{\infty} q^{(12r+6)^2}\Bigg(\sum_{s,t=-\infty}^{\infty} q^{(18s-5)^2+2(18t+2)^2} + \sum_{s,t=-\infty}^{\infty} q^{(18s+1)^2+2(18t-4)^2} \\
&\qquad\qquad + \sum_{s,t=-\infty}^{\infty} q^{(18s+7)^2+2(18t+8)^2}\Bigg) \\
&= 8\sum_{-\infty}^{\infty} q^{(6r+3)^2} \sum_{a+b\equiv 1 \pmod 3} q^{(12a+2)^2+2(6b+2)^2} \\
&+ 8\sum_{-\infty}^{\infty} q^{(12r+6)^2} \sum_{a+b\equiv -1 \pmod 3} q^{(6a+1)^2+2(6b+2)^2}. \qquad (30.9.4)
\end{aligned}
$$

Note that if $a+b\equiv 1 \pmod 3$ then $2a-b\equiv -1 \pmod 3$, we can write $a+b=3k+1$, $2a-b=3l-1$, $a=k+l$, $b=2k-l+1$, while if $a+b\equiv -1 \pmod 3$ then $a-2b\equiv -1 \pmod 3$, we can write $a+b=3k-1$, $a-2b=3l-1$, $a=2k+l-1$, $b=k-l$. Then we have

$$
\begin{aligned}
\sum_{n\geq 0} c(72n+69)q^{72n+69} &= 8\sum_{-\infty}^{\infty} q^{(6r+3)^2} \sum_{k,l=-\infty}^{\infty} q^{(12k+12l+2)^2+2(12k-6l+8)^2} \\
&+ 8\sum_{-\infty}^{\infty} q^{(12r+6)^2} \sum_{k,l=-\infty}^{\infty} q^{(12k+6l-5)^2+2(6k-6l+2)^2} \\
&= 8q^{141} \sum_{r,k,l=-\infty}^{\infty} q^{36r^2+36r+432k^2+432k+216l^2-144l} \\
&+ 8q^{69} \sum_{r,k,l=-\infty}^{\infty} q^{144r^2+144r+216k^2-72k+108l^2-108l}. \qquad (30.9.5)
\end{aligned}
$$

It follows that

$$
\begin{aligned}
\sum_{n\geq 0} c(72n+69)q^{n} &= 8\sum_{r,k,l=-\infty}^{\infty} q^{2r^2+2r+3k^2-k+3(l^2-l)/2} \\
&+ 8q \sum_{r,k,l=-\infty}^{\infty} q^{(r^2+r)/2+6k^2+6k+3l^2-2l} \\
&= 32\psi(q^3)\psi(q^4)\Pi(q^2) + 32q\psi(q)\psi(q^{12})\Omega(q). \qquad (30.9.6)
\end{aligned}
$$

We can write (30.9.6) as follows:

$$\sum_{n\geq 0} c(72n+69)q^n = 32\frac{(q^6;q^6)_\infty^2}{(q^3;q^3)_\infty}\frac{(q^8;q^8)_\infty^2}{(q^4;q^4)_\infty}\frac{(q^4;q^4)_\infty(q^6;q^6)_\infty^2}{(q^2;q^2)_\infty(q^{12};q^{12})_\infty}$$
$$+32q\frac{(q^2;q^2)_\infty^2}{(q;q)_\infty}\frac{(q^{24};q^{24})_\infty^2}{(q^{12};q^{12})_\infty}\frac{(q^2;q^2)_\infty^2(q^3;q^3)_\infty(q^{12};q^{12})_\infty}{(q;q)_\infty(q^4;q^4)_\infty(q^6;q^6)_\infty}$$
$$=32\frac{(q^6;q^6)_\infty^4(q^8;q^8)_\infty^2}{(q^2;q^2)_\infty(q^3;q^3)_\infty(q^{12};q^{12})_\infty}+32q\frac{(q^2;q^2)_\infty^4(q^3;q^3)_\infty(q^{24};q^{24})_\infty^2}{(q;q)_\infty^2(q^4;q^4)_\infty(q^6;q^6)_\infty}$$
$$=32\frac{(q^3;q^3)_\infty(q^4;q^4)_\infty(q^{12};q^{12})_\infty}{(q^2;q^2)_\infty(q^6;q^6)_\infty}$$
$$\times\left(\frac{(q^6;q^6)_\infty^5}{(q^3;q^3)_\infty^2(q^{12};q^{12})_\infty^2}\frac{(q^8;q^8)_\infty^2}{(q^4;q^4)_\infty}+q\frac{(q^2;q^2)_\infty^5}{(q;q)_\infty^2(q^4;q^4)_\infty^2}\frac{(q^{24};q^{24})_\infty^2}{(q^{12};q^{12})_\infty}\right)$$
$$=32\frac{(q^3;q^3)_\infty(q^4;q^4)_\infty(q^{12};q^{12})_\infty}{(q^2;q^2)_\infty(q^6;q^6)_\infty}(\phi(q^3)\psi(q^4)+q\phi(q)\psi(q^{12}))$$
$$=32\frac{(q^3;q^3)_\infty(q^4;q^4)_\infty(q^{12};q^{12})_\infty}{(q^2;q^2)_\infty(q^6;q^6)_\infty}$$
$$\times((\phi(q^{12})+2q^3\psi(q^{24}))\psi(q^4)+q(\phi(q^4)+2q\psi(q^8))\psi(q^{12}))$$
$$=32\frac{(q^3;q^3)_\infty(q^4;q^4)_\infty(q^{12};q^{12})_\infty}{(q^2;q^2)_\infty(q^6;q^6)_\infty}$$
$$\times(\phi(q^{12})\psi(q^4)+2q^2\psi(q^8)\psi(q^{12})+q(\phi(q^4)\psi(q^{12})+2q^2\psi(q^4)\psi(q^{24})))$$
$$=32\frac{(q^3;q^3)_\infty(q^4;q^4)_\infty(q^{12};q^{12})_\infty}{(q^2;q^2)_\infty(q^6;q^6)_\infty}$$
$$\times\left(\left(\frac{(q^8;q^8)_\infty^2(q^{24};q^{24})_\infty^5}{(q^4;q^4)_\infty(q^{12};q^{12})_\infty^2(q^{48};q^{48})_\infty^2}+2q^2\frac{(q^{16};q^{16})_\infty^2(q^{24};q^{24})_\infty^2}{(q^8;q^8)_\infty(q^{12};q^{12})_\infty}\right)\right.$$
$$\left.+q\left(\frac{(q^8;q^8)_\infty^5(q^{24};q^{24})_\infty^2}{(q^4;q^4)_\infty^2(q^{12};q^{12})_\infty(q^{16};q^{16})_\infty^2}+2q^2\frac{(q^8;q^8)_\infty^2(q^{48};q^{48})_\infty^2}{(q^4;q^4)_\infty(q^{24};q^{24})_\infty}\right)\right)$$
$$=32\frac{(q^3;q^3)_\infty(q^4;q^4)_\infty(q^{12};q^{12})_\infty}{(q^2;q^2)_\infty(q^6;q^6)_\infty}$$
$$\times\left(\frac{(q^4;q^4)_\infty^4(q^{16};q^{16})_\infty(q^{24};q^{24})_\infty^3}{(q^8;q^8)_\infty^2(q^{12};q^{12})_\infty^3(q^{48};q^{48})_\infty}\right.$$
$$\times\left(\frac{(q^8;q^8)_\infty^4(q^{12};q^{12})_\infty(q^{24};q^{24})_\infty^2}{(q^4;q^4)_\infty^5(q^{16};q^{16})_\infty(q^{48};q^{48})_\infty}\right.$$
$$\left.+2q^2\frac{(q^8;q^8)_\infty(q^{12};q^{12})_\infty^2(q^{16};q^{16})_\infty(q^{48};q^{48})_\infty}{(q^4;q^4)_\infty^4(q^{24};q^{24})_\infty}\right)$$

$$
\begin{aligned}
&+q\frac{(q^4;q^4)_\infty^3(q^8;q^8)_\infty(q^{48};q^{48})_\infty}{(q^{12};q^{12})_\infty^2(q^{16};q^{16})_\infty}\\
&\times\left(\frac{(q^8;q^8)_\infty^4(q^{12};q^{12})_\infty(q^{24};q^{24})_\infty^2}{(q^4;q^4)_\infty^5(q^{16};q^{16})_\infty(q^{48};q^{48})_\infty}\right.\\
&\qquad\left.\left.+2q^2\frac{(q^8;q^8)_\infty(q^{12};q^{12})_\infty^2(q^{16};q^{16})_\infty(q^{48};q^{48})_\infty}{(q^4;q^4)_\infty^4(q^{24};q^{24})_\infty}\right)\right)\\
&=32\left(\frac{(q^3;q^3)_\infty(q^4;q^4)_\infty^5(q^{16};q^{16})_\infty(q^{24};q^{24})_\infty^3}{(q^2;q^2)_\infty(q^6;q^6)_\infty(q^8;q^8)_\infty^2(q^{12};q^{12})_\infty^2(q^{48};q^{48})_\infty}\right.\\
&\qquad\left.+q\frac{(q^3;q^3)_\infty(q^4;q^4)_\infty^4(q^8;q^8)_\infty(q^{48};q^{48})_\infty}{(q^2;q^2)_\infty(q^6;q^6)_\infty(q^{12};q^{12})_\infty(q^{16};q^{16})_\infty}\right)\\
&\times\left(\frac{(q^8;q^8)_\infty^4(q^{12};q^{12})_\infty(q^{24};q^{24})_\infty^2}{(q^4;q^4)_\infty^5(q^{16};q^{16})_\infty(q^{48};q^{48})_\infty}\right.\\
&\qquad\left.+2q^2\frac{(q^8;q^8)_\infty(q^{12};q^{12})_\infty^2(q^{16};q^{16})_\infty(q^{48};q^{48})_\infty}{(q^4;q^4)_\infty^4(q^{24};q^{24})_\infty}\right)\\
&=32\frac{(q^2;q^2)_\infty(q^3;q^3)_\infty(q^4;q^4)_\infty^4(q^{24};q^{24})_\infty}{(q^6;q^6)_\infty^2(q^8;q^8)_\infty(q^{12};q^{12})_\infty}\\
&\times\left(\frac{(q^4;q^4)_\infty(q^6;q^6)_\infty(q^{16};q^{16})_\infty(q^{24};q^{24})_\infty^2}{(q^2;q^2)_\infty^2(q^8;q^8)_\infty(q^{12};q^{12})_\infty(q^{48};q^{48})_\infty}+q\frac{(q^6;q^6)_\infty(q^8;q^8)_\infty^2(q^{48};q^{48})_\infty}{(q^2;q^2)_\infty^2(q^{16};q^{16})_\infty(q^{24};q^{24})_\infty}\right)\\
&\times\left(\frac{(q^8;q^8)_\infty^4(q^{12};q^{12})_\infty(q^{24};q^{24})_\infty^2}{(q^4;q^4)_\infty^5(q^{16};q^{16})_\infty(q^{48};q^{48})_\infty}\right.\\
&\qquad\left.+2q^2\frac{(q^8;q^8)_\infty(q^{12};q^{12})_\infty^2(q^{16};q^{16})_\infty(q^{48};q^{48})_\infty}{(q^4;q^4)_\infty^4(q^{24};q^{24})_\infty}\right).
\end{aligned}
\tag{30.9.7}
$$

At this point, we need two results, the proofs of which are left to the next section. The results are

$$
\begin{aligned}
&\frac{(q^3;q^3)_\infty}{(q;q)_\infty}\\
&=\frac{(q^4;q^4)_\infty(q^6;q^6)_\infty(q^{16};q^{16})_\infty(q^{24};q^{24})_\infty^2}{(q^2;q^2)_\infty^2(q^8;q^8)_\infty(q^{12};q^{12})_\infty(q^{48};q^{48})_\infty}+q\frac{(q^6;q^6)_\infty(q^8;q^8)_\infty^2(q^{48};q^{48})_\infty}{(q^2;q^2)_\infty^2(q^{16};q^{16})_\infty(q^{24};q^{24})_\infty}
\end{aligned}
\tag{30.9.8}
$$

and

$$
\begin{aligned}
&\frac{(q^3;q^3)_\infty^2}{(q;q)_\infty^2}\\
&=\frac{(q^4;q^4)_\infty^4(q^6;q^6)_\infty(q^{12};q^{12})_\infty^2}{(q^2;q^2)_\infty^5(q^8;q^8)_\infty(q^{24};q^{24})_\infty}+2q\frac{(q^4;q^4)_\infty(q^6;q^6)_\infty^2(q^8;q^8)_\infty(q^{24};q^{24})_\infty}{(q^2;q^2)_\infty^4(q^{12};q^{12})_\infty}.
\end{aligned}
\tag{30.9.9}
$$

If we substitute (30.9.8) and (30.9.9) with q replaced by q^2 into (30.9.7), we find

$$\begin{aligned}\sum_{n\geq 0} c(72n+69)q^n &= 32\frac{(q^2;q^2)_\infty(q^3;q^3)_\infty(q^4;q^4)_\infty^4(q^{24};q^{24})_\infty}{(q^6;q^6)_\infty^2(q^8;q^8)_\infty(q^{12};q^{12})_\infty}\frac{(q^3;q^3)_\infty}{(q;q)_\infty}\frac{(q^6;q^6)_\infty^2}{(q^2;q^2)_\infty^2}\\ &= 32\frac{(q^3;q^3)_\infty^2(q^4;q^4)_\infty^4(q^{24};q^{24})_\infty}{(q;q)_\infty(q^2;q^2)_\infty(q^8;q^8)_\infty(q^{12};q^{12})_\infty}\\ = 32\left(\begin{matrix}q^3,q^4,q^4,q^8,q^9,q^{12},q^{12},q^{12},q^{15},q^{16},q^{20},q^{20},q^{21},q^{24},q^{24},q^{24}\\ q,q^2,q^2,q^5,q^7,q^{10},q^{10}q^{11},q^{13},q^{14},q^{14},q^{17},q^{19},q^{22},q^{22},q^{23}\end{matrix};q^{24}\right)_\infty&,\end{aligned}$$

which is (30.9.1).

30.10 Some Important and Useful 2-Dissections

We prove the following 2-dissections.

$$\begin{aligned}&\frac{(q;q)_\infty}{(q^3;q^3)_\infty}\\ &=\frac{(q^2;q^2)_\infty(q^{16};q^{16})_\infty(q^{24};q^{24})_\infty^2}{(q^6;q^6)_\infty^2(q^8;q^8)_\infty(q^{48};q^{48})_\infty}-q\frac{(q^2;q^2)_\infty(q^8;q^8)_\infty^2(q^{12};q^{12})_\infty(q^{48};q^{48})_\infty}{(q^4;q^4)_\infty(q^6;q^6)_\infty^2(q^{16};q^{16})_\infty(q^{24};q^{24})_\infty},\end{aligned}\tag{30.10.1}$$

$$\begin{aligned}&\frac{(q;q)_\infty^2}{(q^3;q^3)_\infty^2}\\ &=\frac{(q^2;q^2)_\infty(q^4;q^4)_\infty^2(q^{12};q^{12})_\infty^4}{(q^6;q^6)_\infty^5(q^8;q^8)_\infty(q^{24};q^{24})_\infty}-2q\frac{(q^2;q^2)_\infty^2(q^8;q^8)_\infty(q^{12};q^{12})_\infty(q^{24};q^{24})_\infty}{(q^4;q^4)_\infty(q^6;q^6)_\infty^4},\end{aligned}\tag{30.10.2}$$

$$\begin{aligned}&\frac{(q^3;q^3)_\infty}{(q;q)_\infty}\\ &=\frac{(q^4;q^4)_\infty(q^6;q^6)_\infty(q^{16};q^{16})_\infty(q^{24};q^{24})_\infty^2}{(q^2;q^2)_\infty^2(q^8;q^8)_\infty(q^{12};q^{12})_\infty(q^{48};q^{48})_\infty}+q\frac{(q^6;q^6)_\infty(q^8;q^8)_\infty^2(q^{48};q^{48})_\infty}{(q^2;q^2)_\infty^2(q^{16};q^{16})_\infty(q^{24};q^{24})_\infty}\end{aligned}\tag{30.10.3}$$

and

$$\frac{(q^3;q^3)_\infty^2}{(q;q)_\infty^2} = \frac{(q^4;q^4)_\infty^4(q^6;q^6)_\infty(q^{12};q^{12})_\infty^2}{(q^2;q^2)_\infty^5(q^8;q^8)_\infty(q^{24};q^{24})_\infty} + 2q\frac{(q^4;q^4)_\infty(q^6;q^6)_\infty^2(q^8;q^8)_\infty(q^{24};q^{24})_\infty}{(q^2;q^2)_\infty^4(q^{12};q^{12})_\infty}. \tag{30.10.4}$$

We have

$$\begin{aligned}
&\frac{(q;q)_\infty}{(q^3;q^3)_\infty}\\
&= (q,q^2;q^3)_\infty\\
&= (q,q^2,q^4,q^5;q^6)_\infty\\
&= \frac{(q^2;q^2)_\infty}{(q^6;q^6)_\infty^2}(q,q^5,q^6;q^6)_\infty\\
&= \frac{(q^2;q^2)_\infty}{(q^6;q^6)_\infty^2}\sum_{-\infty}^{\infty}(-1)^n q^{3n^2+2n}\\
&= \frac{(q^2z;q^2)_\infty}{(q^6;q^6)_\infty^2}\left(\sum_{-\infty}^{\infty} q^{12n^2+4n} - \sum_{-\infty}^{\infty} q^{12n^2-8n+1}\right)\\
&= \frac{(q^2;q^2)_\infty}{(q^6;q^6)_\infty^2}\left(\Pi(q^8) - q\Omega(q^4)\right)\\
&= \frac{(q^2;q^2)_\infty}{(q^6;q^6)_\infty^2}\left(\frac{(q^{16};q^{16})_\infty(q^{24};q^{24})_\infty^2}{(q^8;q^8)_\infty(q^{48};q^{48})_\infty} - q\frac{(q^8;q^8)_\infty^2(q^{12};q^{12})_\infty(q^{48};q^{48})_\infty}{(q^4;q^4)_\infty(q^{16};q^{16})_\infty(q^{24};q^{24})_\infty}\right)\\
&= \frac{(q^2;q^2)_\infty(q^{16};q^{16})_\infty(q^{24};q^{24})_\infty^2}{(q^6;q^6)_\infty^2(q^8;q^8)_\infty(q^{48};q^{48})_\infty} - q\frac{(q^2;q^2)_\infty(q^8;q^8)_\infty^2(q^{12};q^{12})_\infty(q^{48};q^{48})_\infty}{(q^4;q^4)_\infty(q^6;q^6)_\infty^2(q^{16};q^{16})_\infty(q^{24};q^{24})_\infty},
\end{aligned}$$

which is (30.10.1),
and

$$\begin{aligned}
&\frac{(q;q)_\infty^2}{(q^3;q^3)_\infty^2}\\
&= \frac{(q^2;q^2)_\infty^2}{(q^6;q^6)_\infty^4}\sum_{m,n=-\infty}^{\infty}(-1)^{m+n}q^{3m^2+2m+3n^2+2n}\\
&= \frac{(q^2;q^2)_\infty^2}{(q^6;q^6)_\infty^4}\left(\sum_{r,s=-\infty}^{\infty} q^{3(r+s)^2+2(r+s)+3(r-s)^2+2(r-s)}\right.\\
&\qquad\left. - \sum_{r,s=-\infty}^{\infty} q^{3(r+s+1)^2-2(r+s+1)+3(r-s)^2+2(r-s)}\right)
\end{aligned}$$

$$
\begin{aligned}
&= \frac{(q^2;q^2)_\infty^2}{(q^6;q^6)_\infty^4}\left(\sum_{r,s=-\infty}^{\infty} q^{6r^2+6s^2+4r} - q\sum_{r,s=-\infty}^{\infty} q^{6r^2+6s^2+2r+6s}\right)\\
&= \frac{(q^2;q^2)_\infty^2}{(q^6;q^6)_\infty^4}\left(\phi(q^6)\Omega(q^2) - 2q\psi(q^{12})\Pi(q^4)\right)\\
&= \frac{(q^2;q^2)_\infty^2}{(q^6;q^6)_\infty^4}\left(\frac{(q^{12};q^{12})_\infty^5}{(q^6;q^6)_\infty^2(q^{24};q^{24})_\infty^2}\frac{(q^4;q^4)_\infty^2(q^6;q^6)_\infty(q^{24};q^{24})_\infty}{(q^2;q^2)_\infty(q^8;q^8)_\infty(q^{12};q^{12})_\infty}\right.\\
&\qquad\left. -2q\frac{(q^{24};q^{24})_\infty^2}{(q^{12};q^{12})_\infty}\frac{(q^8;q^8)_\infty(q^{12};q^{12})_\infty^2}{(q^4;q^4)_\infty(q^{24};q^{24})_\infty}\right)\\
&= \frac{(q^2;q^2)_\infty(q^4;q^4)_\infty^2(q^{12};q^{12})_\infty^4}{(q^6;q^6)_\infty^5(q^8;q^8)_\infty(q^{24};q^{24})_\infty} - 2q\frac{(q^2;q^2)_\infty^2(q^8;q^8)_\infty(q^{12};q^{12})_\infty(q^{24};q^{24})_\infty}{(q^4;q^4)_\infty(q^6;q^6)_\infty^4},
\end{aligned}
$$

which is (30.10.2).

In order to prove (30.10.3) and (30.10.4), we use the fact that if the 2-dissection of $f(q)$ is

$$f(q) = g(q^2) + qh(q^2) \tag{30.10.5}$$

then the 2-dissection of $\dfrac{1}{f(q)}$ is

$$\frac{1}{f(q)} = \frac{f(-q)}{f(q)f(-q)} = \frac{1}{f(q)f(-q)}\left(g(q^2) - qh(q^2)\right). \tag{30.10.6}$$

If we apply (30.10.5)–(30.10.6) with $f(q) = \dfrac{(q;q)_\infty}{(q^3;q^3)_\infty}$, we find

$$
\begin{aligned}
f(q)f(-q) &= \frac{(q,q^2;q^2)_\infty}{(q^3,q^6;q^6)_\infty}\frac{(-q,q^2;q^2)_\infty}{(-q^3,q^6;q^6)_\infty}\\
&= \frac{(q^2;q^2)_\infty^2}{(q^6;q^6)_\infty^2}\frac{(q^2;q^4)_\infty}{(q^6;q^{12})_\infty}\\
&= \frac{(q^2;q^2)_\infty^3(q^{12};q^{12})_\infty}{(q^4;q^4)_\infty(q^6;q^6)_\infty^3}
\end{aligned}
$$

and

$$
\begin{aligned}
&\frac{(q^3;q^3)_\infty}{(q;q)_\infty}\\
&= \frac{(q^4;q^4)_\infty(q^6;q^6)_\infty^3}{(q^2;q^2)_\infty^3(q^{12};q^{12})_\infty}
\end{aligned}
$$

$$\times\left(\frac{(q^2;q^2)_\infty(q^{16};q^{16})_\infty(q^{24};q^{24})^2_\infty}{(q^6;q^6)^2_\infty(q^8;q^8)_\infty(q^{48};q^{48})_\infty}+q\frac{(q^2;q^2)_\infty(q^8;q^8)^2_\infty(q^{12};q^{12})_\infty(q^{48};q^{48})_\infty}{(q^4;q^4)_\infty(q^6;q^6)^2_\infty(q^{16};q^{16})_\infty(q^{24};q^{24})_\infty}\right)$$
$$=\frac{(q^4;q^4)_\infty(q^6;q^6)_\infty(q^{16};q^{16})_\infty(q^{24};q^{24})^2_\infty}{(q^2;q^2)^2_\infty(q^8;q^8)_\infty(q^{12};q^{12})_\infty(q^{48};q^{48})_\infty}+q\frac{(q^6;q^6)_\infty(q^8;q^8)^2_\infty(q^{48};q^{48})_\infty}{(q^2;q^2)^2_\infty(q^{16};q^{16})_\infty(q^{24};q^{24})_\infty},$$

which is (30.10.3).

If we apply (30.10.5)–(30.10.6) with $f(q)=\dfrac{(q;q)^2_\infty}{(q^3;q^3)^2_\infty}$, we find

$$f(q)f(-q)=\frac{(q^2;q^2)^6_\infty(q^{12};q^{12})^2_\infty}{(q^4;q^4)^2_\infty(q^6;q^6)^6_\infty}$$

and

$$\frac{(q^3;q^3)^2_\infty}{(q;q)^2_\infty}$$
$$=\frac{(q^4;q^4)^2_\infty(q^6;q^6)^6_\infty}{(q^2;q^2)^6_\infty(q^{12};q^{12})^2_\infty}$$
$$\times\left(\frac{(q^2;q^2)_\infty(q^4;q^4)^2_\infty(q^{12};q^{12})^4_\infty}{(q^6;q^6)^5_\infty(q^8;q^8)_\infty(q^{24};q^{24})_\infty}+2q\frac{(q^2;q^2)^2_\infty(q^8;q^8)_\infty(q^{12};q^{12})_\infty(q^{24};q^{24})_\infty}{(q^4;q^4)_\infty(q^6;q^6)^4_\infty}\right)$$
$$=\frac{(q^4;q^4)^4_\infty(q^6;q^6)_\infty(q^{12};q^{12})^2_\infty}{(q^2;q^2)^5_\infty(q^8;q^8)_\infty(q^{24};q^{24})_\infty}+2q\frac{(q^4;q^4)_\infty(q^6;q^6)^2_\infty(q^8;q^8)_\infty(q^{24};q^{24})_\infty}{(q^2;q^2)^4_\infty(q^{12};q^{12})_\infty},$$

which is (30.10.4).

30.11 The Final Steps in the Proofs of (30.5.5)–(30.5.8)

We show that

$$12\sum_{n\geq 0}c(72n+69)q^n+4\sum_{n\geq 0}b(72n+69)q^n=768\Pi_2 \tag{30.11.1}$$

and

$$12\sum_{n\geq 0}c(72n+69)q^n-4\sum_{n\geq 0}b(72n+69)q^n=768\Pi_3. \tag{30.11.2}$$

We have

$$\begin{aligned}
&12\sum_{n\geq 0} c(72n+69)q^n + 4\sum_{n\geq 0} b(72n+69)q^n \\
&= 384\left(\left(\begin{matrix} q^3,q^4,q^4,q^8,q^9,q^{12},q^{12},q^{12},q^{15},q^{16},q^{20},q^{20},q^{21},q^{24},q^{24},q^{24} \\ q,q^2,q^2,q^5,q^7,q^{10},q^{10},q^{11},q^{13},q^{14},q^{14},q^{17},q^{19},q^{22},q^{22},q^{23} \end{matrix};q^{24}\right)_\infty\right. \\
&\qquad \left. + \left(\begin{matrix} q^4,q^6,q^8,q^{12},q^{12},q^{12},q^{16},q^{18},q^{20},q^{24},q^{24},q^{24} \\ q,q^3,q^5,q^7,q^9,q^{11},q^{13},q^{15},q^{17},q^{19},q^{21},q^{23} \end{matrix};q^{24}\right)_\infty\right) \\
&= 384\left(\begin{matrix} q^4,q^8,q^{12},q^{16},q^{20},q^{24} \\ q,q^2,q^5,q^7,q^{10},q^{11},q^{13},q^{14},q^{17},q^{19},q^{22},q^{23} \end{matrix};q^{24}\right)_\infty \\
&\qquad \times\left(\left(\begin{matrix} q^3,q^4,q^9,q^{12},q^{12},q^{15},q^{20},q^{21},q^{24},q^{24} \\ q^2,q^{10},q^{14},q^{22} \end{matrix};q^{24}\right)_\infty\right. \\
&\qquad\qquad \left. + \left(\begin{matrix} q^2,q^6,q^{10},q^{12},q^{12},q^{14},q^{18},q^{22},q^{24},q^{24} \\ q^3,q^9,q^{15},q^{21} \end{matrix};q^{24}\right)_\infty\right) \\
&= 384\left(\begin{matrix} q^4,q^8,q^{12},q^{16},q^{20},q^{24} \\ q,q^2,q^5,q^7,q^{10},q^{11},q^{13},q^{14},q^{17},q^{19},q^{22},q^{23} \end{matrix};q^{24}\right)_\infty \\
&\qquad \times\left((-q^2,q^3,q^9,-q^{10},q^{12},q^{12};q^{12})_\infty + (q^2,-q^3,-q^9,q^{10},q^{12},q^{12};q^{12})_\infty\right).
\end{aligned} \tag{30.11.3}$$

Similarly,

$$\begin{aligned}
&12\sum_{n\geq 0} c(72n+69)q^n - 4\sum_{n\geq 0} b(72n+69)q^n \\
&= 384\left(\begin{matrix} q^4,q^8,q^{12},q^{16},q^{20},q^{24} \\ q,q^2,q^5,q^7,q^{10},q^{11},q^{13},q^{14},q^{17},q^{19},q^{22},q^{23} \end{matrix};q^{24}\right)_\infty \\
&\qquad \times\left((-q^2,q^3,q^9,-q^{10},q^{12},q^{12};q^{12})_\infty - (q^2,-q^3,-q^9,q^{10},q^{12},q^{12};q^{12})_\infty\right).
\end{aligned} \tag{30.11.4}$$

Now we need only show that

$$\begin{aligned}
(-q^2,q^3,q^9,-q^{10},q^{12},q^{12};q^{12})_\infty &+ (q^2,-q^3,-q^9,q^{10},q^{12},q^{12};q^{12})_\infty \\
&= 2(q^5,q^{11},q^{13},q^{19},q^{24},q^{24};q^{24})_\infty
\end{aligned} \tag{30.11.5}$$

and

$$\begin{aligned}
(-q^2,q^3,q^9,-q^{10},q^{12},q^{12};q^{12})_\infty &- (q^2,-q^3,-q^9,q^{10},q^{12},q^{12};q^{12})_\infty \\
&= 2q^2(q,q^7,q^{17},q^{23},q^{24},q^{24};q^{24})_\infty.
\end{aligned} \tag{30.11.6}$$

Let

$$f(q) = (-q^2,q^3,q^9,-q^{10},q^{12},q^{12};q^{12})_\infty + (q^2,-q^3,-q^9,q^{10},q^{12},q^{12};q^{12})_\infty$$
$$= \sum_{m,n=-\infty}^{\infty} (-1)^n q^{6m^2+4m+6n^2+3n} + \sum_{m,n=-\infty}^{\infty} (-1)^m q^{6m^2+2m+6n^2+3n}.$$

Then

$$\begin{aligned} f(-q) &= \sum_{m,n=-\infty}^{\infty} q^{6m^2+4m+6n^2+3n} + \sum_{m,n=-\infty}^{\infty} (-1)^{m+n} q^{6m^2+4m+6n^2+3n} \\ &= 2\sum_{m+n \text{ even}} q^{6m^2+4m+6n^2+3n} \\ &= 2\sum_{r,s=-\infty}^{\infty} q^{6(r+s)^2+4(r+s)+6(r-s)^2+3(r-s)} \\ &= 2\sum_{r,s=-\infty}^{\infty} q^{12r^2+12s^2+7r+s} \\ &= 2(-q^5,-q^{11},-q^{13},-q^{19},q^{24},q^{24};q^{24})_\infty, \end{aligned}$$

and

$$f(q) = 2(q^5,q^{11},q^{13},q^{19},q^{24},q^{24};q^{24})_\infty.$$

This is (30.11.5). The proof of (30.11.6) is left as an exercise.

If we substitute (30.11.5) into (30.11.3), we obtain

$$\begin{aligned} &12\sum_{n\geq 0} c(72n+69)q^n + 4\sum_{n\geq 0} b(72n+69)q^n \\ &= 768\left(\begin{matrix} q^4,q^8,q^{12},q^{16},q^{20},q^{24},q^{24},q^{24} \\ q,q^2,q^7,q^{10},q^{14},q^{17},q^{22},q^{23} \end{matrix};q^{24}\right)_\infty \\ &= 768\Pi_2, \end{aligned}$$

which is (30.11.1).

If we substitute (30.11.6) into (30.11.4), we obtain

$$\begin{aligned} &12\sum_{n\geq 0} c(72n+69)q^n - 4\sum_{n\geq 0} b(72n+69)q^n \\ &= 768q^2\left(\begin{matrix} q^4,q^8,q^{12},q^{16},q^{20},q^{24},q^{24},q^{24} \\ q^2,q^5,q^{10},q^{11},q^{13},q^{14},q^{19},q^{22} \end{matrix};q^{24}\right)_\infty \\ &= 768\Pi_3, \end{aligned}$$

which is (30.11.2).

Finally, from (30.5.15), (30.6.1) and (30.11.1), we obtain (30.5.5), from (30.5.16), (30.6.1) and (30.11.2), we obtain (30.5.6), from (30.5.17), (30.6.1) and (30.11.2), we obtain (30.5.7) and from (30.5.18), (30.6.1) and (30.11.1) we obtain (30.5.8).

30.12 Our Fifth Proof of the Modular Equation (20.1.1)

We start by obtaining the 2-dissection

$$\begin{aligned}&(q;q)_\infty(q^3;q^3)_\infty\\&=\frac{(q^2;q^2)_\infty(q^8;q^8)_\infty^2(q^{12};q^{12})_\infty^4}{(q^4;q^4)_\infty^2(q^6;q^6)_\infty(q^{24};q^{24})_\infty^2}-q\frac{(q^4;q^4)_\infty^4(q^6;q^6)_\infty(q^{24};q^{24})_\infty^2}{(q^2;q^2)_\infty(q^8;q^8)_\infty^2(q^{12};q^{12})_\infty^2}.\end{aligned} \tag{30.12.1}$$

We have

$$\begin{aligned}&(q;q)_\infty(q^3;q^3)_\infty\\&=(q,q^2,q^3,q^3,q^4,q^5,q^6,q^6;q^6)_\infty\\&=\frac{(q^2;q^2)_\infty}{(q^6;q^6)_\infty}(q,q^5,q^6;q^6)_\infty(q^3,q^3,q^6;q^6)_\infty\\&=\frac{(q^2;q^2)_\infty}{(q^6;q^6)_\infty}\sum_{m,n=-\infty}^{\infty}(-1)^{m+n}q^{3m^2+2m+3n^2}\\&=\frac{(q^2;q^2)_\infty}{(q^6;q^6)_\infty}\left(\sum_{r,s=-\infty}^{\infty}q^{3(r+s)^2+2(r+s)+3(r-s)^2}-\sum_{r,s=-\infty}^{\infty}q^{3(r+s+1)^2-2(r+s+1)+3(r-s)^2}\right)\\&=\frac{(q^2;q^2)_\infty}{(q^6;q^6)_\infty}\left(\sum_{r,s=-\infty}^{\infty}q^{6r^2+6s^2+2r+2s}-\sum_{r,s=-\infty}^{\infty}q^{6r^2+6s^2+4r+4s+1}\right)\\&=\frac{(q^2;q^2)_\infty}{(q^6;q^6)_\infty}\left(\Pi(q^4)^2-q\Omega(q^2)^2\right)\\&=\frac{(q^2;q^2)_\infty}{(q^6;q^6)_\infty}\left(\frac{(q^8;q^8)_\infty^2(q^{12};q^{12})_\infty^4}{(q^4;q^4)_\infty^2(q^{24};q^{24})_\infty^2}-q\frac{(q^4;q^4)_\infty^4(q^6;q^6)_\infty^2(q^{24};q^{24})_\infty^2}{(q^2;q^2)_\infty^2(q^8;q^8)_\infty^2(q^{12};q^{12})_\infty^2}\right)\\&=\frac{(q^2;q^2)_\infty(q^8;q^8)_\infty^2(q^{12};q^{12})_\infty^4}{(q^4;q^4)_\infty^2(q^6;q^6)_\infty(q^{24};q^{24})_\infty^2}-q\frac{(q^4;q^4)_\infty^4(q^6;q^6)_\infty(q^{24};q^{24})_\infty^2}{(q^2;q^2)_\infty(q^8;q^8)_\infty^2(q^{12};q^{12})_\infty^2},\end{aligned}$$

which is (30.12.1).

We now apply (30.10.5)–(30.10.6) to $f(q)=(q;q)_\infty(q^3;q^3)_\infty$.

We have

$$f(q)f(-q) = (q,q^2;q^2)_\infty(q^3,q^6;q^6)_\infty(-q,q^2;q^2)_\infty(-q^3,q^6;q^6)_\infty$$
$$= (q^2;q^2)_\infty^2(q^6;q^6)_\infty^2(q^2;q^4)_\infty(q^6;q^{12})_\infty$$
$$= \frac{(q^2;q^2)_\infty^3(q^6;q^6)_\infty^3}{(q^4;q^4)_\infty(q^{12};q^{12})_\infty} \tag{30.12.2}$$

and we obtain

$$\frac{1}{(q;q)_\infty(q^3;q^3)_\infty}$$
$$= \frac{(q^4;q^4)_\infty(q^{12};q^{12})_\infty}{(q^2;q^2)_\infty^3(q^6;q^6)_\infty^3}$$
$$\times\left(\frac{(q^2;q^2)_\infty(q^8;q^8)_\infty^2(q^{12};q^{12})_\infty^4}{(q^4;q^4)_\infty^2(q^6;q^6)_\infty(q^{24};q^{24})_\infty^2} + q\frac{(q^4;q^4)_\infty^4(q^6;q^6)_\infty(q^{24};q^{24})_\infty^2}{(q^2;q^2)_\infty(q^8;q^8)_\infty^2(q^{12};q^{12})_\infty^2}\right),$$
$$= \frac{(q^8;q^8)_\infty^2(q^{12};q^{12})_\infty^5}{(q^2;q^2)_\infty^2(q^4;q^4)_\infty(q^6;q^6)_\infty^4(q^{24};q^{24})_\infty^2} + q\frac{(q^4;q^4)_\infty^5(q^{24};q^{24})_\infty^2}{(q^2;q^2)_\infty^4(q^6;q^6)_\infty^2(q^8;q^8)_\infty^2(q^{12};q^{12})_\infty}. \tag{30.12.3}$$

If we multiply (30.12.3) by $(q^2;q^2)_\infty^2(q^6;q^6)_\infty^2$, we obtain

$$\frac{(q^2;q^2)_\infty^2(q^6;q^6)_\infty^2}{(q;q)_\infty(q^3;q^3)_\infty}$$
$$= \frac{(q^8;q^8)_\infty^2(q^{12};q^{12})_\infty^5}{(q^4;q^4)_\infty(q^6;q^6)_\infty^2(q^{24};q^{24})_\infty^2} + q\frac{(q^4;q^4)_\infty^5(q^{24};q^{24})_\infty^2}{(q^2;q^2)_\infty^2(q^8;q^8)_\infty^2(q^{12};q^{12})_\infty}. \tag{30.12.4}$$

This, of course, can be written

$$\psi(q)\psi(q^3) = \phi(q^6)\psi(q^4) + q\phi(q^2)\psi(q^{12}),$$

which is (25.2.1), and as we saw in §25.2, implies (20.1.1).

30.13 A Formula for $p_{4\square}(n)$

It is possible to give a formula for $p_{4\square}(n)$ in terms of divisor functions.

We have (30.2.1),

$$\sum_{n\geq 1} p_{4\square}(n)q^n = \frac{1}{384}\left(\phi(q)^4 + 4\phi(q)^3 + 12\phi(q)^2\phi(q^2) + 18\phi(q)^2 + 24\phi(q)\phi(q^2)\right.$$
$$\left. + 32\phi(q)\phi(q^3) + 12\phi(q^2)^2 + 60\phi(q) + 36\phi(q^2) + 32\phi(q^3) + 48\phi(q^4) + 105\right). \tag{30.13.1}$$

Also, from (2.1.2), (2.1.1), (24.1.2) and (23.1.3),

$$\phi(q)^4 = 1 + 8 \sum_{n \geq 1} \left(\sum_{\substack{d|n \\ d \not\equiv 0 \pmod 4}} d \right) q^n, \tag{30.13.2}$$

$$\begin{aligned} \phi(q)^3 &= \phi(q)\phi(q)^2 \\ &= \left(1 + 2\sum_{n\geq 1} q^{n^2}\right)\left(1 + 4\sum_{n\geq 1} e_{\{\{1\},4\}}(n) q^n\right) \\ &= 1 + 2\sum_{n\geq 1} q^{n^2} + 4\sum_{n\geq 1} e_{\{\{1\},4\}}(n) q^n \\ &\quad + 8\sum_{n\geq 1}\left(\sum_{1\leq k<\sqrt{n}} e_{\{\{1\},4\}}(n-k^2)\right) q^n, \end{aligned} \tag{30.13.3}$$

$$\begin{aligned} \phi(q)^2\phi(q^2) &= \left(1 + 2\sum_{n\geq 1} q^{2n^2}\right)\left(1 + 4\sum_{n\geq 1} e_{\{\{1\},4\}}(n) q^n\right) \\ &= 1 + 2\sum_{n\geq 1} q^{2n^2} + 4\sum_{n\geq 1} e_{\{\{1\},4\}}(n) q^n \\ &\quad + 8\sum_{n\geq 1}\left(\sum_{1\leq k<\sqrt{\frac{n}{2}}} e_{\{\{1\},4\}}(n-2k^2)\right), \end{aligned} \tag{30.13.4}$$

$$\phi(q)^2 = 1 + 4\sum_{n\geq 1} e_{\{\{1\},4\}}(n) q^n, \tag{30.13.5}$$

$$\phi(q)\phi(q^2) = 1 + 2\sum_{n\geq 1} e_{\{\{1,3\},8\}}(n) q^n, \tag{30.13.6}$$

$$\phi(q)\phi(q^3) = 1 + 2\sum_{n\geq 1} e_{\{\{1\},3\}}(n) q^n + 4\sum_{n\geq 1} e_{\{\{1\},3\}}(\frac{n}{4}) q^n \tag{30.13.7}$$

and

$$\phi(q^2)^2 = 1 + 4\sum_{n\geq 1} e_{\{\{1\},4\}}(\frac{n}{2}) q^n. \tag{30.13.8}$$

If we substitute (30.13.2)–(30.13.8) into (30.13.1), we find after a little work that for $n \geq 1$,

$$p_{4\square}(n) = \frac{1}{48}\Big(\sum_{\substack{d|n \\ d\not\equiv 0 \pmod 4}} d + 4 \sum_{1\le k<\sqrt{n}} e_{\{\{1\},4\}}(n-k^2) + 12 \sum_{1\le k<\sqrt{\frac{n}{2}}} e_{\{\{1\},4\}}(n-2k^2) + 17e_{\{\{1\},4\}}(n) + 6e_{\{\{1\},4\}}(\frac{n}{2}) + 6e_{\{\{1,3\},8\}}(n) + 8e_{\{\{1\},3\}}(n) + 16e_{\{\{1\},3\}}(\frac{n}{4})\Big) + c(n), \quad (30.13.9)$$

where $c(n) = \frac{1}{3}, \frac{1}{4}, \frac{1}{6}, \frac{7}{12}$ or 0, according as n is an odd square, twice a square, three times a square, an even square or none of these.

Thus we have found an answer to a quest by D.H. Lehmer (1948) [102].

References

40. S. Cooper, M.D. Hirschhorn, J.A. Sellers, Partitions into four squares. Proc. Jangjeon Math. Soc. **8**, 73–94 (2005)
75. M.D. Hirschhorn, Some formulae for partitions into squares. Discret. Math. **211**, 225–228 (2000)
94. M.D. Hirschhorn, J.A. Sellers, Some relations for partitions into squares, in *Special Functions 1999, Proceedings of the International Workshop on Special Functions, Hong Kong* (World Scientific, 2000), pp. 118–124
95. M.D. Hirschhorn, J.A. Sellers, On a problem of Lehmer on partitions into squares. Ramanujan J. **8**, 279–287 (2004)
102. D.H. Lehmer, On the partitions of numbers into squares. Am. Math. Mon. **55**, 476–481 (1948)

Chapter 31
Partitions into Four Distinct Squares of Equal Parity

31.1 Introduction

In Hirschhorn (2013) [87], I wrote the following.

"At the Ramanujan Centenary conference in Illinois in 1987, Bill Gosper gave a plenary address, "Ramanujan as Nemesis", and I made a regrettable error.

"Bill started by saying, and this is often badly misquoted,

"How can we pretend to love this guy [Ramanujan], when he is forever reaching out from the grave to snatch away our neatest results?"

"Bill then proceeded to state a conjecture which he said "even Erdös couldn't do". The conjecture was

If a number is the sum of four distinct odd squares
then it is the sum of four distinct even squares.

"(The converse is false. A number which is congruent to 0 modulo 8 may be the sum of four distinct even squares, but certainly isn't the sum of four odd squares. The converse **does** hold for numbers congruent to 4 modulo 8.)

"The mistake I made was to solve this problem and to show Bill my solution at the end of his talk. The consequence was that the published version of his talk did not contain the conjecture, and my proof has never been published.

"Here it is. I leave it to you to provide the details."

$$\begin{aligned}(4a+1)^2+(4b+1)^2+(4c+1)^2+(4d+1)^2\\ =(2a+2b+2c+2d+2)^2\\ +(2a-2b-2c+2d)^2\\ +(2a+2b-2c-2d)^2\\ +(2a-2b+2c-2d)^2.\end{aligned}$$

M.D. Hirschhorn, *The Power of q*, Developments in Mathematics 49,
DOI 10.1007/978-3-319-57762-3_31

[All you have to note is that if a, b, c and d are distinct, then $2a+2b+2c+2d+2$, $2a-2b+2c-2d$, $2a+2b-2c-2d$ and $2a-2b+2c-2d$ are distinct, and none is the negative of another.]

I continued

"Bill's conjecture led me to further discoveries."

One of these discoveries is

The number $n \equiv 4 \pmod 8$ is the sum of four distinct even squares if and only if it is the sum of four distinct odd squares. Moreover, the number of partitions into four distinct even squares is never less than half, and never more than, the number of partitions of n into four distinct odd squares. (31.1.1)

Let $p_{4o}^{d}(n)$ be the number of partitions of n into four squares of distinct odd non-negative integers, let $p_{4e}^{d}(n)$ be the number of partitions of n into four squares of distinct even non-negative integers, and let $p_{4e}^{d+}(n)$ be the number of partitions of n into four squares of distinct even positive integers.

We will prove the following remarkable result.

For $n \equiv 4 \pmod 8$,

$$p_{4o}^{d}(n) = p_{4e}^{d}(n) + p_{4e}^{d+}(n). \tag{31.1.2}$$

Since $0 \le p_{4e}^{d+}(n) \le p_{4e}^{d}(n)$, (31.1.2) yields

$$p_{4e}^{d}(n) \le p_{4o}^{d}(n) \le 2p_{4e}^{d}(n), \tag{31.1.3}$$

or,

$$\frac{1}{2}p_{4o}^{d}(n) \le p_{4e}^{d}(n) \le p_{4o}^{d}(n), \tag{31.1.4}$$

and our claims in (31.1.1) are proved.

31.2 The Generating Function for $p_{4o}^{d}(8n+4)$

We show that

$$\sum_{n\ge 0} p_{4o}^{d}(8n+4)q^n = \frac{1}{24}\big(\psi(q)^4 - 6\psi(q)^2\psi(q^2) + 8\psi(q)\psi(q^3) + 3\psi(q^2)^2 - 6\psi(q^4)\big). \tag{31.2.1}$$

We start by noting that

$$p_{4o}^{d}(8n+4) = p_{4\Delta}^{d}(n), \tag{31.2.2}$$

where $p^d_{4\triangle}(n)$ is the number of partitions of n into four distinct triangular numbers.

It follows that

$$\sum_{n\geq 0} p^d_{4o}(8n+4)q^n = \sum_{n\geq 0} p^d_{4\triangle}(n)q^n. \tag{31.2.3}$$

We now follow the procedure of §30.2.

We have the following eleven identities.

$$\begin{aligned}(\psi(q)-1)^4 &= \left(\sum_{n\geq 1} q^{\triangle_n}\right)^4\\ &= 24\sum_{n\geq 1} p\{\triangle_a+\triangle_b+\triangle_c+\triangle_d\}(n)q^n + 12\sum_{n\geq 1} p\{\triangle_a+\triangle_a+\triangle_b+\triangle_c\}(n)q^n\\ &+6\sum_{n\geq 1} p\{\triangle_a+\triangle_a+\triangle_b+\triangle_b\}(n)q^n + 4\sum_{n\geq 1} p\{\triangle_a+\triangle_a+\triangle_a+\triangle_b\}(n)q^n\\ &+\sum_{n\geq 1} p\{\triangle_a+\triangle_a+\triangle_a+\triangle_a\}(n)q^n, \end{aligned} \tag{31.2.4}$$

$$\begin{aligned}(\psi(q)-1)^2(\psi(q^2)-1) &= \left(\sum_{n\geq 1} q^{\triangle_n}\right)^2\left(\sum_{n\geq 1} q^{2\triangle_n}\right)\\ &= 2\sum_{n\geq 1} p\{\triangle_a+\triangle_a+\triangle_b+\triangle_c\}(n)q^n + 2\sum_{n\geq 1} p\{\triangle_a+\triangle_a+\triangle_b+\triangle_b\}(n)q^n\\ &+2\sum_{n\geq 1} p\{\triangle_a+\triangle_a+\triangle_a+\triangle_b\}(n)q^n + \sum_{n\geq 1} p\{\triangle_a+\triangle_a+\triangle_a+\triangle_a\}(n)q^n, \end{aligned} \tag{31.2.5}$$

$$\begin{aligned}(\psi(q^2)-1)^2 &= \left(\sum_{n\geq 1} q^{2\triangle_n}\right)^2\\ &= 2\sum_{n\geq 1} p\{\triangle_a+\triangle_a+\triangle_b+\triangle_b\}(n)q^n + \sum_{n\geq 1} p\{\triangle_a+\triangle_a+\triangle_a+\triangle_a\}(n)q^n, \end{aligned} \tag{31.2.6}$$

$$\begin{aligned}(\psi(q)-1)(\psi(q^3)-1) &= \left(\sum_{n\geq 1} q^{\triangle_n}\right)\left(\sum_{n\geq 1} q^{3\triangle_n}\right)\\ &= \sum_{n\geq 1} p\{\triangle_a+\triangle_a+\triangle_a+\triangle_b\}(n)q^n + \sum_{n\geq 1} p\{\triangle_a+\triangle_a+\triangle_a+\triangle_a\}(n)q^n, \end{aligned} \tag{31.2.7}$$

$$\psi(q^4)-1 = \sum_{n\geq 1} q^{4\triangle_n} = \sum_{n\geq 1} p\{\triangle_a+\triangle_a+\triangle_a+\triangle_a\}(n)q^n, \tag{31.2.8}$$

$$\begin{aligned}(\psi(q)-1)^3 &= \left(\sum_{n\geq 1} q^{\triangle_n}\right)^3\\ &= 6\sum_{n\geq 1} p\{\triangle_a+\triangle_b+\triangle_c\}(n)q^n + 3\sum_{n\geq 1} p\{\triangle_a+\triangle_a+\triangle_b\}(n)q^n\end{aligned}$$

$$+\sum_{n\geq 1} p\{\triangle_a+\triangle_a+\triangle_a\}(n)q^n, \tag{31.2.9}$$

$$\left(\psi(q)-1\right)\left(\psi(q^2)-1\right)=\sum_{n\geq 1} q^{\triangle_n}\cdot\sum_{n\geq 1} q^{2\triangle_n}$$
$$=\sum_{n\geq 1} p\{\triangle_a+\triangle_a+\triangle_b\}(n)q^n+\sum_{n\geq 1} p\{\triangle_a+\triangle_a+\triangle_a\}(n)q^n, \tag{31.2.10}$$

$$\psi(q^3)-1=\sum_{n\geq 1} q^{3\triangle_n}=\sum_{n\geq 1} p\{\triangle_a+\triangle_a+\triangle_a\}(n)q^n, \tag{31.2.11}$$

$$\left(\psi(q)-1\right)^2=\left(\sum_{n\geq 1} q^{\triangle_n}\right)^2=2\sum_{n\geq 1} p\{\triangle_a+\triangle_b\}(n)q^n+\sum_{n\geq 1} p\{\triangle_a+\triangle_a\}(n)q^n, \tag{31.2.12}$$

$$\psi(q^2)-1=\sum_{n\geq 1} q^{2\triangle_n}=\sum_{n\geq 1} p\{\triangle_a+\triangle_a\}(n)q^n \tag{31.2.13}$$

and

$$\psi(q)-1=\sum_{n\geq 1} q^{\triangle_n}=\sum_{n\geq 1} p\{\triangle_a\}(n)q^n. \tag{31.2.14}$$

It follows that

$$\sum_{n\geq 1} p\{\triangle_a\}(n)q^n=\psi(q)-1, \tag{31.2.15}$$

$$\sum_{n\geq 1} p\{\triangle_a+\triangle_a\}(n)q^n=\psi(q^2)-1, \tag{31.2.16}$$

$$\sum_{n\geq 1} p\{\triangle_a+\triangle_b\}(n)q^n=\frac{1}{2}\left(\psi(q)^2-2\psi(q)-\psi(q^2)+2\right), \tag{31.2.17}$$

$$\sum_{n\geq 1} p\{\triangle_a+\triangle_a+\triangle_a\}(n)q^n=\psi(q^3)-1, \tag{31.2.18}$$

$$\sum_{n\geq 1} p\{\triangle_a+\triangle_a+\triangle_b\}(n)q^n=\psi(q)\psi(q^2)-\psi(q)-\psi(q^2)-\psi(q^3)+2, \tag{31.2.19}$$

$$\sum_{n\geq 1} p\{\triangle_a+\triangle_b+\triangle_c\}(n)q^n$$
$$=\frac{1}{6}\left(\psi(q)^3-3\psi(q)^2-3\psi(q)\psi(q^2)+6\psi(q)+3\psi(q^2)+2\psi(q^3)-6\right), \tag{31.2.20}$$

$$\sum_{n\geq 1} p\{\triangle_a+\triangle_a+\triangle_a+\triangle_a\}(n)q^n=\psi(q^4)-1, \tag{31.2.21}$$

$$\sum_{n\geq 1} p\{\triangle_a+\triangle_a+\triangle_a+\triangle_b\}(n)q^n=\psi(q)\psi(q^3)-\psi(q)-\psi(q^3)-\psi(q^4)+2, \tag{31.2.22}$$

$$\sum_{n\geq 1} p\{\triangle_a+\triangle_a+\triangle_b+\triangle_b\}(n)q^n = \frac{1}{2}\left(\psi(q^2)^2 - 2\psi(q^2) - \psi(q^4) + 2\right), \tag{31.2.23}$$

$$\sum_{n\geq 1} p\{\triangle_a+\triangle_a+\triangle_b+\triangle_c\}(n)q^n$$
$$= \frac{1}{2}\left(\psi(q)^2\psi(q^2) - \psi(q)^2 - 2\psi(q)\psi(q^2) - 2\psi(q)\psi(q^3) - \psi(q^2)^2\right.$$
$$\left. +4\psi(q) + 3\psi(q^2) + 2\psi(q^3) + 2\psi(q^4) - 6\right) \tag{31.2.24}$$

and

$$\sum_{n\geq 1} p\{\triangle_a+\triangle_b+\triangle_c+\triangle_d\}(n)q^n$$
$$= \frac{1}{24}\left(\psi(q)^4 - 4\psi(q)^3 - 6\psi(q)^2\psi(q^2) + 12\psi(q)^2 + 12\psi(q)\psi(q^2) + 8\psi(q)\psi(q^3)\right.$$
$$\left. +3\psi(q^2)^2 - 24\psi(q) - 12\psi(q^2) - 8\psi(q^3) - 6\psi(q^4) + 24\right). \tag{31.2.25}$$

Thus,

$$\sum_{n\geq 0} p^d_{4o}(8n+4)q^n = \sum_{n\geq 0} p^d_{4\triangle}(n)q^n$$
$$= \sum_{n\geq 1} p\{\triangle_a+\triangle_b+\triangle_c+\triangle_d\}(n)q^n + \sum_{n\geq 1} p\{\triangle_a+\triangle_b+\triangle_c\}(n)q^n$$
$$= \frac{1}{24}\left(\psi(q)^4 - 6\psi(q)^2\psi(q^2) + 8\psi(q)\psi(q^3) + 3\psi(q^2)^2 - 6\psi(q^4)\right),$$

which is (31.2.1).

31.3 The Generating Function for $p^d_{4e}(8n+4) + p^{d+}_{4e}(8n+4)$

We show that

$$\sum_{n\geq 0} (p^d_{4e}(8n+4) + p^{d+}_{4e}(8n+4))q^n$$
$$= \frac{1}{24}(\psi(q)^4 - 6\psi(q)^2\psi(q^2) + 8\psi(q)\psi(q^3) + 3\psi(q^2)^2 - 6\psi(q^4)). \tag{31.3.1}$$

We start by noting that

$$p^d_{4e}(4n) = p^d_{4\square}(n), \quad p^{d+}_{4e}(4n) = p^{d+}_{4\square}(n). \tag{31.3.2}$$

If we add (30.2.3) and (30.2.4) we find

$$\sum_{n\geq 0}(p^{d}_{4e}(4n)+p^{d+}_{4e}(4n))q^n = \sum_{n\geq 0} p^{d}_{4\square}(n)q^n + \sum_{n\geq 0} p^{d+}_{4\square}(n)q^n$$
$$= \frac{1}{192}\left(\phi(q)^4 - 12\phi(q)^2\phi(q^2) + 6\phi(q)^2 + 32\phi(q)\phi(q^3) + 12\phi(q^2)^2\right.$$
$$\left. -24\phi(q) - 12\phi(q^2) - 48\phi(q^4) + 45\right). \tag{31.3.3}$$

We now 2-dissect (31.3.3). We have

$$\sum_{n\geq 0}(p^{d}_{4e}(4n)+p^{d+}_{4e}(4n))q^n$$
$$= \frac{1}{192}((\phi(q^4)^2+4q\psi(q^4)^2)^2 - 12\phi(q^2)(\phi(q^2)^2+4q\psi(q^4)^2)$$
$$+6(\phi(q^2)^2+4q\psi(q^4)^2)+32\phi(q)\phi(q^3)+12\phi(q^2)^2$$
$$-24(\phi(q^4)+2q\psi(q^8)) - 12\phi(q^2) - 48\phi(q^4)+45). \tag{31.3.4}$$

If we extract odd parts making use of (20.1.1), divide by q, replace q^2 by q and use (1.5.13), we find that

$$\sum_{n\geq 0}(p^{d}_{4e}(8n+4)+p^{d+}_{4e}(8n+4))q^n$$
$$= \frac{1}{24}\left(\psi(q)^4 - 6\psi(q)^2\psi(q^2) + 8\psi(q)\psi(q^3) + 3\psi(q^2)^2 - 6\psi(q^4)\right),$$

which is (31.3.1).

31.4 Completing the Proof of (31.1.2)

If we compare (31.2.1) and (31.3.1), we see that we have succeeded in proving (31.1.2).

31.5 Some More Interesting Facts

From the preceding work, we can deduce the following interesting facts.

The number of partitions of $n \equiv 4 \pmod 8$ into four distinct even squares is half the number of partitions of n into four distinct odd squares if and only if n is not the sum of three positive distinct even squares. (31.5.1)

In particular, if we refer to (30.4.1),

The number of partitions of n $\equiv 28 \pmod{32}$ *into four distinct even squares is half the number of partitions of n into four distinct odd squares.* (31.5.2)

The number of partitions of n $\equiv 4 \pmod{8}$ *into four distinct even squares is equal to the number of partitions of n into four distinct odd squares if and only if n is not the sum of four positive distinct even squares.* (31.5.3)

If n is uniquely the sum of four distinct odd squares then n is not the sum of four positive distinct even squares, but is the sum of three positive distinct even squares. (31.5.4)

If n $\equiv 4 \pmod{8}$ *is the sum of four positive distinct even squares then n is not uniquely the sum of four distinct odd squares.* (31.5.5)

Endnotes.

§31.5 These statements have not previously been published by me.

References

52. R.Wm. Gosper, Some identities for your amusement, in *Ramanujan Revisited, Proceedings of the Centenary Conference, University of Illinois at Urbana-Champaign 1–5 June 1987*, pp. 607–610
78. M.D. Hirschhorn, On partitions into four squares of equal parity. Australas. J. Comb. **24**, 285–291 (2001)
87. M.D. Hirschhorn, My Contact with Ramanujan, Journal of the Indian Mathematical Society New Series, Special Volume to Commemorate the 125th Birth Anniversary of Srinivasa Ramanujan and the National Mathematics Year–2012 (2013), pp. 33–43

Chapter 32
Partitions with Odd Parts Distinct

32.1 Introduction

In this chapter, we study $pod(n)$, the number of partitions of n in which the odd parts are distinct. As we shall see, $pod(n)$ is also the number of partitions of n into parts not 2 modulo 4.

In particular, we will prove the following congruences, somewhat analogous to Ramanujan's partition congruences for powers of 5.

For $\alpha \geq 0$,

$$pod\left(3^{2\alpha+3}n + \frac{23 \times 3^{2\alpha+2}+1}{8}\right) \equiv 0 \pmod{3}. \tag{32.1.1}$$

32.2 The Generating Function for $pod(n)$

The generating function for $pod(n)$ is

$$\sum_{n\geq 0} pod(n)q^n = \frac{(-q;q^2)_\infty}{(q^2;q^2)_\infty} = \frac{(q^2;q^4)_\infty}{(q;q)_\infty} = \frac{(q^2;q^2)_\infty}{(q;q)_\infty(q^4;q^4)_\infty}$$

$$= \frac{1}{(q,q^3,q^4;q^4)_\infty} \tag{32.2.1}$$

$$= \frac{1}{\sum_{-\infty}^{\infty}(-1)^n q^{2n^2-n}} = \frac{1}{\psi(-q)}. \tag{32.2.2}$$

M.D. Hirschhorn, *The Power of q*, Developments in Mathematics 49,
DOI 10.1007/978-3-319-57762-3_32

From (32.2.1), we see that

$$pod(n) = p_{\pi \equiv 1,3 \text{ or } 4 \pmod 4}(n),$$

the number of partitions of n into parts not 2 modulo 4.

From (32.2.2), we find that

$$\sum_{n \geq 0} (-1)^n pod(n) q^n = \frac{1}{\psi(q)}. \tag{32.2.3}$$

32.3 Some Congruences

All congruences in this section and in the next are to the modulus 3.

We show that for $\alpha \geq 0$,

$$\sum_{n \geq 0} (-1)^n pod\left(3^{2\alpha+1} n + \frac{5 \times 3^{2\alpha+1}+1}{8}\right) q^n \equiv (-1)^\alpha \psi(q)^5 \tag{32.3.1}$$

and

$$\sum_{n \geq 0} (-1)^n pod\left(3^{2\alpha+2} n + \frac{7 \times 3^{2\alpha+2}+1}{8}\right) q^n \equiv (-1)^\alpha \psi(q)^7. \tag{32.3.2}$$

Our proof is by induction on α.

We have

$$\begin{aligned}
\sum_{n \geq 0} (-1)^n pod(n) q^n &= \frac{1}{\psi(q)} = \frac{\psi(q)^2}{\psi(q)^3} \equiv \frac{\psi(q)^2}{\psi(q^3)} \\
&\equiv \frac{1}{\psi(q^3)} \left(\Pi(q^3) + q\psi(q^9)\right)^2 \\
&\equiv \frac{1}{\psi(q^3)} \left(\Pi(q^3)^2 + 2q\Pi(q^3)\psi(q^9) + \psi(q^9)^2\right).
\end{aligned} \tag{32.3.3}$$

From (32.3.3) we extract

$$\sum_{n \geq 0} (-1)^n pod(3n+2) q^n \equiv \frac{\psi(q^3)^2}{\psi(q)} \equiv \frac{\left(\psi(q)^3\right)^2}{\psi(q)} = \psi(q)^5, \tag{32.3.4}$$

which is the case $\alpha = 0$ of (32.3.1).

Suppose

$$\sum_{n\geq 0}(-1)^n pod\left(3^{2\alpha+1}n+\frac{5\times 3^{2\alpha+1}+1}{8}\right)q^n \equiv (-1)^\alpha \psi(q)^5. \qquad (32.3.5)$$

Then

$$\begin{aligned}
&\sum_{n\geq 0}(-1)^n pod\left(3^{2\alpha+1}n+\frac{5\times 3^{2\alpha+1}+1}{8}\right)q^n \\
&\equiv (-1)^\alpha(\Pi(q^3)+q\psi(q^9))^5 \\
&\equiv (-1)^\alpha(\Pi(q^3)^5+2q\Pi(q^3)^4\psi(q^9)+q^2\Pi(q^3)^3\psi(q^9)^3+q^3\Pi(q^3)^2\psi(q^9)^3 \\
&\qquad +2q^4\Pi(q^3)\psi(q^9)^4+q^5\psi(q^9)^5). \qquad (32.3.6)
\end{aligned}$$

From (32.3.6), we extract

$$\begin{aligned}
&\sum_{n\geq 0}(-1)^{3n+2} pod\left(3^{2\alpha+1}(3n+2)+\frac{5\times 3^{2\alpha+1}+1}{8}\right)q^n \\
&\equiv (-1)^\alpha(\Pi(q)^3\psi(q^3)^2+q\psi(q^3)^5) \\
&= (-1)^\alpha \psi(q^3)^2\frac{\psi(q)^4}{\psi(q^3)} \\
&\equiv (-1)^\alpha\psi(q)^7. \qquad (32.3.7)
\end{aligned}$$

That is,

$$\sum_{n\geq 0}(-1)^n pod\left(3^{2\alpha+2}n+\frac{7\times 3^{2\alpha+2}+1}{8}\right)q^n \equiv (-1)^\alpha \psi(q)^7. \qquad (32.3.8)$$

Now suppose

$$\sum_{n\geq 0}(-1)^n pod\left(3^{2\alpha+2}n+\frac{7\times 3^{2\alpha+2}+1}{8}\right)q^n \equiv (-1)^\alpha \psi(q)^7. \qquad (32.3.9)$$

Then

$$\begin{aligned}
&\sum_{n\geq 0}(-1)^n pod\left(3^{2\alpha+2}n+\frac{7\times 3^{2\alpha+2}+1}{8}\right)q^n \\
&\equiv (-1)^\alpha\left(\Pi(q^3)+q\psi(q^9)\right)^7 \\
&\equiv (-1)^\alpha(\Pi(q^3)^7+q\Pi(q^3)^6\psi(q^9)+2q^3\Pi(q^3)^4\psi(q^9)^3 \\
&\qquad +2q^4\Pi(q^3)^3\psi(q^9)^4+q^6\Pi(q^3)\psi(q^9)^6+q^7\psi(q^9)^7). \qquad (32.3.10)
\end{aligned}$$

From (32.3.10) we extract

$$\begin{aligned}
&\sum_{n\geq 0}(-1)^{3n+1}pod\left(3^{2\alpha+2}(3n+1)+\frac{7\times 3^{2\alpha+2}+1}{8}\right)q^n\\
&\equiv(-1)^{\alpha}(\Pi(q)^6\psi(q^3)+2q\Pi(q)^3\psi(q^3)^4+q^2\psi(q^3)^7)\\
&=(-1)^{\alpha}\psi(q^3)(\frac{\psi(q)^4}{\psi(q^3)})^2\\
&\equiv(-1)^{\alpha}\psi(q)^5.
\end{aligned}\tag{32.3.11}$$

That is,

$$\sum_{n\geq 0}(-1)^n pod\left(3^{2\alpha+3}n+\frac{5\times 3^{2\alpha+3}+1}{8}\right)q^n\equiv(-1)^{\alpha+1}\psi(q)^5.\tag{32.3.12}$$

This completes the proof by induction.

32.4 Proof of the Congruence (32.1.1)

We have (32.3.10),

$$\begin{aligned}
&\sum_{n\geq 0}(-1)^n pod\left(3^{2\alpha+2}n+\frac{7\times 3^{2\alpha+2}+1}{8}\right)q^n\\
&\equiv(-1)^{\alpha}(\Pi(q^3)^7+q\Pi(q^3)^6\psi(q^9)+2q^3\Pi(q^3)^4\psi(q^9)^3\\
&\qquad +2q^4\Pi(q^3)^3\psi(q^9)^4+q^6\Pi(q^3)\psi(q^9)^6+q^7\psi(q^9)^7)
\end{aligned}\tag{32.4.1}$$

from which we extract

$$\sum_{n\geq 0}(-1)^n pod\left(3^{2\alpha+2}(3n+2)+\frac{7\times 3^{2\alpha+2}+1}{8}\right)q^n\equiv 0.\tag{32.4.2}$$

That is,

$$\sum_{n\geq 0}(-1)^n pod\left(3^{2\alpha+3}n+\frac{23\times 3^{2\alpha+2}+1}{8}\right)q^n\equiv 0,\tag{32.4.3}$$

and (32.1.1) follows.

Reference

96. M.D. Hirschhorn, J.A. Sellers, Partitions with odd parts distinct. Ramanujan J. **22**, 273–284 (2010)

Chapter 33
Partitions with Even Parts Distinct

33.1 Introduction

In this chapter, we study $ped(n)$, the number of partitions of n in which the even parts are distinct. As we shall see, $ped(n)$ is also the number of partitions of n into parts not divisible by 4.

In particular, we will prove the following congruences.

For $\alpha \geq 0$,

$$ped\left(3^{2\alpha+2}n + \frac{19 \times 3^{2\alpha+1} - 1}{8}\right) \equiv 0 \pmod 3 \tag{33.1.1}$$

and

$$ped\left(3^{2\alpha+3}n + \frac{17 \times 3^{2\alpha+2} - 1}{8}\right) \equiv 0 \pmod 3. \tag{33.1.2}$$

We will also establish the surprising identity

$$\sum_{n\geq 0} ped(9n+7)q^n = 12\frac{(q^2;q^2)_\infty^4 (q^3;q^3)_\infty^6 (q^4;q^4)_\infty}{(q;q)_\infty^{11}}, \tag{33.1.3}$$

surprising because it shows that $ped(n)$ is divisible by 12 at least 1/9 of the time.

33.2 The Generating Function for $ped(n)$ and Other Preliminaries

We have

M.D. Hirschhorn, *The Power of q*, Developments in Mathematics 49,
DOI 10.1007/978-3-319-57762-3_33

$$\sum_{n\geq 0} ped(n)q^n = \frac{(-q^2;q^2)_\infty}{(q;q^2)_\infty} = \frac{(q^4;q^4)_\infty}{(q^2;q^2)_\infty(q;q^2)_\infty} = \frac{(q^4;q^4)_\infty}{(q;q)_\infty}$$
$$= \frac{1}{(q,q^2,q^3;q^4)_\infty} \qquad (33.2.1)$$

which demonstrates that

$$ped(n) = p_{\pi\equiv 1,2 \text{ or } 3 \pmod 4}(n),$$

the number of partitions of n into parts not 0 modulo 4.

Also,

$$\sum_{n\geq 0} ped(n)q^n = \frac{(-q^2;q^2)_\infty}{(q;q^2)_\infty} = (-q;q)_\infty(-q^2;q^2)_\infty. \qquad (33.2.2)$$

From (33.2.2) it follows that

$$\sum_{n\geq 0} ped(n)q^n = (-q;q)_\infty(-q^2;q^2)_\infty$$
$$= (-q,-q^2,-q^3;q^3)_\infty(-q^2,-q^4,-q^6;q^6)_\infty$$
$$= \frac{(-q^3;q^3)_\infty}{(q^3;q^3)_\infty}\frac{(-q^6;q^6)_\infty}{(q^6;q^6)_\infty}(-q,-q^2,q^3;q^3)_\infty(-q^2,-q^4,q^6;q^6)_\infty$$
$$= \frac{(q^{12};q^{12})_\infty}{(q^3;q^3)^2_\infty(q^6;q^6)_\infty}\sum_{m,n=-\infty}^{\infty} q^{(3m^2+m)/2+3n^2+n}. \qquad (33.2.3)$$

We split the sum on the right of (33.2.3) according to the residue of $m+2n$ modulo 3.

If $m+2n\equiv 0 \pmod 3$, let $m+2n=3r$, $m-n=3s$, $m=r+2s$, $n=r-s$, if $m+2n\equiv -1 \pmod 3$, let $m+2n=3r-1$, $m-n=3s-1$, $m=r+2s-1$, $n=r-s$, and if $m+2n\equiv 1 \pmod 3$, let $m+2n=3r+1$, $m-n=3s+1$, $m=r+2s+1$, $n=r-s$.

Then, the sum becomes

$$\sum_{m,n=-\infty}^{\infty} q^{(3m^2+m)/2+3n^2+n} = \sum_{r,s=-\infty}^{\infty} q^{\left(3(r+2s)^2+(r+2s)\right)/2+3(r-s)^2+(r-s)}$$
$$+\sum_{r,s=-\infty}^{\infty} q^{\left(3(r+2s-1)^2+(r+2s-1)\right)/2+3(r-s)^2+(r-s)}$$
$$+\sum_{r,s=-\infty}^{\infty} q^{\left(3(r+2s+1)^2+(r+2s+1)\right)/2+3(r-s)^2+(r-s)}$$
$$= \sum_{r,s=-\infty}^{\infty} q^{(9r^2+3r)/2+9s^2} + q\sum_{r,s=-\infty}^{\infty} q^{(9r^2-3r)/2+9s^2-6s} + q^2\sum_{r,s=-\infty}^{\infty} q^{(9r^2+9r)/2+9s^2+6s}$$

$$= \Pi(q^3)\phi(q^9) + q\Pi(q^3)\Omega(q^3) + q^2\psi(q^9)\Omega(q^3). \tag{33.2.4}$$

If we substitute (33.2.4) into (33.2.3), we find

$$\sum_{n\geq 0} ped(n)q^n = \frac{(q^{12};q^{12})_\infty}{(q^3;q^3)_\infty^2(q^6;q^6)_\infty}\big(\Pi(q^3)\phi(q^9) + q\Pi(q^3)\Omega(q^3) + q^2\psi(q^9)\Omega(q^3)\big). \tag{33.2.5}$$

From (33.2.5) we extract

$$\begin{aligned}\sum_{n\geq 0} ped(3n+1)q^n &= \frac{(q^4;q^4)_\infty}{(q;q)_\infty^2(q^2;q^2)_\infty}\Pi(q)\Omega(q)\\ &= \frac{(q^4;q^4)_\infty}{(q;q)_\infty^2(q^2;q^2)_\infty}\,\frac{(q^2;q^2)_\infty(q^3;q^3)_\infty^2}{(q;q)_\infty(q^6;q^6)_\infty}\,\frac{(q^2;q^2)_\infty^2(q^3;q^3)_\infty(q^{12};q^{12})_\infty}{(q;q)_\infty(q^4;q^4)_\infty(q^6;q^6)_\infty}\\ &= \frac{(q^2;q^2)_\infty^2(q^3;q^3)_\infty^3(q^{12};q^{12})_\infty}{(q;q)_\infty^4(q^6;q^6)_\infty^2}\\ &= \frac{\phi(-q^3)\psi(-q^3)}{\phi(-q)^2}.\end{aligned} \tag{33.2.6}$$

Now put $-q$ for q, and we obtain

$$\sum_{n\geq 0}(-1)^n ped(3n+1)q^n = \frac{\phi(q^3)\psi(q^3)}{\phi(q)^2}. \tag{33.2.7}$$

33.3 The Derivation of (33.1.3)

From (33.2.7) we find

$$\sum_{n\geq 0}(-1)^n ped(3n+1)q^n = \phi(q^3)\psi(q^3)\left(\frac{\phi(q^9)}{\phi(q^3)^4}\left(\phi(q^9)^2 - 2q\phi(q^9)\Omega(q^3) + 4q^2\Omega(q^3)^2\right)\right)^2. \tag{33.3.1}$$

From (33.3.1) we extract

$$\sum_{n\geq 0}(-1)^n ped(9n+7)q^n = 12\frac{\phi(q^3)^4\psi(q)\Omega(q)^2}{\phi(q)^7}. \tag{33.3.2}$$

If we replace q by $-q$, we find

$$\begin{aligned}\sum_{n\geq 0} ped(9n+7)q^n &= 12\frac{\phi(-q^3)^4\psi(-q)\Omega(-q)^2}{\phi(-q)^7}\\ &= 12\frac{(q^3;q^3)_\infty^8}{(q^6;q^6)_\infty^4}\frac{(q^2;q^2)_\infty^7}{(q;q)_\infty^{14}}\frac{(q;q)_\infty(q^4;q^4)_\infty}{(q^2;q^2)_\infty}\frac{(q;q)_\infty^2(q^6;q^6)_\infty^4}{(q^2;q^2)_\infty^2(q^3;q^3)_\infty^2}\\ &= 12\frac{(q^2;q^2)_\infty^4(q^3;q^3)_\infty^6(q^4;q^4)_\infty}{(q;q)_\infty^{11}},\end{aligned}$$

which is (33.1.3).

Incidentally, (33.1.3) can be written

$$\sum_{n\geq 0} ped(9n+7)q^n = 12\prod_{n\geq 1}(1-q^n)^{-a_n} \tag{33.3.3}$$

where

$$\{a_n\} = \{11, 7, 5, 6, 11, 1, 11, 6, 5, 7, 11, 0, \dots\}$$

is periodic with period 12.

33.4 Some Congruences

All congruences in this and the next section are to the modulus 3.

We show that for $\alpha \geq 0$,

$$\sum_{n\geq 0} ped\left(3^{2\alpha+1}n+\frac{3^{2\alpha+2}-1}{8}\right)q^n \equiv (-1)^\alpha\phi(q)\psi(q^3) \tag{33.4.1}$$

and

$$\sum_{n\geq 0} ped\left(3^{2\alpha+2}n+\frac{3^{2\alpha+2}-1}{8}\right)q^n \equiv (-1)^\alpha\phi(q^3)\psi(q). \tag{33.4.2}$$

Our proof is by induction on α.

We have (33.2.7)

$$\sum_{n\geq 0}(-1)^n ped(3n+1)q^n = \frac{\phi(q^3)\psi(q^3)}{\phi(q)^2} \equiv \phi(q)\psi(q^3), \tag{33.4.3}$$

which is the case $\alpha = 0$ of (33.4.1).

Suppose

$$\sum_{n\geq 0}(-1)^n ped\left(3^{2\alpha+1}n + \frac{3^{2\alpha+2}-1}{8}\right)q^n \equiv (-1)^\alpha \phi(q)\psi(q^3). \tag{33.4.4}$$

Then

$$\sum_{n\geq 0}(-1)^n ped\left(3^{2\alpha+1}n + \frac{3^{2\alpha+2}-1}{8}\right)q^n \equiv (-1)^\alpha \psi(q^3)\left(\phi(q^9)+2q\Omega(q^3)\right) \tag{33.4.5}$$

from which we deduce

$$\sum_{n\geq 0}(-1)^n ped\left(3^{2\alpha+2}n + \frac{3^{2\alpha+2}-1}{8}\right)q^n \equiv (-1)^\alpha \phi(q^3)\psi(q). \tag{33.4.6}$$

Now suppose

$$\sum_{n\geq 0}(-1)^n ped\left(3^{2\alpha+2}n + \frac{3^{2\alpha+2}-1}{8}\right)q^n \equiv (-1)^\alpha \phi(q^3)\psi(q). \tag{33.4.7}$$

Then

$$\sum_{n\geq 0}(-1)^n ped\left(3^{2\alpha+2}n + \frac{3^{2\alpha+2}-1}{8}\right)q^n \equiv (-1)^\alpha \phi(q^3)\left(\Pi(q^3)+q\psi(q^9)\right), \tag{33.4.8}$$

from which we deduce

$$\sum_{n\geq 0}(-1)^{3n+1} ped\left(3^{2\alpha+2}(3n+1) + \frac{3^{2\alpha+2}-1}{8}\right)q^n \equiv (-1)^\alpha \phi(q)\psi(q^3), \tag{33.4.9}$$

or,

$$\sum_{n\geq 0}(-1)^n ped\left(3^{2\alpha+3}n+\frac{3^{2\alpha+4}-1}{8}\right)q^n \equiv (-1)^{\alpha+1}\phi(q)\psi(q^3). \tag{33.4.10}$$

This completes the proof by induction.

33.5 Proofs of the Congruences (33.1.1) and (33.1.2)

We have (33.4.5),

$$\sum_{n\geq 0}(-1)^n ped\left(3^{2\alpha+1}n+\frac{3^{2\alpha+2}-1}{8}\right)q^n \equiv (-1)^{\alpha}\psi(q^3)\left(\phi(q^9)+2q\Omega(q^3)\right) \tag{33.5.1}$$

from which we extract

$$\sum_{n\geq 0}(-1)^n ped\left(3^{2\alpha+1}(3n+2)+\frac{3^{2\alpha+2}-1}{8}\right)q^n \equiv 0, \tag{33.5.2}$$

or,

$$\sum_{n\geq 0}(-1)^n ped\left(3^{2\alpha+2}n+\frac{19\times 3^{2\alpha+1}-1}{8}\right)q^n \equiv 0, \tag{33.5.3}$$

and (33.1.1) follows.

We also have (33.4.8),

$$\sum_{n\geq 0}(-1)^n ped\left(3^{2\alpha+2}n+\frac{3^{2\alpha+2}-1}{8}\right)q^n \equiv (-1)^{\alpha}\phi(q^3)\left(\Pi(q^3)+q\psi(q^9)\right), \tag{33.5.4}$$

from which we extract

$$\sum_{n\geq 0}(-1)^n ped\left(3^{2\alpha+2}(3n+2)+\frac{3^{2\alpha+2}-1}{8}\right)q^n \equiv 0, \tag{33.5.5}$$

or,

$$\sum_{n\geq 0}(-1)^n ped\left(3^{2\alpha+3}n+\frac{17\times 3^{2\alpha+2}-1}{8}\right)q^n \equiv 0, \tag{33.5.6}$$

and (33.1.2) follows.

Endnote.

The arguments presented here are simplified versions of those in Andrews–Hirschhorn–Sellers (2010) [10].

Reference

10. G.E. Andrews, M.D. Hirschhorn, J.A. Sellers, Arithmetic properties of partitions with even parts distinct. Ramanujan J. **23**, 165–181 (2010)

Chapter 34
Some Identities Involving $\phi(q)$ and $\psi(q)$

34.1 Introduction

In this chapter, we prove many identities involving $\phi(q)$ and $\psi(q)$, quite a few of them due to Blecksmith et al. (1987), (1988) [18, 19].

Indeed, we prove

$$\phi(q)+\phi(q^2)=2\left(\begin{matrix}q^3,q^5,q^8\\q,q^4,q^7\end{matrix};q^8\right)_\infty, \tag{34.1.1}$$

$$\phi(q)-\phi(q^2)=2q\left(\begin{matrix}q,q^7,q^8\\q^3,q^4,q^5\end{matrix};q^8\right)_\infty, \tag{34.1.2}$$

$$\phi(-q^2)+\phi(-q)=2\frac{(q;q)_\infty}{(q^4,q^6,q^8,q^{10},q^{22},q^{24},q^{26},q^{28};q^{32})_\infty}, \tag{34.1.3}$$

$$\phi(-q^2)-\phi(-q)=2q\frac{(q;q)_\infty}{(q^2,q^8,q^{12},q^{14},q^{18},q^{20},q^{24},q^{30};q^{32})_\infty}, \tag{34.1.4}$$

$$\phi(q)+\phi(q^3)=2\left(\begin{matrix}q^2,q^6,q^{10},q^{12}\\q,q^3,q^9,q^{11}\end{matrix};q^{12}\right)_\infty, \tag{34.1.5}$$

$$\phi(q)-\phi(q^3)=2q\left(\begin{matrix}q^2,q^6,q^{10},q^{12}\\q^3,q^5,q^7,q^9\end{matrix};q^{12}\right)_\infty, \tag{34.1.6}$$

$$\phi(q)+\phi(q^5)=2\left(\begin{matrix}q^2,q^8,q^{10},q^{12},q^{18},q^{20}\\q,q^4,q^9,q^{11},q^{16},q^{19}\end{matrix};q^{20}\right)_\infty, \tag{34.1.7}$$

The original version of this chapter has been revised. The erratum to this chapter is available at https://doi.org/10.1007/978-3-319-57762-3_44.

M.D. Hirschhorn, *The Power of q*, Developments in Mathematics 49,
DOI 10.1007/978-3-319-57762-3_34

$$\phi(q)-\phi(q^5)=2q\left(\begin{matrix}q^4,q^6,q^{10},q^{14},q^{16},q^{20}\\ q^3,q^7,q^8,q^{12},q^{13},q^{17}\end{matrix};q^{20}\right)_\infty, \tag{34.1.8}$$

$$\psi(q^4)+q\psi(q^{12})=\left(\begin{matrix}q^2,q^5,q^7,q^{12},q^{17},q^{19},q^{22},q^{24}\\ q,q^4,q^6,q^{11},q^{13},q^{18},q^{20},q^{23}\end{matrix};q^{24}\right)_\infty, \tag{34.1.9}$$

$$\psi(q^4)-q\psi(q^{12})=\left(\begin{matrix}q,q^{10},q^{11},q^{12},q^{13},q^{14},q^{23},q^{24}\\ q^4,q^5,q^6,q^7,q^{17},q^{18},q^{19},q^{20}\end{matrix};q^{24}\right)_\infty, \tag{34.1.10}$$

$$\psi(q^2)+q\psi(q^{10})=\left(\begin{matrix}q^3,q^7,q^{10},q^{13},q^{17},q^{20}\\ q,q^6,q^9,q^{11},q^{14},q^{19}\end{matrix};q^{20}\right)_\infty, \tag{34.1.11}$$

$$\psi(q^2)-q\psi(q^{10})=\left(\begin{matrix}q,q^9,q^{10},q^{11},q^{19},q^{20}\\ q^2,q^3,q^7,q^{13},q^{17},q^{18}\end{matrix};q^{20}\right)_\infty, \tag{34.1.12}$$

$$\phi(q)^2-\phi(q^2)^2=4q\psi(q^4)^2, \tag{34.1.13}$$

$$\begin{aligned}\phi(q)^2-\phi(q^3)^2&=4q\left(\begin{matrix}q^2,q^2,q^6,q^6,q^{10},q^{10},q^{12},q^{12}\\ q,q^3,q^3,q^5,q^7,q^9,q^9,q^{11}\end{matrix};q^{12}\right)_\infty\\ &=4q\frac{(q^2;q^2)_\infty^3(q^6;q^6)_\infty(q^{12};q^{12})_\infty^2}{(q;q)_\infty(q^3;q^3)_\infty(q^4;q^4)_\infty^2},\end{aligned} \tag{34.1.14}$$

$$\begin{aligned}\phi(q)^2+\phi(q^3)^2&=2\left(\begin{matrix}q^2,q^6,q^6,q^6,q^6,q^{10},q^{12},q^{12}\\ q,q,q^5,q^5,q^7,q^7,q^{11},q^{11}\end{matrix};q^{12}\right)_\infty\\ &=2\frac{(q^2;q^2)_\infty^3(q^3;q^3)_\infty^2(q^6;q^6)_\infty}{(q;q)_\infty^2(q^4;q^4)_\infty(q^{12};q^{12})_\infty},\end{aligned} \tag{34.1.15}$$

$$\begin{aligned}&\phi(q)^4-\phi(q^3)^4\\ &\quad=8q\left(\begin{matrix}q^2,q^2,q^2,q^6,q^6,q^6,q^6,q^6,q^6,q^{10},q^{10},q^{10},q^{12},q^{12},q^{12},q^{12}\\ q,q,q,q^3,q^3,q^5,q^5,q^5,q^7,q^7,q^7,q^9,q^9,q^{11},q^{11},q^{11}\end{matrix};q^{12}\right)_\infty\\ &\quad=8q\frac{(q^2;q^2)_\infty^6(q^3;q^3)_\infty(q^6;q^6)_\infty^2(q^{12};q^{12})_\infty}{(q;q)_\infty^3(q^4;q^4)_\infty^3}\\ &\quad=8q\phi(q^3)\Omega(q)^3,\end{aligned} \tag{34.1.16}$$

$$\begin{aligned}\psi(q^2)^2-q\psi(q^6)^2&=\left(\begin{matrix}q,q^5,q^6,q^6,q^7,q^{11},q^{12},q^{12}\\ q^2,q^2,q^3,q^3,q^9,q^9,q^{10},q^{10}\end{matrix};q^{12}\right)_\infty\\ &=\frac{(q;q)_\infty(q^4;q^4)_\infty^2(q^6;q^6)_\infty^7}{(q^2;q^2)_\infty^3(q^3;q^3)_\infty^3(q^{12};q^{12})_\infty^2},\end{aligned} \tag{34.1.17}$$

$$\begin{aligned}\psi(q^2)^2+q\psi(q^6)^2&=\left(\begin{matrix}q^3,q^3,q^9,q^9,q^{12},q^{12}\\ q,q^2.q^5,q^7,q^{10},q^{11}\end{matrix};q^{12}\right)_\infty\\ &=\frac{(q^3;q^3)_\infty^3(q^4;q^4)_\infty(q^{12};q^{12})_\infty}{(q;q)_\infty(q^6;q^6)_\infty^2},\end{aligned}\tag{34.1.18}$$

$$\begin{aligned}\psi(q)^4-q\psi(q^3)^4&=\left(\begin{matrix}q^3,q^3,q^6,q^6,q^6,q^6\\ q,q,q,q^5,q^5,q^5\end{matrix};q^6\right)_\infty\\ &=\frac{(q^2;q^2)_\infty^3(q^3;q^3)_\infty^5}{(q;q)_\infty^3(q^6;q^6)_\infty}\\ &=\psi(q^3)\Pi(q)^3,\end{aligned}\tag{34.1.19}$$

$$\begin{aligned}\phi(q)^2-\phi(q^5)^2&=4q\left(\begin{matrix}q^2,q^6,q^{10},q^{10},q^{14},q^{18},q^{20},q^{20}\\ q,q^3,q^7,q^9,q^{11},q^{13},q^{17},q^{19}\end{matrix};q^{20}\right)_\infty\\ &=4q\frac{(q^2;q^2)_\infty^2(q^5;q^5)_\infty(q^{20};q^{20})_\infty}{(q;q)_\infty(q^4;q^4)_\infty}\\ &=4q(-q,-q^3,-q^7,-q^9,q^{10},q^{10};q^{10})_\infty,\end{aligned}\tag{34.1.20}$$

$$\begin{aligned}\psi(q)^2-q\psi(q^5)^2&=\left(\begin{matrix}q^5,q^5,q^{10},q^{10}\\ q,q^3,q^7,q^9\end{matrix};q^{10}\right)_\infty\\ &=\frac{(q^2;q^2)_\infty(q^5;q^5)_\infty^3}{(q;q)_\infty(q^{10};q^{10})_\infty}\\ &=(-q,-q^2,-q^3,-q^4,q^5,q^5;q^5)_\infty,\end{aligned}\tag{34.1.21}$$

$$\begin{aligned}5\phi(q^5)^2-\phi(q)^2&=4(q,q^2,q^3,q^4,q^4,q^6,q^7,q^8,q^8,q^9,q^{10},q^{10},\\ &\qquad q^{11},q^{12},q^{12},q^{13},q^{14},q^{16},q^{16},q^{17},q^{18},q^{19},q^{20},q^{20};q^{20})_\infty\\ &=4\frac{(q;q)_\infty(q^4;q^4)_\infty(q^{10};q^{10})_\infty^2}{(q^5;q^5)_\infty(q^{20};q^{20})_\infty},\end{aligned}\tag{34.1.22}$$

$$\begin{aligned}\psi(q)^2-5q\psi(q^5)^2&=(q,q,q,q^2,q^2,q^3,q^3,q^3,q^4,q^4,q^5,q^5,\\ &\qquad q^6,q^6,q^7,q^7,q^7,q^8,q^8,q^9,q^9,q^9,q^{10},q^{10};q^{10})_\infty\\ &=\frac{(q;q)_\infty^3(q^{10};q^{10})_\infty}{(q^2;q^2)_\infty(q^5;q^5)_\infty}.\end{aligned}\tag{34.1.23}$$

$$3\phi(q^3)^2+\phi(q)^2=4\frac{(q^2;q^2)_\infty(q^4;q^4)_\infty^2(q^6;q^6)_\infty^3}{(q;q)_\infty(q^3;q^3)_\infty(q^{12};q^{12})_\infty^2} \tag{34.1.24}$$

and

$$3\phi(q^3)^2-\phi(q)^2=2\frac{(q;q)_\infty^2(q^2;q^2)_\infty(q^6;q^6)_\infty^3}{(q^3;q^3)_\infty^2(q^4;q^4)_\infty(q^{12};q^{12})_\infty}. \tag{34.1.25}$$

The proofs of all these occupy §§34.2–34.9, and in §34.10 I present a conjecture.

34.2 Proofs of (34.1.1) and (34.1.2)

We have

$$\begin{aligned}
&\left(\begin{matrix}q^3,q^5,q^8\\ q,q^4,q^7\end{matrix};q^8\right)_\infty+q\left(\begin{matrix}q,q^7,q^8\\ q^3,q^4,q^5\end{matrix};q^8\right)_\infty\\
&=\frac{1}{(q,q^3,q^4,q^5,q^7,q^8;q^8)_\infty}((q^3,q^5,q^8;q^8)_\infty^2+q(q,q^7,q^8;q^8)_\infty^2)\\
&=\frac{(q^2;q^2)_\infty}{(q;q)_\infty(q^4;q^4)_\infty}f(q),
\end{aligned} \tag{34.2.1}$$

where

$$f(q)=\sum_{m,n=-\infty}^{\infty}(-1)^{m+n}q^{4m^2+m+4n^2+n}+q\sum_{m,n=-\infty}^{\infty}(-1)^{m+n}q^{4m^2-3m+4n^2-3n},$$

$$f(-q)=\sum_{m,n=-\infty}^{\infty}q^{4m^2+m+4n^2+n}-q\sum_{m,n=-\infty}^{\infty}q^{4m^2-3m+4n^2-3n},$$

$$\begin{aligned}
q^2f(-q^{16})&=\sum_{m,n=-\infty}^{\infty}q^{(8m+1)^2+(8n+1)^2}-\sum_{m,n=-\infty}^{\infty}q^{(8m-3)^2+(8n-3)^2}\\
&=\sum_{m,n\text{ even}}q^{(4m+1)^2+(4n+1)^2}-\sum_{m,n\text{ odd}}q^{(4m+1)^2+(4n+1)^2}\\
&=\sum_{m\equiv n\pmod 2}(-1)^m q^{(4m+1)^2+(4n+1)^2}\\
&=\sum_{r,s=-\infty}^{\infty}(-1)^{r+s}q^{(4(r+s)+1)^2+(4(r-s)+1)^2}\\
&=\sum_{r,s=-\infty}^{\infty}(-1)^{r+s}q^{32r^2+16r+32s^2+2},
\end{aligned}$$

$$f(-q) = \sum_{r,s=-\infty}^{\infty} (-1)^{r+s} q^{2r^2+r+2s^2}$$
$$= \phi(-q^2)\psi(-q)$$

and

$$\begin{aligned} f(q) &= \phi(-q^2)\psi(q) \\ &= \frac{(q^2;q^2)_\infty^2}{(q^4;q^4)_\infty} \frac{(q^2;q^2)_\infty^2}{(q;q)_\infty} \\ &= \frac{(q^2;q^2)_\infty^4}{(q;q)_\infty (q^4;q^4)_\infty}. \end{aligned} \tag{34.2.2}$$

If we substitute (34.2.2) into (34.2.1), we find

$$\left(\begin{matrix} q^3, q^5, q^8 \\ q, q^4, q^7 \end{matrix}; q^8\right)_\infty + q\left(\begin{matrix} q, q^7, q^8 \\ q^3, q^4, q^5 \end{matrix}; q^8\right)_\infty = \frac{(q^2;q^2)_\infty^5}{(q;q)_\infty^2 (q^4;q^4)_\infty^2} = \phi(q),$$

which is (34.1.1).

Similarly,

$$\left(\begin{matrix} q^3, q^5, q^8 \\ q, q^4, q^7 \end{matrix}; q^8\right)_\infty - q\left(\begin{matrix} q, q^7, q^8 \\ q^3, q^4, q^5 \end{matrix}; q^8\right)_\infty = \frac{(q^2;q^2)_\infty}{(q;q)_\infty (q^4;q^4)_\infty} g(q), \tag{34.2.3}$$

where

$$g(q) = \sum_{m,n=-\infty}^{\infty} (-1)^{m+n} q^{4m^2+m+4n^2+n} - q \sum_{m,n=-\infty}^{\infty} (-1)^{m+n} q^{4m^2-3m+4n^2-3n},$$

$$g(-q) = \sum_{m,n=-\infty}^{\infty} q^{4m^2+m+4n^2+n} + q \sum_{m,n=-\infty}^{\infty} q^{4m^2-3m+4n^2-3n},$$

$$\begin{aligned} q^2 g(-q^{16}) &= \sum_{m,n=-\infty}^{\infty} q^{(8m+1)^2+(8n+1)^2} + \sum_{m,n=-\infty}^{\infty} q^{(8m-3)^2+(8n-3)^2} \\ &= \sum_{m,n \text{ even}} q^{(4m+1)^2+(4n+1)^2} + \sum_{m,n \text{ odd}} q^{(4m+1)^2+(4n+1)^2} \\ &= \sum_{m \equiv n \pmod 2} q^{(4m+1)^2+(4n+1)^2} \end{aligned}$$

$$= \sum_{r,s=-\infty}^{\infty} q^{(4(r+s)+1)^2+(4(r-s)+1)^2}$$
$$= \sum_{r,s=-\infty}^{\infty} q^{32r^2+16r+32s^2+2},$$

$$g(-q) = \sum_{r,s=-\infty}^{\infty} q^{2r^2+r+2s^2}$$
$$= \phi(q^2)\psi(q)$$

and

$$\begin{aligned} g(q) &= \phi(q^2)\psi(-q) \\ &= \frac{(q^2;q^2)_\infty^5}{(q;q)_\infty^2(q^4;q^4)_\infty^2}\,\frac{(q;q)_\infty(q^4;q^4)_\infty}{(q^2;q^2)_\infty} \\ &= \frac{(q;q)_\infty(q^4;q^4)_\infty^6}{(q^2;q^2)_\infty^3(q^8;q^8)_\infty^2}. \end{aligned} \tag{34.2.4}$$

If we substitute (34.2.4) into (34.2.3), we find

$$\left(\begin{matrix} q^3,q^5,q^8 \\ q,q^4,q^7 \end{matrix};q^8\right)_\infty - q\left(\begin{matrix} q,q^7,q^8 \\ q^3,q^4,q^5 \end{matrix};q^8\right)_\infty = \frac{(q^4;q^4)_\infty^5}{(q^2;q^2)_\infty^2(q^8;q^8)_\infty^2} = \phi(q^2),$$

which is (34.1.2).

34.3 Proofs of (34.1.3) and (34.1.4)

In order to prove (34.1.3) and (34.1.4), we require the following 2-dissections:

$$\left(\begin{matrix} q^3,q^5,q^8 \\ q,q^4,q^7 \end{matrix};q^8\right)_\infty = \left(\begin{matrix} q^6,q^{10},q^{16} \\ q^2,q^8,q^{14} \end{matrix};q^{16}\right)_\infty + q\left(\begin{matrix} q^{16} \\ q^8 \end{matrix};q^{16}\right)_\infty, \tag{34.3.1}$$

$$\left(\begin{matrix} q,q^7,q^8 \\ q^3,q^4,q^5 \end{matrix};q^8\right)_\infty = \left(\begin{matrix} q^{16} \\ q^8 \end{matrix};q^{16}\right)_\infty - q\left(\begin{matrix} q^2,q^{14},q^{16} \\ q^6,q^8,q^{10} \end{matrix};q^{16}\right)_\infty \tag{34.3.2}$$

and

$$\begin{aligned} (q;q)_\infty &= (q^2,q^{12},q^{14},q^{16},q^{18},q^{20},q^{30},q^{32};q^{32})_\infty \\ &\quad - q(q^4,q^6,q^{10},q^{16},q^{22},q^{26},q^{28},q^{32};q^{32})_\infty. \end{aligned} \tag{34.3.3}$$

We have from (34.1.1) and (34.1.2),

$$
\begin{aligned}
2\begin{pmatrix} q^3,q^5,q^8 \\ q,q^4,q^7 \end{pmatrix};q^8\Big)_\infty &= \phi(q)+\phi(q^2) \\
&= \left(\phi(q^4)+2q\psi(q^8)\right)+\phi(q^2) \\
&= \left(\phi(q^2)+\phi(q^4)\right)+2q\psi(q^8) \\
&= 2\left(\begin{matrix} q^6,q^{10},q^{16} \\ q^2,q^8,q^{14} \end{matrix};q^{16}\right)_\infty + 2q\left(\begin{matrix} q^{16} \\ q^8 \end{matrix};q^{16}\right)_\infty,
\end{aligned}
$$

which yields (34.3.1) on division by 2, and

$$
\begin{aligned}
2q\left(\begin{matrix} q,q^7,q^8 \\ q^3,q^4,q^5 \end{matrix};q^8\right)_\infty &= \phi(q)-\phi(q^2) \\
&= \left(\phi(q^4)+2q\psi(q^8)\right)-\phi(q^2) \\
&= 2q\psi(q^8)-\left(\phi(q^2)-\phi(q^4)\right) \\
&= 2q\left(\begin{matrix} q^{16} \\ q^8 \end{matrix};q^{16}\right)_\infty - 2q^2\left(\begin{matrix} q^2,q^{14},q^{16} \\ q^6,q^8,q^{10} \end{matrix};q^{16}\right)_\infty,
\end{aligned}
$$

which yields (34.3.2) on division by $2q$.

Also,

$$
\begin{aligned}
(q;q)_\infty &= (q,q^2,q^3,q^4;q^4)_\infty \\
&= (q^2;q^4)_\infty(q,q^3,q^4;q^4)_\infty \\
&= (q^2;q^4)_\infty \sum_{-\infty}^{\infty}(-1)^n q^{2n^2+n} \\
&= (q^2;q^4)_\infty\left(\sum_{-\infty}^{\infty} q^{8n^2+2n} - q\sum_{-\infty}^{\infty} q^{8n^2+6n}\right) \\
&= (q^2;q^4)_\infty\left((-q^6,-q^{10},q^{16};q^{16})_\infty - q(-q^2,-q^{14},q^{16};q^{16})_\infty\right) \\
&= (q^2;q^4)_\infty\left(\left(\begin{matrix} q^{12},q^{16},q^{20},q^{32} \\ q^6,q^{10},q^{22},q^{26} \end{matrix};q^{32}\right)_\infty - q\left(\begin{matrix} q^4,q^{16},q^{28},q^{32} \\ q^2,q^{14},q^{18},q^{30} \end{matrix};q^{32}\right)_\infty\right) \\
&= (q^2,q^{12},q^{14},q^{16},q^{18},q^{20},q^{30},q^{32};q^{32})_\infty \\
&\qquad - q(q^4,q^6,q^{10},q^{16},q^{22},q^{26},q^{28},q^{32};q^{32})_\infty,
\end{aligned}
$$

which is (34.3.3).

We have, by (34.1.1), (34.3.2) and (34.3.3),

$$
\begin{aligned}
&\phi(-q^2)+\phi(-q)=(\phi(q^8)-2q^2\psi(q^{16}))+(\phi(q^4)-2q\psi(q^8))\\
&=(\phi(q^4)+\phi(q^8))-2q^2\psi(q^{16})-2q\psi(q^8)\\
&=2\left(\begin{matrix}q^{12},q^{20},q^{32}\\ q^4,q^{16},q^{28}\end{matrix};q^{32}\right)-2q^2\left(\begin{matrix}q^{32}\\ q^{16}\end{matrix};q^{32}\right)_\infty-2q\left(\begin{matrix}q^{16}\\ q^{8}\end{matrix};q^{16}\right)_\infty\\
&=2\left(\begin{matrix}q^{12},q^{20}\\ q^4.q^{28}\end{matrix};q^{32}\right)_\infty\left(\left(\begin{matrix}q^{32}\\ q^{16}\end{matrix};q^{32}\right)_\infty-q^2\left(\begin{matrix}q^4,q^{28},q^{32}\\ q^{12},q^{16},q^{20}\end{matrix};q^{32}\right)_\infty\right)-2q\left(\begin{matrix}q^{16}\\ q^{8}\end{matrix};q^{16}\right)_\infty\\
&=2\left(\begin{matrix}q^{12},q^{20}\\ q^4.q^{28}\end{matrix};q^{32}\right)_\infty\left(\begin{matrix}q^2,q^{14},q^{16}\\ q^6,q^8,q^{10}\end{matrix};q^{16}\right)_\infty-2q\left(\begin{matrix}q^{16}\\ q^{8}\end{matrix};q^{16}\right)_\infty\\
&=2\left(\begin{matrix}q^2,q^{12},q^{14},q^{16},q^{18},q^{20},q^{30},q^{32}\\ q^4,q^6,q^8,q^{10},q^{22},q^{24},q^{26},q^{28}\end{matrix};q^{32}\right)_\infty-2q\left(\begin{matrix}q^{16},q^{32}\\ q^8,q^{24}\end{matrix};q^{32}\right)_\infty\\
&=2\frac{(q^2,q^{12},q^{14},q^{16},q^{18},q^{20},q^{30},q^{32};q^{32})_\infty-q(q^4,q^6,q^{10},q^{16},q^{22},q^{26},q^{28},q^{32};q^{32})_\infty}{(q^4,q^6,q^8,q^{10},q^{22},q^{24},q^{26},q^{28};q^{32})_\infty}\\
&=2\frac{(q;q)_\infty}{(q^4,q^6,q^8,q^{10},q^{22},q^{24},q^{26},q^{28};q^{32})_\infty},
\end{aligned}
$$

which is (34.1.3), and by (34.1.2), (34.3.1) and (34.3.3),

$$
\begin{aligned}
&\phi(-q^2)-\phi(-q)\\
&=(\phi(q^8)-2q^2\psi(q^{16}))-(\phi(q^4)-2q\psi(q^8))\\
&=2q\psi(q^8)-2q^2\psi(q^{16})-(\phi(q^4)-\phi(q^8))\\
&=2q\left(\begin{matrix}q^{16}\\ q^{8}\end{matrix};q^{16}\right)_\infty-2q^2\left(\begin{matrix}q^{32}\\ q^{16}\end{matrix};q^{32}\right)_\infty-2q^4\left(\begin{matrix}q^4,q^{28},q^{32}\\ q^{12},q^{16},q^{20}\end{matrix};q^{32}\right)_\infty\\
&=2q\left(\begin{matrix}q^{16}\\ q^{8}\end{matrix};q^{16}\right)_\infty-2q^2\left(\begin{matrix}q^4,q^{28}\\ q^{12},q^{20}\end{matrix};q^{32}\right)_\infty\left(\left(\begin{matrix}q^{12},q^{20},q^{32}\\ q^4,q^{16},q^{28}\end{matrix};q^{32}\right)_\infty+q^2\left(\begin{matrix}q^{32}\\ q^{16}\end{matrix};q^{32}\right)_\infty\right)\\
&=2q\left(\begin{matrix}q^{16}\\ q^{8}\end{matrix};q^{16}\right)_\infty-2q^2\left(\begin{matrix}q^4,q^{28}\\ q^{12},q^{20}\end{matrix};q^{32}\right)_\infty\left(\begin{matrix}q^6,q^{10},q^{16}\\ q^2,q^8,q^{14}\end{matrix};q^{16}\right)_\infty\\
&=2q\left(\begin{matrix}q^{16},q^{32}\\ q^8,q^{24}\end{matrix};q^{32}\right)_\infty-2q^2\left(\begin{matrix}q^4,q^6,q^{10},q^{16},q^{22},q^{24},q^{26},q^{32}\\ q^2,q^8,q^{12},q^{14},q^{18},q^{20},q^{24},q^{28}\end{matrix};q^{32}\right)_\infty\\
&=2q\frac{(q^2,q^{12},q^{14},q^{16},q^{18},q^{20},q^{30},q^{32};q^{32})_\infty-q(q^4,q^6,q^{10},q^{16},q^{22},q^{26},q^{28},q^{32};q^{32})_\infty}{(q^2,q^8,q^{12},q^{14},q^{18},q^{20},q^{24},q^{28};q^{32})_\infty}\\
&=2q\frac{(q;q)_\infty}{(q^2,q^8,q^{12},q^{14},q^{18},q^{20},q^{24},q^{28};q^{32})_\infty},
\end{aligned}
$$

which is (34.1.4).

34.4 Proofs of (34.1.5) and (34.1.6)

We start by obtaining the 3-dissections

$$\left(\begin{matrix} q^2,q^6,q^{10},q^{12} \\ q,q^3,q^9,q^{12} \end{matrix};q^{12}\right)_\infty = \left(\begin{matrix} q^6,q^{18},q^{30},q^{36} \\ q^3,q^9,q^{27},q^{33} \end{matrix};q^{36}\right)_\infty + q\Omega(q^3) \tag{34.4.1}$$

and

$$\left(\begin{matrix} q^2,q^6,q^{10},q^{12} \\ q^3,q^5,q^7,q^9 \end{matrix};q^{12}\right)_\infty = \Omega(q^3) - q^2\left(\begin{matrix} q^6,q^{18},q^{30},q^{36} \\ q^9,q^{15},q^{21},q^{27} \end{matrix};q^{36}\right)_\infty. \tag{34.4.2}$$

We have

$$\begin{aligned}
&\left(\begin{matrix} q^2,q^6,q^{10},q^{12} \\ q,q^3,q^9.q^{11} \end{matrix};q^{12}\right)_\infty \\
&= \frac{(q^6;q^6)_\infty}{(q^3;q^6)_\infty}\left(\begin{matrix} q^2,q^{10} \\ q,q^{11} \end{matrix};q^{12}\right)_\infty \\
&= \frac{(q^6;q^6)_\infty^2}{(q^3;q^3)_\infty}\frac{(q^2,q^{10},q^{14},q^{22};q^{24})_\infty}{(q,q^{11};q^{12})_\infty} \\
&= \frac{(q^6;q^6)_\infty^2}{(q^3;q^3)_\infty}(-q,-q^{11};q^{12})_\infty(q^{10},q^{14};q^{24})_\infty \\
&= \frac{(q^6;q^6)_\infty^2}{(q^3;q^3)_\infty(q^{12};q^{12})_\infty(q^{24};q^{24})_\infty}(-q,-q^{11},q^{12};q^{12})_\infty(q^{10},q^{14},q^{24};q^{24})_\infty \\
&= \frac{(q^6;q^6)_\infty^2}{(q^3;q^3)_\infty(q^{12};q^{12})_\infty(q^{24};q^{24})_\infty}\sum_{m,n=-\infty}^{\infty}(-1)^n q^{6m^2-5m+12n^2-2n}.
\end{aligned} \tag{34.4.3}$$

We now 3-dissect the sum, according to the residue modulo 3 of $5m+2n \equiv -(m+n)$.

If $m+n \equiv 0 \pmod 3$, let $m+n = 3r$, $m-2n = 3s$, $m = 2r+s$, $n = r-s$, if $m+n = 3r+1$, $m-2n = 3s+1$, $m = 2r+s+1$, $n = r-s$, if $m+n = 3r-1$, $m-2n = 3s-1$, $m = 2r+s-1$, $n = r-s$, and the sum becomes

$$\begin{aligned}
&\sum_{m,n=-\infty}^{\infty}(-1)^n q^{6m^2-5m+12n^2-2n} \\
&= \sum_{r,s=-\infty}^{\infty}(-1)^{r-s} q^{(6(2r+s)^2-5(2r+s)+12(r-s)^2-2(r-s)} \\
&+ \sum_{r,s=-\infty}^{\infty}(-1)^{r-s} q^{6(2r+s+1)^2-5(2r+s+1)+12(r-s)^2-2(r-s)} \\
&+ \sum_{r,s=-\infty}^{\infty}(-1)^{r-s} q^{6(2r+s-1)^2-5(2r+s-1)+12(r-s)^2-2(r-s)}
\end{aligned}$$

$$\begin{aligned}
&= \sum_{r,s=-\infty}^{\infty} (-1)^{r+s} q^{36r^2-12r+18s^2-3s} \\
&+ q \sum_{r,s=-\infty}^{\infty} (-1)^{r+s} q^{36r^2+12r+18s^2+9s} \\
&+ q^{11} \sum_{r,s=-\infty}^{\infty} (-1)^{r+s} q^{36r^2-36r+18s^2-15s} \\
&= (q^{24};q^{24})_\infty (q^{15},q^{21},q^{36};q^{36})_\infty + q(q^{24};q^{24})_\infty (q^{9},q^{27},q^{36};q^{36})_\infty. \qquad (34.4.4)
\end{aligned}$$

If we substitute (34.4.4) into (34.4.3), we find

$$\begin{aligned}
&\left(\begin{matrix} q^2 q^6, q^{10}, q^{12} \\ q, q^3, q^9, q^{11} \end{matrix} ; q^{12} \right)_\infty \\
&= \frac{(q^6;q^6)_\infty^2}{(q^3;q^3)_\infty (q^{12};q^{12})_\infty} \left((q^{15},q^{21},q^{36};q^{36})_\infty + q(q^9,q^{27},q^{36};q^{36})_\infty \right) \\
&= \left(\begin{matrix} q^6, q^{18}, q^{30}, q^{36} \\ q^3, q^9, q^{27}, q^{33} \end{matrix} ; q^{36} \right)_\infty + q \left(\begin{matrix} q^6, q^{18}, q^{30}, q^{36} \\ q^3, q^{15}, q^{21}, q^{33} \end{matrix} ; q^{36} \right)_\infty \\
&= \left(\begin{matrix} q^6, q^{18}, q^{30}, q^{36} \\ q^3, q^9, q^{27}, q^{33} \end{matrix} ; q^{36} \right)_\infty + q(-q^3, -q^{15}, q^{18}; q^{18})_\infty \\
&= \left(\begin{matrix} q^6, q^{12}, q^{30}, q^{36} \\ q^3, q^9, q^{27}, q^{33} \end{matrix} ; q^{36} \right)_\infty + q\Omega(q^3),
\end{aligned}$$

which is (34.4.1).

The proof of (34.4.2) is left as an exercise.

If we let

$$P(q) = \left(\begin{matrix} q^2, q^6, q^{10}, q^{12} \\ q, q^3, q^9, q^{11} \end{matrix} ; q^{12} \right)_\infty,$$

then (34.4.1) can be written

$$P(q) - \frac{1}{2} \left(\phi(q) + \phi(q^3) \right) = P(q^3) - \frac{1}{2} \left(\phi(q^3) + \phi(q^9) \right). \qquad (34.4.5)$$

By iteration,

$$P(q) - \frac{1}{2} (\phi(q) + \phi(q^3)) = P(q^{3^n}) - \frac{1}{2} (\phi(q^{3^n}) + \phi(q^{3^{n+1}})). \qquad (34.4.6)$$

If we let $n \to \infty$ in (34.4.6), we obtain

$$P(q) - \frac{1}{2} \left(\phi(q) + \phi(q^3) \right) = 0, \qquad (34.4.7)$$

which is (34.1.5).

The proof of (34.1.6) from (34.4.2) is left as an exercise.

34.5 Proofs of (34.1.7) and (34.1.8)

We start by establishing the following 5-dissections.

$$\begin{aligned}\left(\begin{matrix}q^2,q^8,q^{10},q^{12},q^{18},q^{20}\\ q,q^4,q^9,q^{11},q^{16},q^{19}\end{matrix};q^{20}\right)_\infty &= \left(\begin{matrix}q^{10},q^{40},q^{50},q^{60},q^{90},q^{100}\\ q^5,q^{20},q^{45},q^{55},q^{80},q^{95}\end{matrix};q^{100}\right)_\infty \\ &\quad+q\left(\begin{matrix}q^{30},q^{50},q^{70},q^{100}\\ q^{15},q^{35},q^{65},q^{85}\end{matrix};q^{100}\right)_\infty \\ &\quad+q^4\left(\begin{matrix}q^{10},q^{50},q^{90},q^{100}\\ q^5,q^{45},q^{55},q^{95}\end{matrix};q^{100}\right)_\infty \end{aligned} \tag{34.5.1}$$

and

$$\begin{aligned}\left(\begin{matrix}q^4,q^6,q^{10},q^{14},q^{16},q^{20}\\ q^3,q^7,q^8,q^{12},q^{13},q^{17}\end{matrix};q^{20}\right)_\infty &= \left(\begin{matrix}q^{30},q^{50},q^{70},q^{100}\\ q^{15},q^{35},q^{65},q^{85}\end{matrix};q^{100}\right)_\infty \\ &\quad+q^3\left(\begin{matrix}q^{10},q^{50},q^{90},q^{100}\\ q^5,q^{45},q^{55},q^{95}\end{matrix};q^{100}\right)_\infty \\ &\quad-q^4\left(\begin{matrix}q^{20},q^{30},q^{50},q^{70},q^{80},q^{100}\\ q^{15},q^{35},q^{40},q^{60},q^{65},q^{85}\end{matrix};q^{100}\right)_\infty . \end{aligned} \tag{34.5.2}$$

We make use of the quintuple product identity (10.1.2) to find

$$\begin{aligned}&\left(\begin{matrix}q^2,q^8,q^{10},q^{12},q^{18},q^{20}\\ q,q^4,q^9,q^{11},q^{16},q^{19}\end{matrix};q^{20}\right)_\infty \\ &= \frac{1}{(q^{20};q^{20})_\infty}\left(\begin{matrix}q^8,q^{12},q^{20}\\ q^4,q^{16}\end{matrix};q^{20}\right)_\infty \left(\begin{matrix}q^2,q^{10},q^{18},q^{20}\\ q,q^9,q^{11},q^{19}\end{matrix};q^{20}\right)_\infty \\ &= \frac{1}{(q^{20};q^{20})_\infty}\left(\begin{matrix}q^8,q^{12},q^{20}\\ q^4,q^{16}\end{matrix};q^{20}\right)_\infty (-q,-q^9,q^{10};q^{10})_\infty \\ &= \frac{1}{(q^{20};q^{20})_\infty}\left(\sum_{-\infty}^{\infty}(-1)^m q^{30m^2-2m}+\sum_{-\infty}^{\infty}(-1)^m q^{30m^2-22m+4}\right)\sum_{-\infty}^{\infty} q^{5n^2-4n} \\ &= \frac{1}{(q^{20};q^{20})_\infty}\left(\sum_{m,n=-\infty}^{\infty}(-1)^m q^{30m^2-2m+5n^2-4n}\right. \\ &\qquad\qquad \left.+\sum_{m,n=-\infty}^{\infty}(-1)^m q^{30m^2-22m+5n^2-4n+4}\right). \end{aligned} \tag{34.5.3}$$

We now split the sums into five, according to the residue modulo 5 of $m+2n$.

If $m+2n \equiv 0$, $3m+n \equiv 0$, $2m-n \equiv 0$, let $3m+n = 5r$, $2m-n = 5s$, $m = r+s$, $n = 2r-3s$,
if $m+2n \equiv 1$, $3m+n \equiv -2$, $2m-n \equiv 2$, $3m+n = 5r-2$, $2m-n = 5s+2$, $m = r+s$, $n = 2r-3s-2$,
if $m+2n \equiv 2$, $3m+n \equiv 1$, $2m-n \equiv -1$, $3m+n = 5r+1$, $2m-n = 5s-1$, $m = r+s$, $n = 2r-3s+1$,
if $m+2n \equiv -1$, $3m+n \equiv 2$, $2m-n \equiv -2$, $3m+n = 5r+2$, $2m-n = 5s-2$, $m = r+s$, $n = 2r-3s+2$,
and if $m+2n \equiv -2$, $3m+n \equiv -1$, $2m-n \equiv 1$, $3m+n = 5r-1$, $2m-n = 5s+1$, $m = r+s$, $n = 2r-3s-1$.

The sum in (34.5.3) becomes

$$\sum_{r,s=-\infty}^{\infty} (-1)^{r+s} q^{30(r+s)^2-2(r+s)+5(2r-3s)^2-4(2r-3s)}$$
$$+ \sum_{r,s=-\infty}^{\infty} (-1)^{r+s} q^{30(r+s)^2-22(r+s)+5(2r-3s)^2-4(2r-3s)+4}$$
$$+ \sum_{r,s=-\infty}^{\infty} (-1)^{r+s} q^{30(r+s)^2-2(r+s)+5(2r-3s-2)^2-4(2r-3s-2)}$$
$$+ \sum_{r,s=-\infty}^{\infty} (-1)^{r+s} q^{30(r+s)^2-22(r+s)+5(2r-3s-2)^2-4(2r-3s-2)+4}$$
$$+ \sum_{r,s=-\infty}^{\infty} (-1)^{r+s} q^{30(r+s)^2-2(r+s)+5(2r-3s+1)^2-4(2r-3s+1)}$$
$$+ \sum_{r,s=-\infty}^{\infty} (-1)^{r+s} q^{30(r+s)^2-22(r+s)+5(2r-3s+1)^2-4(2r-3s+1)+4}$$
$$+ \sum_{r,s=-\infty}^{\infty} (-1)^{r+s} q^{30(r+s)^2-2(r+s)+5(2r-3s+2)^2-4(2r-3s+2)}$$
$$+ \sum_{r,s=-\infty}^{\infty} (-1)^{r+s} q^{30(r+s)^2-22(r+s)+5(2r-3s+2)^2-4(2r-3s+2)+4}$$
$$+ \sum_{r,s=-\infty}^{\infty} (-1)^{r+s} q^{30(r+s)^2-2(r+s)+5(2r-3s-1)^2-4(2r-3s-1)}$$
$$+ \sum_{r,s=-\infty}^{\infty} (-1)^{r+s} q^{30(r+s)^2-22(r+s)+5(2r-3s-1)^2-4(2r-3s-1)+4}$$
$$= \sum_{r,s=-\infty}^{\infty} (-1)^{r+s} q^{50r^2-10r+75s^2+10s} + \sum_{r,s=-\infty}^{\infty} (-1)^{r+s} q^{50r^2-30r+75s^2-10s+4}$$
$$+ \sum_{r,s=-\infty}^{\infty} (-1)^{r+s} q^{50r^2-50r+75s^2+70s+28} + \sum_{r,s=-\infty}^{\infty} (-1)^{r+s} q^{50r^2-70r+75s^2+50s+32}$$

$$+\sum_{r,s=-\infty}^{\infty}(-1)^{r+s}q^{50r^2+10r+75s^2-20s+1}+\sum_{r,s=-\infty}^{\infty}(-1)^{r+s}q^{50r^2-10r+75s^2-40s+5}$$
$$+\sum_{r,s=-\infty}^{\infty}(-1)^{r+s}q^{50r^2+30r+75s^2-50s+12}+\sum_{r,s=-\infty}^{\infty}(-1)^{r+s}q^{50r^2+10r+75s^2-70s+16}$$
$$+\sum_{r,s=-\infty}^{\infty}(-1)^{r+s}q^{50r^2-30r+75s^2+40s+9}+\sum_{r,s=-\infty}^{\infty}(-1)^{r+s}q^{50r^2-50r+75s^2+20s+13}. \tag{34.5.4}$$

Now, the third and last sums are both zero, and the fourth,

$$\sum_{r,s=-\infty}^{\infty}(-1)^{r+s}q^{50r^2-70r+75s^2+50s+32}=\sum_{r,s=-\infty}^{\infty}(-1)^{r+s}q^{50(r-1)^2+30(r-1)+75s^2+50s+12}$$
$$=-\sum_{r,s=-\infty}^{\infty}(-1)^{r+s}q^{50r^2+30r+75s^2+50s+12}$$

cancels with the seventh, so the sum in (34.5.4) becomes

$$\sum_{r,s=-\infty}^{\infty}(-1)^{r+s}q^{50r^2-10r+75s^2+10s}+\sum_{r,s=-\infty}^{\infty}(-1)^{r+s}q^{50r^2-30r+75s^2-10s+4}$$
$$+\sum_{r,s=-\infty}^{\infty}(-1)^{r+s}q^{50r^2+10r+75s^2-20s+1}+\sum_{r,s=-\infty}^{\infty}(-1)^{r+s}q^{50r^2-10r+75s^2-40s+5}$$
$$+\sum_{r,s=-\infty}^{\infty}(-1)^{r+s}q^{50r^2+10r+75s^2-70s+16}+\sum_{r,s=-\infty}^{\infty}(-1)^{r+s}q^{50r^2-30r+75s^2+40s+9}$$

$$=(q^{40},q^{60},q^{100};q^{100})_\infty((q^{65},q^{85},q^{150};q^{150})_\infty+q^5(q^{35},q^{115},q^{150};q^{150})_\infty)$$
$$+q(q^{40},q^{60},q^{100};q^{100})_\infty((q^{55},q^{95},q^{150};q^{150})_\infty+q^{15}(q^{5},q^{145},q^{150};q^{150})_\infty)$$
$$+q^4(q^{20},q^{80},q^{100};q^{100})_\infty((q^{65},q^{85},q^{150};q^{150})_\infty+q^5(q^{35},q^{115},q^{150};q^{150})_\infty)$$
$$=(q^{40},q^{60},q^{100};q^{100})_\infty\left(\begin{matrix}q^{10},q^{40},q^{50}\\q^5,q^{45}\end{matrix};q^{50}\right)_\infty$$
$$+q(q^{40},q^{60},q^{100};q^{100})_\infty\left(\begin{matrix}q^{20},q^{30},q^{50}\\q^{15},q^{35}\end{matrix};q^{50}\right)_\infty$$
$$+q^4(q^{20},q^{80},q^{100};q^{100})_\infty\left(\begin{matrix}q^{10},q^{40},q^{50}\\q^5,q^{45}\end{matrix};q^{50}\right)_\infty$$

$$=\left(\begin{matrix}q^{10},q^{40},q^{40},q^{50},q^{60},q^{60},q^{90},q^{100},q^{100}\\q^5,q^{45},q^{55},q^{95}\end{matrix};q^{100}\right)_\infty$$

$$+q\left(\begin{matrix}q^{20},q^{30},q^{40},q^{50},q^{60},q^{70},q^{80},q^{100},q^{100}\\ q^{15},q^{35},q^{65},q^{85}\end{matrix};q^{100}\right)_\infty$$
$$+q^4\left(\begin{matrix}q^{10},q^{20},q^{40},q^{50},q^{60},q^{80},q^{90},q^{100},q^{100}\\ q^{5},q^{45},q^{55},q^{95}\end{matrix};q^{100}\right)_\infty$$

$$=(q^{20};q^{20})_\infty\left(\left(\begin{matrix}q^{10},q^{40},q^{50},q^{60},q^{90},q^{100}\\ q^{5},q^{20},q^{45},q^{55},q^{80},q^{95}\end{matrix};q^{100}\right)_\infty\right.$$
$$+q\left(\begin{matrix}q^{30},q^{50},q^{70},q^{100}\\ q^{15},q^{35},q^{65},q^{85}\end{matrix};q^{100}\right)_\infty$$
$$\left.+q^4\left(\begin{matrix}q^{10},q^{50},q^{90},q^{100}\\ q^{5},q^{45},q^{55},q^{95}\end{matrix};q^{100}\right)_\infty\right). \tag{34.5.5}$$

If we substitute (34.5.5) into (34.5.3), we obtain (34.5.1).
The proof of (34.5.2) is left as an exercise.
Now,

$$\begin{aligned}q\left(\begin{matrix}q^{30},q^{50},q^{70},q^{100}\\ q^{15},q^{35},q^{65},q^{85}\end{matrix};q^{100}\right)_\infty &= q(-q^{15},-q^{35},q^{50};q^{50})_\infty\\ &= q\sum_{-\infty}^{\infty} q^{25n^2+10n}\\ &= \sum_{-\infty}^{\infty} q^{(5n+1)^2}\\ &= \frac{1}{2}\sum_{n\equiv\pm1 \pmod 5} q^{n^2},\end{aligned} \tag{34.5.6}$$

and

$$\begin{aligned}q^4\left(\begin{matrix}q^{10},q^{50},q^{90},q^{100}\\ q^{5},q^{45},q^{55},q^{95}\end{matrix};q^{100}\right)_\infty &= q^4(-q^{5},-q^{45},q^{50};q^{50})_\infty\\ &= q^4\sum_{-\infty}^{\infty} q^{25n^2+20n}\\ &= \sum_{-\infty}^{\infty} q^{(5n+2)^2}\\ &= \frac{1}{2}\sum_{n\equiv\pm2 \pmod 5} q^{n^2}.\end{aligned} \tag{34.5.7}$$

So if we let

$$P(q)=\left(\begin{matrix}q^{2},q^{8},q^{10},q^{12},q^{18},q^{20}\\ q,q^{4},q^{9},q^{11},q^{16},q^{19}\end{matrix};q^{20}\right)_\infty, \tag{34.5.8}$$

then (34.5.1) can be written

$$P(q) = P(q^5) + \frac{1}{2} \sum_{n \not\equiv 0 \pmod 5} q^{n^2}$$
$$= P(q^5) + \frac{1}{2}(\phi(q) - \phi(q^{25})), \tag{34.5.9}$$

or,

$$P(q) - \frac{1}{2}(\phi(q) + \phi(q^5)) = P(q^5) - \frac{1}{2}(\phi(q^5) + \phi(q^{25})). \tag{34.5.10}$$

By iteration,

$$P(q) - \frac{1}{2}(\phi(q) + \phi(q^5)) = P(q^{5^n}) - \frac{1}{2}(\phi(q^{5^n}) + \phi(q^{5^{n+1}})). \tag{34.5.11}$$

If we let $n \to \infty$ in (34.5.11), we find

$$P(q) - \frac{1}{2}(\phi(q) + \phi(q^5)) = 0,$$

which is (34.1.7).

The proof of (34.1.8) from (34.5.2) is left as an exercise.

34.6 Proofs of (34.1.9) and (34.1.10)

We start with the 2-dissection of the product on the left of (34.4.1),

$$\begin{aligned}\left(\begin{matrix} q^2, q^6, q^{10}, q^{12} \\ q, q^3, q^9, q^{11} \end{matrix}; q^{12}\right)_\infty &= \left(\begin{matrix} q^8, q^{24}, q^{40}, q^{48} \\ q^4, q^{12}, q^{36}, q^{44} \end{matrix}; q^{48}\right)_\infty \\ &+ q\left(\begin{matrix} q^4, q^{10}, q^{14}, q^{24}, q^{34}, q^{38}, q^{44}, q^{48} \\ q^2, q^8, q^{12}, q^{22}, q^{26}, q^{36}, q^{40}, q^{46} \end{matrix}; q^{48}\right)_\infty.\end{aligned} \tag{34.6.1}$$

We have

$$\begin{aligned}&(-q, -q^3, -q^9, -q^{11}, q^{12}, q^{12}; q^{12})_\infty \\ &= \sum_{m,n=-\infty}^{\infty} q^{6m^2-5m+6n^2-3n} \\ &= \sum_{m \equiv n \pmod 2} q^{6m^2-5m+6n^2-3n} + \sum_{m \not\equiv n \pmod 2} q^{6m^2-5m+6n^2-3n}\end{aligned}$$

$$
\begin{aligned}
&= \sum_{r,s=-\infty}^{\infty} q^{6(r+s)^2-5(r+s)+6(r-s)^2-3(r-s)} \\
&\quad + \sum_{r,s=-\infty}^{\infty} q^{6(r+s+1)^2-5(r+s+1)+6(r-s)^2-3(r-s)} \\
&= \sum_{r,s=-\infty}^{\infty} q^{12r^2-8r+12s^2-2s} + q \sum_{r,s=-\infty}^{\infty} q^{12r^2+4r+12s^2+10s} \\
&= (-q^4,-q^{10},-q^{14},-q^{20},q^{24},q^{24};q^{24})_\infty \\
&\quad + q(-q^2,-q^8,-q^{16},-q^{22},q^{24},q^{24};q^{24})_\infty. \qquad (34.6.2)
\end{aligned}
$$

We can write (34.6.2)

$$
\begin{aligned}
&\left(\begin{matrix} q^2,q^6,q^{12},q^{12},q^{18},q^{22},q^{24},q^{24} \\ q,q^3,q^9,q^{11},q^{13},q^{15},q^{21},q^{23} \end{matrix};q^{24}\right)_\infty \\
&= \left(\begin{matrix} q^8,q^{24},q^{24},q^{40},q^{48},q^{48} \\ q^4,q^{10},q^{14},q^{34},q^{38},q^{44} \end{matrix};q^{48}\right)_\infty + q\left(\begin{matrix} q^4,q^{24},q^{24},q^{44},q^{48},q^{48} \\ q^2,q^8,q^{22},q^{26},q^{40},q^{46} \end{matrix};q^{48}\right)_\infty. \qquad (34.6.3)
\end{aligned}
$$

If we multiply (34.6.3) by $\left(\begin{matrix} q^{10},q^{14} \\ q^{12},q^{24} \end{matrix};q^{24}\right)_\infty$, we obtain

$$
\begin{aligned}
\left(\begin{matrix} q^2,q^6,q^{10},q^{12} \\ q,q^3,q^9,q^{11} \end{matrix};q^{12}\right)_\infty &= \left(\begin{matrix} q^8,q^{24},q^{40},q^{48} \\ q^4,q^{12},q^{36},q^{44} \end{matrix};q^{48}\right)_\infty \\
&\quad + q\left(\begin{matrix} q^4,q^{10},q^{14},q^{24},q^{34},q^{38},q^{44},q^{48} \\ q^2,q^8,q^{12},q^{22},q^{26},q^{36},q^{40},q^{46} \end{matrix};q^{48}\right)_\infty, \qquad (34.6.4)
\end{aligned}
$$

which is (34.6.1).

If we now multiply by 2 and use (34.1.5) with q for q and q^4 for q, (34.6.4) yields

$$
\phi(q)+\phi(q^3) = \phi(q^4)+\phi(q^{12})+2q\left(\begin{matrix} q^4,q^{10},q^{14},q^{24},q^{34},q^{38},q^{44},q^{48} \\ q^2,q^8,q^{12},q^{22},q^{26},q^{36},q^{40},q^{46} \end{matrix};q^{48}\right)_\infty. \qquad (34.6.5)
$$

If we take the odd parts in (34.6.5), divide by $2q$ then replace q by $q^{\frac{1}{2}}$, we obtain

$$
\psi(q^4)+q\psi(q^{12}) = \left(\begin{matrix} q^2,q^5,q^7,q^{12},q^{17},q^{19},q^{22},q^{24} \\ q,q^4,q^6,q^{11},q^{13},q^{18},q^{20},q^{23} \end{matrix};q^{24}\right)_\infty,
$$

which is (34.1.9).

We have

$$\psi(q^4)+q\psi(q^{12})=\left(\begin{matrix}q^2,q^{12},q^{22},q^{24}\\ q^4,q^6,q^{18},q^{20}\end{matrix};q^{24}\right)_\infty\left(\begin{matrix}q^5,q^7\\ q,q^{11}\end{matrix};q^{12}\right)_\infty. \tag{34.6.6}$$

If we replace q by $-q$ in (34.6.6), we find that

$$\begin{aligned}\psi(q^4)-q\psi(q^{12})&=\left(\begin{matrix}q^2,q^{12},q^{22},q^{24}\\ q^4,q^6,q^{18},q^{20}\end{matrix};q^{24}\right)_\infty\left(\begin{matrix}-q^5,-q^7\\ -q,-q^{11}\end{matrix};q^{12}\right)_\infty\\ &=\left(\begin{matrix}q^2,q^{12},q^{22},q^{24}\\ q^4,q^6,q^{18},q^{20}\end{matrix};q^{24}\right)_\infty\left(\begin{matrix}q,q^{10},q^{11},q^{13},q^{14},q^{23}\\ q^2,q^5,q^7,q^{17},q^{19},q^{22}\end{matrix};q^{24}\right)_\infty\\ &=\left(\begin{matrix}q,q^{10},q^{11},q^{12},q^{13},q^{14},q^{23},q^{24}\\ q^4,q^5,q^6,q^7,q^{17},q^{18},q^{19},q^{20}\end{matrix};q^{24}\right)_\infty,\end{aligned}$$

which is (34.1.10).

34.7 Proofs of (34.1.11) and (34.1.12)

We start with the 2-dissection of the product that appears on the left of (34.5.1),

$$\begin{aligned}&\left(\begin{matrix}q^2,q^8,q^{10},q^{12},q^{18},q^{20}\\ q,q^4,q^9,q^{11},q^{16},q^{19}\end{matrix};q^{20}\right)_\infty\\ &=\left(\begin{matrix}q^8,q^{22},q^{40},q^{48},q^{72},q^{80}\\ q^4,q^{16},q^{36},q^{44},q^{64},q^{76}\end{matrix};q^{80}\right)_\infty+q\left(\begin{matrix}q^{12},q^{28},q^{40},q^{52},q^{68},q^{80}\\ q^4,q^{24},q^{36},q^{44},q^{56},q^{76}\end{matrix};q^{80}\right)_\infty\end{aligned} \tag{34.7.1}$$

We have

$$\begin{aligned}(-q,-q^9,q^{10};q^{10})_\infty&=\sum_{-\infty}^{\infty}q^{5n^2-4n}\\ &=\sum_{-\infty}^{\infty}q^{20n^2-8n}+q\sum_{-\infty}^{\infty}q^{20n^2+12n}\\ &=(-q^{12},-q^{28},q^{40};q^{40})_\infty+q(-q^8,-q^{32},q^{40};q^{40})_\infty.\end{aligned} \tag{34.7.2}$$

We can write (34.7.2) as follows:

$$\left({q^2,q^{10},q^{18},q^{20} \atop q,q^9,q^{11},q^{19}};q^{20}\right)_\infty = \left({q^{24},q^{40},q^{56},q^{80} \atop q^{12},q^{28},q^{52},q^{68}};q^{80}\right)_\infty + q\left({q^{16},q^{40},q^{64},q^{80} \atop q^8,q^{32},q^{48},q^{72}};q^{80}\right)_\infty. \tag{34.7.3}$$

If we multiply (34.7.3) by $\left({q^8,q^{12} \atop q^4,q^{16}};q^{20}\right)_\infty$, we obtain

$$\begin{aligned}\left({q^2,q^8,q^{10},q^{12},q^{18},q^{20} \atop q,q^4,q^9,q^{11},q^{16},q^{19}};q^{20}\right)_\infty &= \left({q^8,q^{32},q^{40},q^{48},q^{72},q^{80} \atop q^4,q^{16},q^{36},q^{44},q^{64},q^{76}};q^{80}\right)_\infty \\ &+ q\left({q^{12},q^{28},q^{40},q^{52},q^{68},q^{80} \atop q^4,q^{24},q^{36},q^{44},q^{56},q^{76}};q^{80}\right)_\infty,\end{aligned} \tag{34.7.4}$$

which is (34.7.1).

If we multiply by 2 and use (34.1.7) with q for q and q^4 for q, (34.7.4) yields

$$\phi(q)+\phi(q^5) = \phi(q^4)+\phi(q^{20}) + 2q\left({q^{12},q^{28},q^{40},q^{52},q^{68},q^{80} \atop q^4,q^{24},q^{30},q^{44},q^{56},q^{76}};q^{80}\right)_\infty. \tag{34.7.5}$$

If we take the odd parts in (34.7.5), divide by $2q$ and replace q by $q^{\frac{1}{4}}$, we obtain

$$\psi(q^2)+q\psi(q^{10}) = \left({q^3,q^7,q^{10},q^{13},q^{17},q^{20} \atop q,q^6,q^9,q^{11},q^{14},q^{19}};q^{20}\right)_\infty,$$

which is (34.1.11).

We have

$$\psi(q^2)+q\psi(q^{10}) = \left({q^{10},q^{20} \atop q^6,q^{14}};q^{20}\right)_\infty \left({q^3,q^7 \atop q,q^9};q^{10}\right)_\infty. \tag{34.7.6}$$

If we replace q by $-q$ in (34.7.6), we find that

$$\begin{aligned}\psi(q^2)-q\psi(q^{10}) &= \left({q^{10},q^{20} \atop q^6,q^{14}};q^{20}\right)_\infty \left({-q^3,-q^7 \atop -q,-q^9};q^{10}\right)_\infty \\ &= \left({q^{10},q^{20} \atop q^6,q^{14}};q^{20}\right)_\infty \left({q,q^6,q^9,q^{11},q^{14},q^{19} \atop q^2,q^3,q^7,q^{13},q^{17},q^{18}};q^{20}\right)_\infty \\ &= \left({q,q^9,q^{10},q^{11},q^{19},q^{20} \atop q^2,q^3,q^7,q^{13},q^{17},q^{18}};q^{20}\right)_\infty,\end{aligned}$$

which is (34.1.12).

34.8 Proofs of (34.1.13)–(34.1.23)

If we multiply (34.1.1) and (34.1.2), we obtain

$$\phi(q)^2 - \phi(q^2)^2 = 4q \begin{pmatrix} q^8 \\ q^4 ; q^8 \end{pmatrix}_\infty^2 = 4q\psi(q^4)^2, \tag{34.8.1}$$

which is (34.1.13) or (1.10.1). So the pair (34.1.1) and (34.1.2) constitute a factorisation of (1.10.1).

If we multiply (34.1.5) and (34.1.6), we obtain

$$\phi(q)^2 - \phi(q^3)^2 = 4q\frac{(q^2;q^2)_\infty^3 (q^6;q^6)_\infty (q^{12};q^{12})_\infty^2}{(q;q)_\infty (q^3;q^3)_\infty (q^4;q^4)_\infty^2},$$

which is (34.1.14).

Now,

$$\frac{\phi(q^3)^4}{\phi(q^9)} = \phi(q)\phi(\omega q)\phi(\omega^2 q) = \phi(q^9)^3 + 8q^3\Omega(q^3)^3, \tag{34.8.2}$$

It follows that

$$\phi(q)^4 - \phi(q^3)^4 = 8q\phi(q^3)\Omega(q)^3,$$

which is (34.1.16).

If we divide (34.1.16) by (34.1.14), we obtain (34.1.15).

If we multiply (34.1.9) and (34.1.10) then replace q by $q^{\frac{1}{2}}$, we obtain (34.1.17).

If we replace q by $-q$ in (34.1.17), we obtain (34.1.18).

If we multiply (34.1.17) and (34.1.18) and replace q by $q^{\frac{1}{2}}$, we obtain (34.1.19).

Exercise: Show that (34.1.18) and (34.1.17) follow from the 2-dissection of $q^{-\frac{1}{3}}c(q)$, (22.1.14).

If we multiply (34.1.7) and (34.1.8), we obtain (34.1.20), and if we multiply (34.1.11) and (34.1.12) and replace q by $q^{\frac{1}{2}}$, we obtain (34.1.21).

If we replace q by $-q$ in (34.1.20), we obtain

$$\phi(-q^5)^2 - \phi(-q)^2 = 4q\frac{(q;q)_\infty (q^{10};q^{10})_\infty^3}{(q^2;q^2)_\infty (q^5;q^5)_\infty},$$

or,

$$\frac{(q^5;q^5)_\infty^4}{(q^{10};q^{10})_\infty^2} - \frac{(q;q)_\infty^4}{(q^2;q^2)_\infty^2} = 4q\frac{(q;q)_\infty (q^{10};q^{10})_\infty^3}{(q^2;q^2)_\infty (q^5;q^5)_\infty}. \tag{34.8.3}$$

If we multiply (34.8.3) by $\dfrac{(q^2;q^2)_\infty(q^{10};q^{10})_\infty}{(q;q)_\infty(q^5;q^5)_\infty}$, we obtain

$$\frac{(q^2;q^2)_\infty(q^5;q^5)^3_\infty}{(q;q)_\infty(q^{10};q^{10})_\infty}-\frac{(q;q)^3_\infty(q^{10};q^{10})_\infty}{(q^2;q^2)_\infty(q^5;q^5)_\infty}=4q\frac{(q^{10};q^{10})^4_\infty}{(q^5;q^5)^2_\infty},$$

or, by (34.1.21),

$$\psi(q)^2-q\psi(q^5)^2-\frac{(q;q)^3_\infty(q^{10};q^{10})_\infty}{(q^2;q^2)_\infty(q^5;q^5)_\infty}=4q\psi(q^5)^2,$$

which yields (34.1.23).

In similar fashion, (34.1.21) can be written

$$\frac{(q^2;q^2)^4_\infty}{(q;q)^2_\infty}-q\frac{(q^{10};q^{10})^4_\infty}{(q^5;q^5)^2_\infty}=\frac{(q^2;q^2)_\infty(q^5;q^5)^3_\infty}{(q;q)_\infty(q^{10};q^{10})_\infty}. \tag{34.8.4}$$

If we multiply (34.8.4) by $4\dfrac{(q;q)_\infty(q^5;q^5)_\infty}{(q^2;q^2)_\infty(q^{10};q^{10})_\infty}$, we obtain

$$4\frac{(q^2;q^2)^3_\infty(q^5;q^5)_\infty}{(q;q)_\infty(q^{10};q^{10})_\infty}-4q\frac{(q;q)_\infty(q^{10};q^{10})^3_\infty}{(q^2;q^2)_\infty(q^5;q^5)_\infty}=4\frac{(q^5;q^5)^4_\infty}{(q^{10};q^{10})^2_\infty},$$

or, by (34.8.3),

$$4\frac{(q^2;q^2)^3_\infty(q^5;q^5)_\infty}{(q;q)_\infty(q^{10};q^{10})_\infty}-(\phi(-q^5)^2-\phi(-q)^2)=4\phi(-q^5)^2. \tag{34.8.5}$$

If we now replace q by $-q$, we obtain

$$5\phi(q^5)^2-\phi(q)^2=4\frac{(q;q)_\infty(q^4;q^4)_\infty(q^{10};q^{10})^2_\infty}{(q^5;q^5)_\infty(q^{20},q^{20})_\infty},$$

which is (34.1.22).

We note the following neat results.

$$(\phi(q)^2-\phi(q^5)^2)(5\phi(q^5)^2-\phi(q)^2)=16q(q^2;q^2)^2_\infty(q^{10};q^{10})^2_\infty \tag{34.8.6}$$

and

$$(\psi(q)^2-q\psi(q^5)^2)(\psi(q)^2-5q\psi(q^5)^2)=(q;q)^2_\infty(q^5;q^5)^2_\infty. \tag{34.8.7}$$

34.9 Proofs of (34.1.24) and (34.1.25)

We have, making use of (34.1.15), (34.1.14) and (22.7.5),

$$
\begin{aligned}
3\phi(q^3)^2+\phi(q)^2 &= 2\left(\phi(q)^2+\phi(q^3)^2\right)-\left(\phi(q)^2-\phi(q^3)^2\right)\\
&= 4\frac{(q^2;q^2)_\infty^3(q^3;q^3)_\infty^2(q^6;q^6)_\infty}{(q;q)_\infty^2(q^4;q^4)_\infty(q^{12};q^{12})_\infty}-4q\frac{(q^2;q^2)_\infty^3(q^6;q^6)_\infty(q^{12};q^{12})_\infty^2}{(q;q)_\infty(q^3;q^3)_\infty(q^4;q^4)_\infty^2}\\
&= 4\frac{(q^2;q^2)_\infty^3(q^6;q^6)_\infty}{(q;q)_\infty(q^3;q^3)_\infty(q^4;q^4)_\infty(q^{12};q^{12})_\infty}\left(\frac{(q^3;q^3)_\infty^3}{(q;q)_\infty}-q\frac{(q^{12};q^{12})_\infty^3}{(q^4;q^4)_\infty}\right)\\
&= 4\frac{(q^2;q^2)_\infty^3(q^6;q^6)_\infty}{(q;q)_\infty(q^3;q^3)_\infty(q^4;q^4)_\infty(q^{12};q^{12})_\infty}\cdot\frac{(q^4;q^4)_\infty^3(q^6;q^6)_\infty^2}{(q^2;q^2)_\infty^2(q^{12};q^{12})_\infty}\\
&= 4\frac{(q^2;q^2)_\infty(q^4;q^4)_\infty^2(q^6;q^6)_\infty^3}{(q;q)_\infty(q^3;q^3)_\infty(q^{12};q^{12})_\infty^2}
\end{aligned}
$$

which is (34.1.24) and, making use of (34.1.15), (34.1.14) and (22.7.6),

$$
\begin{aligned}
3\phi(q^3)^2-\phi(q)^2 &= \left(\phi(q)^2+\phi(q^3)^2\right)-2\left(\phi(q)^2-\phi(q^3)^2\right)\\
&= 2\frac{(q^2;q^2)_\infty^3(q^3;q^3)_\infty^2(q^6;q^6)_\infty}{(q;q)_\infty^2(q^4;q^4)_\infty(q^{12};q^{12})_\infty}-8q\frac{(q^2;q^2)_\infty^3(q^6;q^6)_\infty(q^{12};q^{12})_\infty^2}{(q;q)_\infty(q^3;q^3)_\infty(q^4;q^4)_\infty^2}\\
&= 2\frac{(q^2;q^2)_\infty^3(q^6;q^6)_\infty}{(q;q)_\infty(q^3;q^3)_\infty(q^4;q^4)_\infty(q^{12};q^{12})_\infty}\left(\frac{(q^3;q^3)_\infty^3}{(q;q)_\infty}-4q\frac{(q^{12};q^{12})_\infty^3}{(q^4;q^4)_\infty}\right)\\
&= 2\frac{(q^2;q^2)_\infty^3(q^6;q^6)_\infty}{(q;q)_\infty(q^3;q^3)_\infty(q^4;q^4)_\infty(q^{12};q^{12})_\infty}\cdot\frac{(q;q)_\infty^3(q^6;q^6)_\infty^2}{(q^2;q^2)_\infty^2(q^3;q^3)_\infty}\\
&= 2\frac{(q;q)_\infty^2(q^2;q^2)_\infty(q^6;q^6)_\infty^3}{(q^3;q^3)_\infty^2(q^4;q^4)_\infty(q^{12};q^{12})_\infty},
\end{aligned}
$$

which is (34.1.25).

34.10 The 2^n-dissection of Euler's Product

We saw (34.3.3) that the 2-dissection of Euler's product is

$$
\begin{aligned}
(q;q)_\infty &= (q^2,q^{12},q^{14},q^{16},q^{18},q^{20},q^{30},q^{32};q^{32})_\infty\\
&\quad -q(q^4,q^6,q^{10},q^{16},q^{22},q^{26},q^{28},q^{32};q^{32})_\infty. \qquad (34.10.1)
\end{aligned}
$$

This can be written (somewhat strangely),

$$(q;q)_\infty = -q(q^4,q^{28};q^{32})_\infty(q^6,q^{10},q^{16};q^{16})_\infty + (q^{12},q^{20};q^{32})_\infty(q^2,q^{14},q^{16};q^{16})_\infty. \tag{34.10.2}$$

If we 4-dissect the products $(q^6,q^{10},q^{16};q^{16})_\infty$ and $(q^2,q^{14},q^{16};q^{16})_\infty$, we (eventually) obtain the 4-dissection of Euler's product, which can be written

$$\begin{aligned}(q;q)_\infty &= q^7(q^8,q^{120};q^{128})_\infty(q^{28},q^{36},q^{64};q^{64})_\infty\\ &\quad - q^2(q^{24},q^{104};q^{128})_\infty(q^{20},q^{44},q^{64};q^{64})_\infty\\ &\quad + (q^{40},q^{88};q^{128})_\infty(q^{12};q^{52},q^{64};q^{64})_\infty\\ &\quad - q(q^{56},q^{72};q^{128})_\infty(q^4,q^{60},q^{64};q^{64})_\infty.\end{aligned} \tag{34.10.3}$$

We can continue in the same manner and find the 8-dissection, the 16-dissection, and so on, of Euler's product.

I conjecture that the 2^n-dissection is

$$\begin{aligned}(q;q)_\infty = \sum_{k=1}^{2^n}(-1)^{n+k+1}q^{c_k}&(q^{(2k-1)2^{n+1}},q^{2^{2n+3}-(2k-1)2^{n+1}};q^{2^{2n+3}})_\infty\\ &\times (q^{2^{2n+1}-(2k-1)2^n},q^{2^{2n+1}+(2k-1)2^n},q^{2^{2n+2}};q^{2^{2n+2}})_\infty,\end{aligned} \tag{34.10.4}$$

where

$$\text{if } n \text{ is odd,}\quad c_k = P\left(\frac{2^{n+1}-1}{3}-(k-1)\right),\quad k=1,\ \dots\ ,2^n,$$

$$\text{if } n \text{ is even,}\quad c_k = P\left(-\frac{2^{n+1}-2}{3}+(k-1)\right),\quad k=1,\ \dots\ ,2^n,$$

and

$$P(n) = \frac{3n^2-n}{2}.$$

Endnote.

I have found (January, 2016) a much simpler proof of (34.1.3) and (34.1.4).

We have

$$\phi(-q^2)\pm\phi(-q) = (q;q)_\infty\left((-q;q^2)_\infty \pm (q;q^2)_\infty\right).$$

Now use the exercise at the end of §19.4.

References

18. R. Blecksmith, J. Brillhart, I. Gerst, Parity results for certain partition functions and identities similar to theta function identities. Math. Comput. **48**, 29–38 (1987)
19. R. Blecksmith, J. Brillhart, I. Gerst, Some infinite product identities. Math. Comput. **51**, 301–314 (1988)
36. S. Cooper, M.D. Hirschhorn, On some sum-to-product identities. Bull. Aust. Math. Soc. **63**, 353–365 (2001)
37. S. Cooper, M.D. Hirschhorn, On some infinite product identities. Rocky Mountain J. Math. **31**, 131–139 (2001)

Chapter 35
Some Useful Parametrisations

35.1 Introduction

In this chapter, we establish the following useful parametrisations:

$$(q;q)_\infty = s^{\frac{1}{2}} t^{\frac{1}{24}} (1+4qt)^{\frac{1}{8}} (1+2qt)^{\frac{1}{6}} (1+qt)^{\frac{1}{8}} (1-2qt)^{\frac{1}{2}}, \tag{35.1.1}$$

$$(q^2;q^2)_\infty = s^{\frac{1}{2}} t^{\frac{1}{12}} (1+4qt)^{\frac{1}{4}} (1+2qt)^{\frac{1}{12}} (1+qt)^{\frac{1}{4}} (1-2qt)^{\frac{1}{4}}, \tag{35.1.2}$$

$$(q^3;q^3)_\infty = s^{\frac{1}{2}} t^{\frac{1}{8}} (1+4qt)^{\frac{1}{24}} (1+2qt)^{\frac{1}{2}} (1+qt)^{\frac{1}{24}} (1-2qt)^{\frac{1}{6}}, \tag{35.1.3}$$

$$(q^4;q^4)_\infty = s^{\frac{1}{2}} t^{\frac{1}{6}} (1+4qt)^{\frac{1}{8}} (1+2qt)^{\frac{1}{24}} (1+qt)^{\frac{1}{2}} (1-2qt)^{\frac{1}{8}}, \tag{35.1.4}$$

$$(q^6;q^6)_\infty = s^{\frac{1}{2}} t^{\frac{1}{4}} (1+4qt)^{\frac{1}{12}} (1+2qt)^{\frac{1}{4}} (1+qt)^{\frac{1}{12}} (1-2qt)^{\frac{1}{12}} \tag{35.1.5}$$

and

$$(q^{12};q^{12})_\infty = s^{\frac{1}{2}} t^{\frac{1}{2}} (1+4qt)^{\frac{1}{24}} (1+2qt)^{\frac{1}{8}} (1+qt)^{\frac{1}{6}} (1-2qt)^{\frac{1}{24}}, \tag{35.1.6}$$

where

$$s = \frac{\phi(q^3)^3}{\phi(q)} = \frac{(q;q)_\infty^2 (q^4;q^4)_\infty^2 (q^6;q^6)_\infty^{15}}{(q^2;q^2)_\infty^5 (q^3;q^3)_\infty^6 (q^{12};q^{12})_\infty^6} \tag{35.1.7}$$

and

$$t = \frac{\phi(q)^2 - \phi(q^3)^2}{4q\phi(q^3)^2} = \frac{(q^2;q^2)_\infty^3 (q^3;q^3)_\infty^3 (q^{12};q^{12})_\infty^6}{(q;q)_\infty (q^4;q^4)_\infty^2 (q^6;q^6)_\infty^9}. \tag{35.1.8}$$

The original version of this chapter has been revised. The erratum to this chapter is available at https://doi.org/10.1007/978-3-319-57762-3_44.

M.D. Hirschhorn, *The Power of q*, Developments in Mathematics 49,
DOI 10.1007/978-3-319-57762-3_35

This idea is entirely due to A. Alaca, S. Alaca and K.S. Williams (2006) [1].

35.2 The Derivations

We begin by defining

$$r=\frac{\phi(q)^2}{\phi(q^3)^2}=\frac{(q^2;q^2)_\infty^{10}(q^3;q^3)_\infty^4(q^{12};q^{12})_\infty^4}{(q;q)_\infty^4(q^4;q^4)_\infty^4(q^6;q^6)_\infty^{10}}. \tag{35.2.1}$$

We now observe that (34.1.15), (34.1.14), (34.1.24) and (34.1.25) yield

$$r+1=2\frac{(q^2;q^2)_\infty^3(q^3;q^3)_\infty^6(q^{12};q^{12})_\infty^3}{(q;q)_\infty^2(q^4;q^4)_\infty(q^6;q^6)_\infty^9}, \tag{35.2.2}$$

$$r-1=4q\frac{(q^2;q^2)_\infty^3(q^3;q^3)_\infty^3(q^{12};q^{12})_\infty^6}{(q;q)_\infty(q^4;q^4)_\infty^2(q^6;q^6)_\infty^9}, \tag{35.2.3}$$

$$3+r=4\frac{(q^2;q^2)_\infty(q^3;q^3)_\infty^3(q^4;q^4)_\infty^2(q^{12};q^{12})_\infty^2}{(q;q)_\infty(q^6;q^6)_\infty^7} \tag{35.2.4}$$

and

$$3-r=2\frac{(q;q)_\infty^2(q^2;q^2)_\infty(q^3;q^3)_\infty^2(q^{12};q^{12})_\infty^3}{(q^4;q^4)_\infty(q^6;q^6)_\infty^7}. \tag{35.2.5}$$

Note that (35.2.3) gives

$$r=1+4qt, \tag{35.2.6}$$

where t is defined in (35.1.8).

If we substitute (35.2.6) into (35.2.2), (35.2.4) and (35.2.5), we obtain

$$\frac{(q^2;q^2)_\infty^3(q^3;q^3)_\infty^6(q^{12};q^{12})_\infty^3}{(q;q)_\infty^2(q^4;q^4)_\infty(q^6;q^6)_\infty^9}=1+2qt, \tag{35.2.7}$$

$$\frac{(q^2;q^2)_\infty(q^3;q^3)_\infty^3(q^4;q^4)_\infty^2(q^{12};q^{12})_\infty^2}{(q;q)_\infty(q^6;q^6)_\infty^7}=1+qt \tag{35.2.8}$$

and

$$\frac{(q;q)_\infty^2(q^2;q^2)_\infty(q^3;q^3)_\infty^2(q^{12};q^{12})_\infty^3}{(q^4;q^4)_\infty(q^6;q^6)_\infty^7}=1-2qt. \tag{35.2.9}$$

From (35.1.7), (35.1.8), (35.2.6), (35.2.1) and (35.2.7)–(35.2.9) we have

$$
\begin{aligned}
&s^{\alpha}t^{\beta}(1+4qt)^{\gamma}(1+2qt)^{\delta}(1+qt)^{\mu}(1-2qt)^{\nu} \\
&= \left(\frac{(q;q)_\infty^2(q^4;q^4)_\infty^2(q^6;q^6)_\infty^{15}}{(q^2;q^2)_\infty^5(q^3;q^3)_\infty^6(q^{12};q^{12})_\infty^6}\right)^{\alpha}\left(\frac{(q^2;q^2)_\infty^3(q^3;q^3)_\infty^3(q^{12};q^{12})_\infty^6}{(q;q)_\infty(q^4;q^4)_\infty^2(q^6;q^6)_\infty^9}\right)^{\beta} \\
&\times\left(\frac{(q^2;q^2)_\infty^{10}(q^3;q^3)_\infty^4(q^{12};q^{12})_\infty^4}{(q;q)_\infty^4(q^4;q^4)_\infty^4(q^6;q^6)_\infty^{10}}\right)^{\gamma}\left(\frac{(q^2;q^2)_\infty^3(q^3;q^3)_\infty^6(q^{12};q^{12})_\infty^3}{(q;q)_\infty^2(q^4;q^4)_\infty(q^6;q^6)_\infty^9}\right)^{\delta} \\
&\times\left(\frac{(q^2;q^2)_\infty(q^3;q^3)_\infty^3(q^4;q^4)_\infty^2(q^{12};q^{12})_\infty^2}{(q;q)_\infty(q^6;q^6)_\infty^7}\right)^{\mu} \\
&\qquad\qquad\left(\frac{(q;q)_\infty^2(q^2;q^2)_\infty(q^3;q^3)_\infty^2(q^{12};q^{12})_\infty^3}{(q^4;q^4)_\infty(q^6;q^6)_\infty^7}\right)^{\nu} \\
&= (q;q)_\infty^{\pi}(q^2;q^2)_\infty^{\rho}(q^3;q^3)_\infty^{\sigma}(q^4;q^4)_\infty^{\tau}(q^6;q^6)_\infty^{\upsilon}(q^{12};q^{12})_\infty^{\omega},
\end{aligned} \tag{35.2.10}
$$

where

$$
\begin{aligned}
\pi &= 2\alpha-\beta-4\gamma-2\delta-\mu+2\nu, \\
\rho &= -5\alpha+3\beta+10\gamma+3\delta+\mu+\nu, \\
\sigma &= -6\alpha+3\beta+4\gamma+6\delta+3\mu+2\nu, \\
\tau &= 2\alpha-2\beta-4\gamma-\delta+2\mu-\nu, \\
\upsilon &= 15\alpha-9\beta-10\gamma-9\delta-7\mu-7\nu
\end{aligned}
$$

and

$$
\omega = -6\alpha+6\beta+4\gamma+3\delta+2\mu+3\nu. \tag{35.2.11}
$$

If we invert (35.2.11). We find

$$
\begin{aligned}
\alpha &= \frac{1}{2}\pi+\frac{1}{2}\rho+\frac{1}{2}\sigma+\frac{1}{2}\tau+\frac{1}{2}\upsilon+\frac{1}{2}\omega, \\
\beta &= \frac{1}{24}\pi+\frac{1}{12}\rho+\frac{1}{8}\sigma+\frac{1}{6}\tau+\frac{1}{4}\upsilon+\frac{1}{2}\omega, \\
\gamma &= \frac{1}{8}\pi+\frac{1}{4}\rho+\frac{1}{24}\sigma+\frac{1}{8}\tau+\frac{1}{12}\upsilon+\frac{1}{24}\omega, \\
\delta &= \frac{1}{6}\pi+\frac{1}{12}\rho+\frac{1}{2}\sigma+\frac{1}{24}\tau+\frac{1}{4}\upsilon+\frac{1}{8}\omega, \\
\mu &= \frac{1}{8}\pi+\frac{1}{4}\rho+\frac{1}{24}\sigma+\frac{1}{2}\tau+\frac{1}{12}\upsilon+\frac{1}{6}\omega
\end{aligned}
$$

and

$$\nu = \frac{1}{2}\pi + \frac{1}{4}\rho + \frac{1}{6}\sigma + \frac{1}{8}\tau + \frac{1}{12}\upsilon + \frac{1}{24}\omega. \tag{35.2.12}$$

If $(\pi, \rho, \sigma, \tau, \upsilon, \omega) = (1,0,0,0,0,0)$ then $(\alpha, \beta, \gamma, \delta, \mu, \nu) = (\frac{1}{2}, \frac{1}{24}, \frac{1}{8}, \frac{1}{6}, \frac{1}{8}, \frac{1}{2})$ and

$$(q;q)_\infty = s^{\frac{1}{2}} t^{\frac{1}{24}} (1+4qt)^{\frac{1}{8}} (1+2qt)^{\frac{1}{6}} (1+qt)^{\frac{1}{8}} (1-2qt)^{\frac{1}{2}},$$

which is (35.1.1). In the same way, we obtain (35.1.2)–(35.1.6).

Exercise: Show that

$$s = \frac{\eta(\tau)^2 \eta(4\tau)^2 \eta(6\tau)^{15}}{\eta(2\tau)^5 \eta(3\tau)^6 \eta(12\tau)^6}$$

and

$$qt = \frac{\eta(2\tau)^3 \eta(3\tau)^3 \eta(12\tau)^6}{\eta(\tau)\eta(4\tau)^2 \eta(6\tau)^9}.$$

Reference

1. A. Alaca, S. Alaca, K.S. Williams, On the two-dimensional theta functions of the Borweins. Acta Arith. **124**, 177–195 (2006)

Chapter 36
Overpartitions

36.1 Introduction

In this chapter, we make a brief study of $\overline{p}(n)$, the number of overpartitions of n, that is, partitions of n in two colours, where the parts in one colour are distinct, and the parts in the other colour are unrestricted. As we will see, $\overline{p}(n)$ is also the number of partitions of n where odd parts come in two colours. (Compare this with Chapter 14.)

We will show that $\overline{p}(n)$ satisfies the following congruences.
For $n > 0$,

$$\overline{p}(n) \equiv 0 \pmod{2} \tag{36.1.1}$$

and for $n \geq 0$,

$$\overline{p}(4n+3) \equiv 0 \pmod{8}, \tag{36.1.2}$$

$$\overline{p}(8n+7) \equiv 0 \pmod{64} \tag{36.1.3}$$

and

$$\overline{p}(40n+35) \equiv 0 \pmod{40}. \tag{36.1.4}$$

(36.1.1)–(36.1.3) are proved with ease in §36.2.

The congruence (36.1.4) was conjectured by Hirschhorn and Sellers (2005) [97], and was first proved by W.Y.C. Chen and E.X.W. Xia (2014) [31]. Their proof made use of the parametrisations (35.1.1), (35.1.2) and (35.1.4).

We give two different proofs of (36.1.4), the first inspired by that of Chen and Xia, but avoiding the use of the parametrisation results.

The original version of this chapter has been revised. The erratum to this chapter is available at https://doi.org/10.1007/978-3-319-57762-3_44.

M.D. Hirschhorn, *The Power of q*, Developments in Mathematics 49,
DOI 10.1007/978-3-319-57762-3_36

36.2 Congruences for $\overline{p}(n)$ Modulo Powers of 2

The generating function for the $\overline{p}(n)$ is

$$\sum_{n\geq 0}\overline{p}(n)q^n=\frac{(-q;q)_\infty}{(q;q)_\infty}=\frac{1}{(q;q^2)_\infty(q;q)_\infty}=\frac{1}{(q;q^2)_\infty^2(q^2;q^2)_\infty}=\frac{1}{\phi(-q)}. \quad (36.2.1)$$

Since

$$\phi(-q)=1+2\sum_{n\geq 1}(-1)^n q^{n^2}\equiv 1 \pmod 2,$$

we have

$$\sum_{n\geq 0}\overline{p}(n)q^n\equiv 1 \pmod 2, \quad (36.2.2)$$

and (36.1.1) follows.

Also,

$$\begin{aligned}\sum_{n\geq 0}\overline{p}(n)q^n&=\frac{1}{\phi(-q)}\\&=\frac{\phi(q)}{\phi(-q^2)^2}\\&=\frac{\phi(q^4)+2q\psi(q^8)}{\phi(-q^2)^2}, \quad (36.2.3)\\ \sum_{n\geq 0}\overline{p}(2n+1)q^n&=2\frac{\psi(q^4)}{\phi(-q)^2}\\&=2\frac{\psi(q^4)\phi(q)^2}{\phi(-q)^2\phi(q)^2}\\&=2\frac{\psi(q^4)\left(\phi(q^2)^2+4q\psi(q^4)^2\right)}{\phi(-q^2)^4}, \quad (36.2.4)\\ \sum_{n\geq 0}\overline{p}(4n+3)q^n&=8\frac{\psi(q^2)^3}{\phi(-q)^4}\\&=8\frac{\psi(q^2)^3}{\phi(-q^2)^8}\phi(q)^4\\&=8\frac{\psi(q^2)^3}{\phi(-q^2)^8}\left(\phi(q)^2\right)^2\\&=8\frac{\psi(q^2)^3}{\phi(-q^2)^8}\left(\phi(q^2)^2+4q\psi(q^4)^2\right)^2 \quad (36.2.5)\end{aligned}$$

and

$$\sum_{n\geq 0}\overline{p}(8n+7)q^n = 64\frac{\psi(q)^3}{\phi(-q)^8}\phi(q)^2\psi(q^2)^2$$
$$= 64\frac{\psi(q)^7}{\phi(-q)^8}. \tag{36.2.6}$$

(36.1.2) and (36.1.3) follow from (36.2.5) and (36.2.6).

36.3 Our First Proof that $\overline{p}(40n+35) \equiv 0 \pmod{40}$

We have, modulo 5,

$$\sum_{n\geq 0}(-1)^n\overline{p}(n)q^n = \frac{1}{\phi(q)} = \frac{\phi(q)^4}{\phi(q)^5} \equiv \frac{\phi(q)^4}{\phi(q^5)}. \tag{36.3.1}$$

Now, the 5-dissection of $\phi(q)$ is

$$\phi(q) = a + 2qb + 2q^4c \tag{36.3.2}$$

where

$$a = \phi(q^{25}),\ \ b = (-q^{15}, -q^{35}, q^{50}; q^{50})_\infty,\ \ c = (-q^5, -q^{45}, q^{50}; q^{50})_\infty. \tag{36.3.3}$$

Note that by (34.1.20) with q^5 for q,

$$a^2 + 4q^5bc = \phi(q^5)^2. \tag{36.3.4}$$

We have

$$\sum_{n\geq 0}(-1)^n\overline{p}(n)q^n \equiv \frac{(a+2qb+2q^4c)^4}{\phi(q^5)}$$
$$\equiv \frac{1}{\phi(q^5)}(a^4 + 8qa^3b + 24q^2a^2b^2 + 32q^3ab^3 + 8q^4a^3c + 16q^4b^4$$
$$+ 48q^5a^2bc + 96q^6ab^2c + 64q^7b^3c + 24q^8a^2c^2 + 96q^9abc^2$$
$$+ 96q^{10}b^2c^2 + 32q^{12}ac^3 + 64q^{13}bc^3 + 16q^{16}c^4). \tag{36.3.5}$$

From (36.3.5) we extract

$$\begin{aligned}\sum_{n\geq 0}(-1)^n\overline{p}(5n)q^{5n} &\equiv \frac{a^4+8q^5a^2bc+16q^{10}b^2c^2}{\phi(q^5)}\\ &= \frac{(a^2+4q^5bc)^2}{\phi(q^5)}\\ &= \frac{(\phi(q^5)^2)^2}{\phi(q^5)}\\ &= \phi(q^5)^3, \end{aligned} \tag{36.3.6}$$

and

$$\begin{aligned}\sum_{n\geq 0}(-1)^n\overline{p}(5n)q^{n} &\equiv \phi(q)^3\\ &= \left(\phi(q^4)+2q\psi(q^8)\right)^3. \end{aligned} \tag{36.3.7}$$

From (36.3.7) we extract

$$\sum_{n\geq 0}\overline{p}(20n+15)q^n \equiv 2\psi(q^2)^3. \tag{36.3.8}$$

From (36.3.8) we extract

$$\sum_{n\geq 0}\overline{p}(40n+15)q^n \equiv 2\psi(q)^3 \tag{36.3.9}$$

and

$$\sum_{n\geq 0}\overline{p}(40n+35)q^n \equiv 0. \tag{36.3.10}$$

Thus

$$\overline{p}(40n+35) \equiv 0 \pmod 5 \tag{36.3.11}$$

and, by (36.1.2),

$$\overline{p}(40n+35) = \overline{p}(4(10n+8)+3) \equiv 0 \pmod 8. \tag{36.3.12}$$

So

$$\overline{p}(40n+35) \equiv 0 \pmod{40},$$

which is (36.1.4).

36.4 Our Second Proof that $\overline{p}(40n+35) \equiv 0 \pmod{40}$

We saw in (36.2.5) that

$$\sum_{n\geq 0} \overline{p}(4n+3)q^n = 8\frac{\psi(q^2)^3}{\phi(-q^2)^8}\left(\phi(q^2)^2 + 4q\psi(q^4)^2\right)^2. \tag{36.4.1}$$

It follows that

$$\begin{aligned}\sum_{n\geq 0} \overline{p}(8n+3)q^n &= 8\frac{\psi(q)^3}{\phi(-q)^8}\left(\phi(q)^4 + 16q\psi(q^2)^4\right)\\ &= 8\frac{(q^2;q^2)_\infty^{14}}{(q;q)_\infty^{19}}\left(\frac{(q^2;q^2)_\infty^{20}}{(q;q)_\infty^8(q^4;q^4)_\infty^8} + 16q\frac{(q^4;q^4)_\infty^8}{(q^2;q^2)_\infty^4}\right)\\ &= 8\frac{(q^2;q^2)_\infty^{34}}{(q;q)_\infty^{27}(q^4;q^4)_\infty^8} + 128q\frac{(q^2;q^2)_\infty^{10}(q^4;q^4)_\infty^8}{(q;q)_\infty^{19}}.\end{aligned} \tag{36.4.2}$$

Hence, modulo 5,

$$\begin{aligned}\sum_{n\geq 0} \overline{p}(8n+3)q^n &\equiv 3\frac{\left((q^2;q^2)_\infty^5\right)^7}{\left((q;q)_\infty^5\right)^6\left((q^4;q^4)_\infty^5\right)^2}(q;q)_\infty^3\frac{(q^4;q^4)_\infty^2}{(q^2;q^2)_\infty}\\ &\quad + 3q\frac{\left((q^2;q^2)_\infty^5\right)^2\left((q^4;q^4)_\infty^5\right)}{\left((q;q)_\infty^5\right)^4}(q;q)_\infty(q^4;q^4)_\infty^3\\ &\equiv 3\frac{(q^{10};q^{10})_\infty^7}{(q^5;q^5)_\infty^6(q^{20};q^{20})_\infty}J(q)\psi(q^2)\\ &\quad + 3q\frac{(q^{10};q^{10})_\infty^2(q^{20};q^{20})_\infty}{(q^5;q^5)_\infty^4}E(q)J(q^4).\end{aligned} \tag{36.4.3}$$

Now,

$$J(q) = (q;q)_\infty^3 = \sum_{n\geq 0}(-1)^n(2n+1)q^{(n^2+n)/2} \equiv J_0 + J_1,$$

$$\psi(q^2) = \frac{(q^4;q^4)_\infty^2}{(q^2;q^2)_\infty} = \sum_{n\geq 0} q^{n^2+n} = \psi_0^* + \psi_1^* + \psi_2^*,$$

$$E(q) = (q;q)_\infty = \sum_{-\infty}^{\infty} q^{(3n^2+n)/2} = E_0 + E_1 + E_2$$

and

$$J(q^4) = (q^4;q^4)_\infty^3 = \sum_{n\geq 0}(-1)^n(2n+1)q^{2(n^2+n)} \equiv J_0^* + J_4^*. \tag{36.4.4}$$

So

$$\begin{aligned}
&\sum_{n\geq 0}\overline{p}(8n+3)q^n \\
&\equiv 3\frac{(q^{10};q^{10})_\infty^7}{(q^5;q^5)_\infty^6(q^{20};q^{20})_\infty}(J_0+J_1)(\psi_0^*+\psi_1^*+\psi_2^*) \\
&\qquad + 3q\frac{(q^{10};q^{10})_\infty^2(q^{20};q^{20})_\infty}{(q^5;q^5)_\infty^4}(E_0+E_1+E_2)(J_0^*+J_4^*).
\end{aligned} \tag{36.4.5}$$

There are no terms on the right side of (36.4.5) in which the exponent is 4 modulo 5, so

$$\sum_{n\geq 0}\overline{p}(8(5n+4)+3)q^n \equiv 0, \tag{36.4.6}$$

or,

$$\sum_{n\geq 0}\overline{p}(40n+35)q^n \equiv 0. \tag{36.4.7}$$

So

$$\overline{p}(40n+35) \equiv 0 \pmod 5,$$

and, as we saw in §36.3, this gives us (36.1.4).

Exercise: Show that

$$\overline{p}(88n+55) \equiv 0 \pmod{704}. \tag{36.4.8}$$

Endnote.

There is a vast literature on the subject of overpartitions. One of the major contributors is J. Lovejoy.

References

31. W.Y.C. Chen, E.X.W. Xia, Proof of a conjecture of Hirschhorn and Sellers on overpartitions. Acta Arith. **163**, 59–63 (2014)
97. M.D. Hirschhorn, J.A. Sellers, Arithmetic relations for overpartitions. J. Comb. Math. Comb. Comput. **53**, 65–73 (2005)

Chapter 37
Bipartitions with Odd Parts Distinct

37.1 Introduction

W.Y.C. Chen and B.L.S. Lin (2011) [30] introduced what I will denote $bipod(n)$, the number of bipartitions of n with odd parts distinct, that is, the number of partitions of n into parts of two colours in which the odd parts in each colour are distinct.

We will show, among other things, that
modulo 3,
for $\alpha \geq 1$,

$$bipod\left(3^{2\alpha+1}n+\frac{7\times 3^{2\alpha}+1}{4}\right)\equiv 0 \tag{37.1.1}$$

and

$$bipod\left(3^{2\alpha+1}n+\frac{11\times 3^{2\alpha}+1}{4}\right)\equiv 0, \tag{37.1.2}$$

and
modulo 5,
for $\alpha \geq 1$,

$$bipod\left(5^{\alpha+1}n+\frac{11\times 5^{\alpha}+1}{4}\right)\equiv 0 \tag{37.1.3}$$

and

$$bipod\left(5^{\alpha+1}n+\frac{19\times 5^{\alpha}+1}{4}\right)\equiv 0. \tag{37.1.4}$$

The original version of this chapter has been revised. The erratum to this chapter is available at https://doi.org/10.1007/978-3-319-57762-3_44.

M.D. Hirschhorn, *The Power of q*, Developments in Mathematics 49,
DOI 10.1007/978-3-319-57762-3_37

37.2 The Mod 3 Results

In this section, all congruences are to the modulus 3.

The generating function of *bipod*(*n*) is

$$\sum_{n\geq 0} bipod(n)q^n = \frac{(-q;q^2)_\infty^2}{(q^2;q^2)_\infty^2}. \tag{37.2.1}$$

It follows that

$$\sum_{n\geq 0}(-1)^n bipod(n)q^n = \frac{(q;q^2)_\infty^2}{(q^2;q^2)_\infty^2} = \frac{(q;q)_\infty^2}{(q^2;q^2)_\infty^4} = \frac{1}{\psi(q)^2}. \tag{37.2.2}$$

We now 3-dissect (37.2.2).

We have

$$\begin{aligned}
&\sum_{n\geq 0}(-1)^n bipod(n)q^n \\
&= \left(\frac{\psi(q^9)}{\psi(q^3)^4}\left(\Pi(q^3)^2 - q\Pi(q^3)\psi(q^9) + q^2\psi(q^9)^2\right)\right)^2 \\
&= \frac{\psi(q^9)^2}{\psi(q^3)^8}\left(\Pi(q^3)^4 - 2q\Pi(q^3)^3\psi(q^9) + 3q^2\Pi(q^3)^2\psi(q^9)^2\right. \\
&\qquad\qquad \left. -2q^3\Pi(q^3)\psi(q^9)^3 + q^4\psi(q^9)^4\right).
\end{aligned} \tag{37.2.3}$$

It follows that

$$\begin{aligned}
\sum_{n\geq 0}(-1)^n bipod(3n+2)q^n &= 3\frac{\psi(q^3)^4\Pi(q)^2}{\psi(q)^8} \\
&= 3\frac{(q;q)_\infty^6(q^6;q^6)_\infty^6}{(q^2;q^2)_\infty^{14}},
\end{aligned} \tag{37.2.4}$$

or,

$$\sum_{n\geq 0} bipod(3n+2)q^n = 3\frac{(q^2;q^2)_\infty^4(q^6;q^6)_\infty^6}{(q;q)_\infty^6(q^4;q^4)_\infty^6}. \tag{37.2.5}$$

Also,

$$\sum_{n\geq 0}(-1)^n bipod(n)q^n = \frac{1}{\psi(q)^2} = \frac{\psi(q)}{\psi(q)^3} \equiv \frac{\psi(q)}{\psi(q^3)}$$
$$\equiv \frac{\Pi(q^3)+q\psi(q^9)}{\psi(q^3)} \tag{37.2.6}$$

and

$$\sum_{n\geq 0}(-1)^n bipod(3n+1)q^n \equiv -\frac{\psi(q^3)}{\psi(q)} \equiv -\frac{\psi(q)^2\psi(q^3)}{\psi(q)^3} \equiv -\psi(q)^2$$
$$\equiv -\left(\Pi(q^3)+q\psi(q^9)\right)^2. \tag{37.2.7}$$

From (37.2.7) it follows by induction on α that for $\alpha \geq 0$,

$$\sum_{n\geq 0}(-1)^n bipod\left(3^{2\alpha+1}n+\frac{3^{2\alpha+1}+1}{4}\right)q^n \equiv -\psi(q)^2$$
$$\equiv -\left(\Pi(q^3)+q\psi(q^9)\right)^2 \tag{37.2.8}$$

and

$$\sum_{n\geq 0}(-1)^n bipod\left(3^{2\alpha+2}n+\frac{3^{2\alpha+3}+1}{4}\right)q^n \equiv -\psi(q^3)^2. \tag{37.2.9}$$

From (37.2.9) we deduce that for $\alpha \geq 0$,

$$\sum_{n\geq 0}(-1)^n bipod\left(3^{2\alpha+3}n+\frac{7\times 3^{2\alpha+2}+1}{4}\right)q^n \equiv 0 \tag{37.2.10}$$

and

$$\sum_{n\geq 0}(-1)^n bipod\left(3^{2\alpha+3}n+\frac{11\times 3^{2\alpha+2}+1}{4}\right)q^n \equiv 0. \tag{37.2.11}$$

From (37.2.10) and (37.2.11) we deduce (37.1.1) and (37.1.2).

37.3 The Mod 5 Results

In this section, all congruences are to the modulus 5.

We have

$$\sum_{n\geq 0}(-1)^n bipod(n)q^n = \frac{1}{\psi(q)^2} \equiv \frac{\psi(q)^3}{\psi(q^5)}. \tag{37.3.1}$$

Now, the 5-dissection of $\psi(q)$ is

$$\psi(q) = a + qb + q^3c \tag{37.3.2}$$

where

$$a = (-q^{10}, -q^{15}, q^{25}; q^{25})_\infty, \quad b = (-q^5, -q^{20}, q^{25}; q^{25})_\infty, \quad c = \psi(q^{25}). \tag{37.3.3}$$

Note that by (34.1.21) with q^5 for q,

$$ab + q^5c^2 = \psi(q^5)^2. \tag{37.3.4}$$

We substitute (37.3.2) into (37.3.1), and extract

$$\begin{aligned}\sum_{n\geq 0}(-1)^n bipod(5n+4)q^{5n} &\equiv \frac{abc + q^5c^3}{\psi(q^5)} \\ &= \frac{c(ab + q^5c^2)}{\psi(q^5)} \\ &= \frac{\psi(q^{25})\psi(q^5)^2}{\psi(q^5)} \\ &= \psi(q^{25})\psi(q^5)\end{aligned} \tag{37.3.5}$$

and

$$\sum_{n\geq 0}(-1)^n bipod(5n+4)q^n \equiv \psi(q^5)\psi(q) = \psi(q^5)(a + qb + q^3\psi(q^{25})).$$

It follows by induction that for $\alpha \geq 1$,

$$\begin{aligned}\sum_{n\geq 0}(-1)^n bipod\left(5^\alpha n + \frac{3\times 5^\alpha + 1}{4}\right)q^n &\equiv (-1)^{\alpha-1}\psi(q)\psi(q^5) \\ &= (-1)^{\alpha-1}\psi(q^5)(a + qb + q^3\psi(q^{25}))\end{aligned} \tag{37.3.6}$$

from which (37.1.3) and (37.1.4) follow directly.

Reference

30. W.Y.C. Chen, B.L.S. Lin, Congruences for bipartitions with odd parts distinct. Ramanujan J. **25**, 277–293 (2011)

Chapter 38
Overcubic Partitions

38.1 Introduction

B. Kim (2010) [100] introduced $\overline{a}(n)$, the number of overcubic partitions of n, that is, the number of partitions of n into odd parts in two colours and even parts in four, in which odd parts in one colour and even parts in two are distinct. He showed, using deeper methods than ours, that

$$\sum_{n\geq 0}\overline{a}(3n+2)q^n = 6\frac{(q^3;q^3)_\infty^6(q^4;q^4))_\infty^3}{(q;q)_\infty^8(q^2;q^2)_\infty^3}. \tag{38.1.1}$$

We prove (38.1.1), as well as

$$\sum_{n\geq 0}\overline{a}(3n)q^n = \frac{(q^4;q^4)_\infty^4(q^6;q^6)_\infty^7}{(q^2;q^2)_\infty^8(q^3;q^3)_\infty^2(q^{12};q^{12})_\infty^3} + 12q\frac{(q^3;q^3)_\infty^3(q^4;q^4)_\infty^3(q^6;q^6)_\infty^3}{(q;q)_\infty^7(q^2;q^2)_\infty^4} \tag{38.1.2}$$

and

$$\sum_{n\geq 0}\overline{a}(3n+1)q^n = 2\frac{(q^3;q^3)_\infty^9(q^4;q^4)_\infty^3}{(q;q)_\infty^9(q^2;q^2)_\infty^2(q^6;q^6)_\infty^3} + 8q\frac{(q^4;q^4)_\infty^3(q^6;q^6)_\infty^6}{(q;q)_\infty^6(q^2;q^2)_\infty^5}. \tag{38.1.3}$$

The original version of this chapter has been revised. The erratum to this chapter is available at https://doi.org/10.1007/978-3-319-57762-3_44.

M.D. Hirschhorn, *The Power of q*, Developments in Mathematics 49,
DOI 10.1007/978-3-319-57762-3_38

38.2 The Generating Function and First Steps

The generating function for the $\{\overline{a}(n)\}$ is

$$\sum_{n\geq 0}\overline{a}(n)q^n = \frac{(-q;q^2)_\infty}{(q;q^2)_\infty}\frac{(-q^2;q^2)_\infty^2}{(q^2;q^2)_\infty^2}$$
$$= \frac{(-q;q)_\infty(-q^2;q^2)_\infty}{(q;q)_\infty(q^2;q^2)_\infty}$$
$$= \frac{(q^2;q^2)_\infty}{(q;q)_\infty^2}\frac{(q^4;q^4)_\infty}{(q^2;q^2)_\infty^2}$$
$$= \frac{1}{\phi(-q)\phi(-q^2)}. \tag{38.2.1}$$

We now 3-dissect (38.2.1).

We have

$$\sum_{n\geq 0}\overline{a}(n)q^n = \frac{1}{\phi(-q)\phi(-q^2)}$$
$$= \frac{\phi(-\omega q)\phi(-\omega^2 q)\phi(-\omega q^2)\phi(-\omega^2 q^2)}{\phi(-q)\phi(-\omega q)\phi(-\omega^2 q)\phi(-q^2)\phi(-\omega q^2)\phi(-\omega^2 q^2)}$$
$$= \frac{\phi(-q^9)\phi(-q^{18})}{\phi(-q^3)^4\phi(-q^6)^4}\left(\phi(-q^9)-2\omega q\Omega(-q^3)\right)\left(\phi(-q^9)-2\omega^2 q\Omega(-q^3)\right)$$
$$\times(\phi(-q^{18})-2\omega q^2\Omega(-q^6))(\phi(-q^{18})-2\omega^2 q^2\Omega(-q^6))$$
$$= \frac{\phi(-q^9)\phi(-q^{18})}{\phi(-q^3)^4\phi(-q^6)^4}\left(\phi(-q^9)^2+2q\phi(-q^9)\Omega(-q^3)+4q^2\Omega(-q^3)^2\right)$$
$$\times(\phi(-q^{18})^2+2q^2\phi(-q^{18})\Omega(-q^6)+4q^4\Omega(-q^6)^2). \tag{38.2.2}$$

It follows that

$$\sum_{n\geq 0}\overline{a}(3n)q^n = \frac{\phi(-q^3)\phi(-q^6)}{\phi(-q)^4\phi(-q^2)^4}(\phi(-q^3)^2\phi(-q^6)^2$$
$$+4q\phi(-q^3)\phi(-q^6)\Omega(-q)\Omega(-q^2)+16q^2\Omega(-q)^2\Omega(-q^2)^2), \tag{38.2.3}$$

$$\sum_{n\geq 0}\overline{a}(3n+1)q^n = \frac{\phi(-q^3)\phi(-q^6)}{\phi(-q)^4\phi(-q^2)^4}(2\phi(-q^3)\phi(-q^6)^2\Omega(-q)$$
$$+4q\phi(-q^3)^2\Omega(-q^2)^2+8q\phi(-q^6)\Omega(-q)^2\Omega(-q^2)) \tag{38.2.4}$$

and

$$\sum_{n\geq 0}\overline{a}(3n+2)q^n = \frac{\phi(-q^3)\phi(-q^6)}{\phi(-q)^4\phi(-q^2)^4}\Big(2\phi(-q^3)^2\phi(-q^6)\Omega(-q^2) \\ +4\phi(-q^6)^2\Omega(-q^2)^2 + 8q\phi(-q^3)\Omega(-q)\Omega(-q^2)^2\Big). \quad (38.2.5)$$

In product form, these are

$$\sum_{n\geq 0}\overline{a}(3n)q^n = \frac{(q^3;q^3)_\infty^6(q^4;q^4)_\infty^4(q^6;q^6)_\infty^3}{(q;q)_\infty^8(q^2;q^2)_\infty^4(q^{12};q^{12})_\infty^3} + 4q\frac{(q^3;q^3)_\infty^3(q^4;q^4)_\infty^3(q^6;q^6)_\infty^3}{(q;q)_\infty^7(q^2;q^2)_\infty^4} \\ + 16q^2\frac{(q^4;q^4)_\infty^2(q^6;q^6)_\infty^3(q^{12};q^{12})_\infty^3}{(q;q)_\infty^6(q^2;q^2)_\infty^4}, \quad (38.2.6)$$

$$\sum_{n\geq 0}\overline{a}(3n+1)q^n = 2\frac{(q^3;q^3)_\infty^3(q^4;q^4)_\infty^4(q^6;q^6)_\infty^6}{(q;q)_\infty^7(q^2;q^2)_\infty^5(q^{12};q^{12})_\infty^3} + 4q\frac{(q^3;q^3)_\infty^6(q^4;q^4)_\infty^2(q^{12};q^{12})_\infty^3}{(q;q)_\infty^8(q^2;q^2)_\infty^2(q^6;q^6)_\infty^3} \\ + 8q\frac{(q^4;q^4)_\infty^3(q^6;q^6)_\infty^6}{(q;q)_\infty^6(q^2;q^2)_\infty^5} \quad (38.2.7)$$

and

$$\sum_{n\geq 0}\overline{a}(3n+2)q^n = 2\frac{(q^3;q^3)_\infty^6(q^4;q^4)_\infty^3}{(q;q)_\infty^8(q^2;q^2)_\infty^3} + 4\frac{(q^4;q^4)_\infty^4(q^6;q^6)_\infty^9}{(q;q)_\infty^6(q^2;q^2)_\infty^6(q^{12};q^{12})_\infty^3} \\ + 8q\frac{(q^3;q^3)_\infty^3(q^4;q^4)_\infty^2(q^{12};lq^{12})_\infty^3}{(q;q)_\infty^7(q^2;q^2)_\infty^3}. \quad (38.2.8)$$

38.3 Completion

We now reduce the number of terms on the right sides of (38.2.6)–(38.2.8).

We require the following two identities:

$$\frac{(q^3;q^3)_\infty^3}{(q;q)_\infty} - 4q\frac{(q^{12};q^{12})_\infty^3}{(q^4;q^4)_\infty} = \frac{(q;q)_\infty^3(q^6;q^6)_\infty^2}{(q^2;q^2)_\infty^2(q^3;q^3)_\infty} \quad (38.3.1)$$

and

$$\frac{(q;q)_\infty}{(q^3;q^3)_\infty^3}+2q\frac{(q^2;q^2)_\infty^3(q^{12};q^{12})_\infty^6}{(q^4;q^4)_\infty^2(q^6;q^6)_\infty^9}=\frac{(q^2;q^2)_\infty^3(q^3;q^3)_\infty^3(q^{12};q^{12})_\infty^3}{(q;q)_\infty(q^4;q^4)_\infty(q^6;q^6)_\infty^9}. \tag{38.3.2}$$

(38.3.1) is simply (22.7.6).

In order to prove (38.3.2), note that if the 2-dissection of $f(q)$ is

$$f(q)=g(q^2)+qh(q^2) \tag{38.3.3}$$

then

$$f(-q)+2qh(q^2)=f(q) \tag{38.3.4}$$

and

$$\frac{1}{f(q)}+2q\frac{h(q^2)}{f(q)f(-q)}=\frac{f(q)}{f(q)f(-q)}\left(=\frac{1}{f(-q)}\right). \tag{38.3.5}$$

With

$$f(q)=\frac{(q^3;q^3)_\infty^3}{(q;q)_\infty}, \tag{38.3.6}$$

we have (22.7.5),

$$f(q)=\frac{(q^4;q^4)_\infty^3(q^6;q^6)_\infty^2}{(q^2;q^2)_\infty^2(q^{12};q^{12})_\infty}+q\frac{(q^{12};q^{12})_\infty^3}{(q^4;q^4)_\infty} \tag{38.3.7}$$

and

$$f(q)f(-q)=\frac{(q^4;q^4)_\infty(q^6;q^6)_\infty^9}{(q^2;q^2)_\infty^3(q^{12};q^{12})_\infty^3}. \tag{38.3.8}$$

(38.3.1) and (38.3.5) now yield

$$\frac{(q;q)_\infty}{(q^3;q^3)_\infty^3}+2q\frac{(q^2;q^2)_\infty^3(q^{12};q^{12})_\infty^6}{(q^4;q^4)_\infty^2(q^6;q^6)_\infty^9}=\frac{(q^2;q^2)_\infty^3(q^3;q^3)_\infty^3(q^{12};q^{12})_\infty^3}{(q;q)_\infty(q^4;q^4)_\infty(q^6;q^6)_\infty^9},$$

which is (38.3.2).

So now we have

$$\sum_{n\geq 0}\overline{a}(3n)q^n=\frac{(q^3;q^3)_\infty^6(q^4;q^4)_\infty^4(q^6;q^6)_\infty^3}{(q;q)_\infty^8(q^2;q^2)_\infty^4(q^{12};q^{12})_\infty^3}+4q\frac{(q^3;q^3)_\infty^3(q^4;q^4)_\infty^3(q^6;q^6)_\infty^3}{(q;q)_\infty^7(q^2;q^2)_\infty^4}$$
$$+16q^2\frac{(q^4;q^4)_\infty^2(q^6;q^6)_\infty^3(q^{12};q^{12})_\infty^3}{(q;q)_\infty^6(q^2;q^2)_\infty^4}$$

$$= \frac{(q^4;q^4)_\infty^4(q^6;q^6)_\infty^3}{(q;q)_\infty^6(q^2;q^2)_\infty^4(q^{12};q^{12})_\infty^3}\left(\frac{(q^3;q^3)_\infty^3}{(q;q)_\infty} - 4q\frac{(q^{12};q^{12})_\infty^3}{(q^4;q^4)_\infty}\right)^2$$
$$+ 12q\frac{(q^3;q^3)_\infty^3(q^4;q^4)_\infty^3(q^6;q^6)_\infty^3}{(q;q)_\infty^7(q^2;q^2)_\infty^4}$$
$$= \frac{(q^4;q^4)_\infty^4(q^6;q^6)_\infty^3}{(q;q)_\infty^6(q^2;q^2)_\infty^4(q^{12};q^{12})_\infty^3}\left(\frac{(q;q)_\infty^3(q^6;q^6)_\infty^2}{(q^2;q^2)_\infty^2(q^3;q^3)_\infty}\right)^2$$
$$+ 12q\frac{(q^3;q^3)_\infty^3(q^4;q^4)_\infty^3(q^6;q^6)_\infty^3}{(q;q)_\infty^7(q^2;q^2)_\infty^4}$$
$$= \frac{(q^4;q^4)_\infty^4(q^6;q^6)_\infty^7}{(q^2;q^2)_\infty^8(q^3;q^3)_\infty^2(q^{12};q^{12})_\infty^3} + 12q\frac{(q^3;q^3)_\infty^3(q^4;q^4)_\infty^3(q^6;q^6)_\infty^3}{(q;q)_\infty^7(q^2;q^2)_\infty^4},$$

which is (38.1.2),

$$\sum_{n\geq 0}\overline{a}(3n+1)q^n = 2\frac{(q^3;q^3)_\infty^3(q^4;q^4)_\infty^4(q^6;q^6)_\infty^6}{(q;q)_\infty^7(q^2;q^2)_\infty^5(q^{12};q^{12})_\infty^3} + 4q\frac{(q^3;q^3)_\infty^6(q^4;q^4)_\infty^2(q^{12};q^{12})_\infty^3}{(q;q)_\infty^8(q^2;q^2)_\infty^2(q^6;q^6)_\infty^3}$$
$$+ 8q\frac{(q^4;q^4)_\infty^3(q^6;q^6)_\infty^6}{(q;q)_\infty^6(q^2;q^2)_\infty^5}$$
$$= 2\frac{(q^3;q^3)_\infty^6(q^4;q^4)_\infty^4(q^6;q^6)_\infty^6}{(q;q)_\infty^8(q^2;q^2)_\infty^5(q^{12};q^{12})_\infty^3}\left(\frac{(q;q)_\infty}{(q^3;q^3)_\infty^3} + 2q\frac{(q^2;q^2)_\infty^3(q^{12};q^{12})_\infty^6}{(q^4;q^4)_\infty^2(q^6;q^6)_\infty^9}\right)$$
$$+ 8q\frac{(q^4;q^4)_\infty^3(q^6;q^6)_\infty^6}{(q;q)_\infty^6(q^2;q^2)_\infty^5}$$
$$= 2\frac{(q^3;q^3)_\infty^6(q^4;q^4)_\infty^4(q^6;q^6)_\infty^6}{(q;q)_\infty^8(q^2;q^2)_\infty^5(q^{12};q^{12})_\infty}\cdot\frac{(q^2;q^2)_\infty^3(q^3;q^3)_\infty^3(q^{12};q^{12})_\infty^3}{(q;q)_\infty(q^4;q^4)_\infty(q^6;q^6)_\infty^9}$$
$$+ 8q\frac{(q^4;q^4)_\infty^3(q^6;q^6)_\infty^6}{(q;q)_\infty^6(q^2;q^2)_\infty^5}$$
$$= 2\frac{(q^3;q^3)_\infty^9(q^4;q^4)_\infty^3}{(q;q)_\infty^9(q^2;q^2)_\infty^2(q^6;q^6)_\infty^3} + 8q\frac{(q^4;q^4)_\infty^3(q^6;q^6)_\infty^6}{(q;q)_\infty^6(q^2;q^2)_\infty^5},$$

which is (38.1.3), and

$$\sum_{n\geq 0}\overline{a}(3n+2)q^n = 2\frac{(q^3;q^3)_\infty^6(q^4;q^4)_\infty^3}{(q;q)_\infty^8(q^2;q^2)_\infty^3} + 4\frac{(q^4;q^4)_\infty^4(q^6;q^6)_\infty^9}{(q;q)_\infty^6(q^2;q^2)_\infty^6(q^{12};q^{12})_\infty^3}$$
$$+ 8q\frac{(q^3;q^3)_\infty^3(q^4;q^4)_\infty^2(q^{12};q^{12})_\infty^2}{(q;q)_\infty^7(q^2;q^2)_\infty^3}$$
$$= 2\frac{(q^3;q^3)_\infty^6(q^4;q^4)_\infty^3}{(q;q)_\infty^8(q^2;q^2)_\infty^3}$$

$$
\begin{aligned}
&+4\frac{(q^3;q^3)_\infty^3(q^4;q^4)_\infty^4(q^6;q^6)_\infty^9}{(q;q)_\infty^7(q^2;q^2)_\infty^6(q^{12};q^{12})_\infty^3}\left(\frac{(q;q)_\infty}{(q^3;q^3)_\infty^3}+2q\frac{(q^2;q^2)_\infty^3(q^{12};q^{12})_\infty^6}{(q^4;q^4)_\infty^2(q^6;q^6)_\infty^9}\right)\\
&=2\frac{(q^3;q^3)_\infty^6(q^4;q^4)_\infty^3}{(q;q)_\infty^8(q^2;q^2)_\infty^3}\\
&+4\frac{(q^3;q^3)_\infty^3(q^4;q^4)_\infty^4(q^6;q^6)_\infty^9}{(q;q)_\infty^7(q^2;q^2)_\infty^6(q^{12};q^{12})_\infty^3}\cdot\frac{(q^2;q^2)_\infty^3(q^3;q^3)_\infty^3(q^{12};q^{12})_\infty^3}{(q;q)_\infty(q^4;q^4)_\infty(q^6;q^6)_\infty^9}\\
&=6\frac{(q^3;q^3)_\infty^6(q^4;q^4)_\infty^3}{(q;q)_\infty^8(q^2;q^2)_\infty^3},
\end{aligned}
$$

which is (38.1.1), and we are done.

Exercise: Show using (38.2.6) and (38.3.1) that

$$
\begin{aligned}
\sum_{n\geq 0}\overline{a}(3n)q^n=&\frac{(q^3;q^3)_\infty^{10}(q^4;q^4)_\infty^4(q^6;q^6)_\infty}{(q;q)_\infty^{12}(q^2;q^2)_\infty^2(q^{12};q^{12})_\infty^3}\\
&-64q^3\frac{(q^3;q^3)_\infty(q^4;q^4)_\infty(q^6;q^6)_\infty(q^{12};q^{12})_\infty^6}{(q;q)_\infty^9(q^2;q^2)_\infty^2}.
\end{aligned}
\tag{38.3.9}
$$

References

86. M.D. Hirschhorn, A note on overcubic partitions. N. Z. J. Math. **42**, 229–234 (2012)
100. B. Kim, The overcubic partition function mod 3, *Ramanujan rediscovered, proceedings of a Conference on Elliptic Functions, Partitions, and q-Series in Memory of K. Venkatachaliengar, Bangalore*, 1–5 June, 2009 Ramanujan Math. Soc. Lect. Notes **14**, 157–163 (2010)
118. J.A. Sellers, Elementary proofs of congruences for the cubic and overcubic partition functions. Australas. J. Comb. **66**(2), 191–197 (2014)

Chapter 39
Generalised Frobenius Partitions

39.1 Introduction

A partition of n can be represented by a Ferrers diagram. The Ferrers diagram consists of left-justified rows of π_i dots, where the π_i are the parts of the partition π of n, $\pi_1 \geq \pi_2 \geq \cdots \geq \pi_m \geq 1$, $m = p(n)$, the number of partitions of n and $\pi_1 + \pi_2 + \cdots + \pi_m = n$.

The Ferrers diagram of a partition contains a maximal square of dots in the top left corner, known as the Durfee square.

We define the Frobenius symbol of the partition as follows. Suppose the Durfee square is $r \times r$ for some $r > 0$.

We define the a_i by $a_i = \pi_i - i$, $i = 1, \cdots, r$.

Now consider the partition π' of n, where the Ferrers diagram of π is transposed. π' is called the conjugate of π. The Durfee square of π' is $r \times r$ also.

We define b_i by $b_i = \pi'_i - i$, $i = 1, \cdots, r$.

Then the Frobenius symbol for the partition π is the array

$$\begin{pmatrix} a_1 \; a_2 \cdots a_r \\ b_1 \; b_2 \cdots b_r \end{pmatrix}.$$

Note that

$$n = a_1 + b_1 + \cdots + a_r + b_r + r$$

and that

$$a_1 > a_2 > \cdots > a_r \geq 0, \quad b_1 > b_2 > \cdots > b_r \geq 0.$$

Following Andrews in his seminal paper [5], we define a generalised Frobenius partition of n to be an array of the form

$$\begin{pmatrix} a_1 \; a_2 \; \cdots \; a_r \\ b_1 \; b_2 \; \cdots \; b_r \end{pmatrix}$$

M.D. Hirschhorn, *The Power of q*, Developments in Mathematics 49,
DOI 10.1007/978-3-319-57762-3_39

with

$$a_1 \geq a_2 \geq \cdots \geq a_r \geq 0, \quad b_1 \geq b_2 \geq \cdots \geq b_r \geq 0$$

and

$$a_1 + b_1 \cdots + a_r + b_r + r = n.$$

If we now allow the a_i and b_i to be coloured from an ordered set of k colours, with the additional condition that if two a's (or two b's) are equal, then they are of different colours, we have a generalised Frobenius partition of n in k colours.

A number of papers have appeared recently on the subject of $c\phi_k(n)$, the number of k-coloured generalised Frobenius partitions of n, which is shown by Andrews [5], Theorem 5.2 to be given by

$$\sum_{n\geq 0} c\phi_k(n)q^n = \frac{1}{(q;q)_\infty^k} \sum_{m_1, \quad , m_k=-\infty}^{\infty} q^Q.$$

where

$$Q = \sum_{i=1}^{k-1} m_i^2 + \sum_{1\leq i<j\leq k-1} m_i m_j.$$

In particular, we consider the case $k = 6$, and prove the following congruences:

$$c\phi_6(3n+1) \equiv 0 \pmod 9, \tag{39.1.1}$$

$$c\phi_6(3n+2) \equiv 0 \pmod{27}, \tag{39.1.2}$$

$$c\phi_6(9n+7) \equiv 0 \pmod{27} \tag{39.1.3}$$

and

$$c\phi_6(27n+16) \equiv 0 \pmod{243}. \tag{39.1.4}$$

Our proof of (39.1.4) follows Gu, Wang and Xia (2016), who prove rather more.

39.2 Proofs of (39.1.1) and (39.1.2)

We showed in §21.2 that

$$\left(\sum_{-\infty}^{\infty} a^r q^{r^2}\right)^3 = a(q^2)\sum_{-\infty}^{\infty} a^{3r} q^{3r^2} + q^{\frac{1}{3}} c(q^2)\left(\sum_{-\infty}^{\infty} a^{3r+1} q^{3r^2+2r} + \sum_{-\infty}^{\infty} a^{-3r-1} q^{3r^2+2r}\right). \tag{39.2.1}$$

If we square (39.2.1) and extract the constant term, we find

$$\sum_{m_1+\cdots+m_6=0} q^{m_1^2+\cdots+m_6^2} = a(q^2)^2\sum_{-\infty}^{\infty} q^{6r^2} + 2q^{\frac{2}{3}}c(q^2)^2\sum_{-\infty}^{\infty} q^{6r^2+4r}$$
$$= a(q^2)^2\phi(q^6) + 2q^{\frac{2}{3}}c(q^2)^2\Omega(q^2). \tag{39.2.2}$$

That is,

$$\sum_{m_1,\ldots,m_5=-\infty}^{\infty} q^{2Q} = a(q^2)^2\phi(q^6) + 2q^{\frac{2}{3}}c(q^2)^2\Omega(q^2), \tag{39.2.3}$$

so

$$\sum_{n\geq 0} c\phi_6(n)q^n = \frac{1}{(q;q)_\infty^6}\left(a(q)^2\phi(q^3) + 2q^{\frac{1}{3}}c(q)^2\Omega(q)\right). \tag{39.2.4}$$

We require the 3-dissections (22.1.4), (22.1.5) and (26.1.1),

$$a(q) = a(q^3) + 2c(q^3), \tag{39.2.5}$$
$$b(q) = a(q^3) - c(q^3), \tag{39.2.6}$$
$$\phi(q) = \phi(q^9) + 2q\Omega(q^3) \tag{39.2.7}$$

as well as

$$\frac{1}{b(q)} = \frac{b(\omega q)b(\omega^2 q)}{b(q)b(\omega q)b(\omega^2 q)}$$
$$= \frac{a(q^3)^2 + a(q^3)c(q^3) + c(q^3)^2}{b(q^3)^3}. \tag{39.2.8}$$

To obtain (39.2.8) we use (39.2.6) and the fact that

$$b(q)b(\omega q)b(\omega^2 q) = \frac{1}{(q^3;q^3)_\infty^3}\left(\frac{(q^3;q^3)_\infty^4}{(q^9;q^9)_\infty}\right)^3 = \frac{(q^3;q^3)_\infty^9}{(q^9;q^9)_\infty^3} = b(q^3)^3. \tag{39.2.9}$$

We can write (39.2.4) as follows:

$$\sum_{n\geq 0} c\phi_6(n)q^n$$
$$= \frac{\phi(q^3)}{(q^3;q^3)_\infty^2}\left(\frac{(q^3;q^3)_\infty}{(q;q)_\infty^3}\right)^2 a(q)^2 + 18q\frac{(q^3;q^3)_\infty^4(q^{12};q^{12})_\infty}{(q^6;q^6)_\infty}\left(\frac{(q^3;q^3)_\infty}{(q;q)_\infty^3}\right)^3\phi(-q^2)$$
$$= \frac{\phi(q^3)}{(q^3;q^3)_\infty^2}\frac{a(q)^2}{b(q)^2} + 18q\frac{(q^3;q^3)_\infty^4(q^{12};q^{12})_\infty}{(q^6;q^6)_\infty}\frac{\phi(-q^2)}{b(q)^3}$$
$$= \frac{\phi(q^3)}{(q^3;q^3)_\infty^2}\left(a(q^3)+2c(q^3)\right)^2\frac{\left(a(q^3)^2 + a(q^3)c(q^3) + c(q^3)^2\right)^2}{b(q^3)^6}$$

$$+18q\frac{(q^3;q^3)_\infty^4(q^{12};q^{12})_\infty}{(q^6;q^6)_\infty}(\phi(-q^{18})-2q^2\Omega(-q^6))$$
$$\times\frac{\left(a(q^3)^2+a(q^3)c(q^3)+c(q^3)^2\right)^3}{b(q^3)^9}$$
$$=\frac{\phi(q^3)}{(q^3;q^3)_\infty^2 b(q^3)^6}(a(q^3)^6+6a(q^3)^5c(q^3)+15a(q^3)^4c(q^3)^2$$
$$+22a(q^3)^3c(q^3)^3+21a(q^3)^2c(q^3)^4+12a(q^3)c(q^3)^5+4c(q^3)^6)$$
$$+18q\frac{(q^3;q^3)_\infty^4(q^{12};q^{12})_\infty}{(q^6;q^6)_\infty b(q^3)^9}(\phi(-q^{18})-2q^2\Omega(-q^6))(a(q^3)^6+3a(q^3)^5c(q^3)$$
$$+6a(q^3)^4c(q^3)^2+7a(q^3)^3c(q^3)^3+6a(q^3)^2c(q^3)^4+3a(q^3)c(q^3)^5+c(q^3)^6)$$

$$=\frac{(q^6;q^6)_\infty^5(q^9;q^9)_\infty^6}{(q^3;q^3)_\infty^{22}(q^{12};q^{12})_\infty^2}\left(a(q^3)^6+18qa(q^3)^5\frac{(q^9;q^9)_\infty^3}{(q^3;q^3)_\infty}+135q^2a(q^3)^4\frac{(q^9;q^9)_\infty^6}{(q^3;q^3)_\infty^2}\right.$$
$$+594q^3a(q^3)^3\frac{(q^9;q^9)_\infty^9}{(q^3;q^3)_\infty^6}+1701q^4a(q^3)^2\frac{(q^9;q^9)_\infty^{12}}{(q^3;q^3)_\infty^4}$$
$$\left.+2916q^5a(q^3)\frac{(q^9;q^9)_\infty^{15}}{(q^3;q^3)_\infty^5}+2916q^6\frac{(q^9;q^9)_\infty^{18}}{(q^3;q^3)_\infty^6}\right)$$
$$+18q\frac{(q^9;q^9)_\infty^9(q^{12};q^{12})_\infty(q^{18};q^{18})_\infty^2}{(q^3;q^3)_\infty^{23}(q^6;q^6)_\infty(q^{36};q^{36})_\infty}\left(a(q^3)^6+9qa(q^3)^5\frac{(q^9;q^9)_\infty^3}{(q^3;q^3)_\infty}\right.$$
$$+54q^2a(q^3)^4\frac{(q^9;q^9)_\infty^6}{(q^3;q^3)_\infty^2}+189q^3a(q^3)^3\frac{(q^9;q^9)_\infty^9}{(q^3;q^3)_\infty^3}$$
$$+486q^4a(q^3)^2\frac{(q^9;q^9)_\infty^{12}}{(q^3;q^3)_\infty^4}+729q^5a(q^3)\frac{(q^9;q^9)_\infty^{15}}{(q^3;q^3)_\infty^5}$$
$$\left.+729q^6\frac{(q^9;q^9)_\infty^{18}}{(q^3;q^3)_\infty^6}\right)$$
$$-36q^3\frac{(q^9;q^9)_\infty^9(q^{36};q^{36})_\infty^2}{(q^3;q^3)_\infty^{23}(q^{18};q^{18})_\infty}\left(a(q^3)^6+9qa(q^3)^5\frac{(q^9;q^9)_\infty^3}{(q^3;q^3)_\infty}+54q^2a(q^3)^4\frac{(q^9;q^9)_\infty^6}{(q^3;q^3)_\infty^2}\right.$$
$$+189q^3a(q^3)^3\frac{(q^9;q^9)_\infty^9}{(q^3;q^3)_\infty^3}+486q^4a(q^3)^2\frac{(q^9;q^9)_\infty^{12}}{(q^3;q^3)_\infty^4}$$
$$\left.+729q^5a(q^3)\frac{(q^9;q^9)_\infty^{15}}{(q^3;q^3)_\infty^5}+729q^6\frac{(q^9;q^9)_\infty^{18}}{(q^3;q^3)_\infty^6}\right). \tag{39.2.10}$$

It follows that

$$\begin{aligned}
&\sum_{n\geq 0} c\phi_6(3n+1)q^n \\
&= \frac{(q^2;q^2)_\infty^5 (q^3;q^3)_\infty^6}{(q;q)_\infty^{22}(q^4;q^4)_\infty^2}\left(18a(q)^5\frac{(q^3;q^3)_\infty^3}{(q;q)_\infty}+1701qa(q)^2\frac{(q^3;q^3)_\infty^{12}}{(q;q)_\infty^4}\right) \\
&+18\frac{(q^3;q^3)_\infty^9(q^4;q^4)_\infty(q^6;q^6)_\infty^2}{(q;q)_\infty^{23}(q^2;q^2)_\infty(q^{12};q^{12})_\infty}\left(a(q)^6+189qa(q)^3\frac{(q^3;q^3)_\infty^9}{(q;q)_\infty^3}+729q^2\frac{(q^3;q^3)_\infty^{18}}{(q;q)_\infty^6}\right) \\
&-36q\frac{(q^3;q^3)_\infty^9(q^{12};q^{12})_\infty^2}{(q;q)_\infty^{23}(q^6;q^6)_\infty}\left(9a(q)^5\frac{(q^3;q^3)_\infty^3}{(q;q)_\infty}+486qa(q)^2\frac{(q^3;q^3)_\infty^{12}}{(q;q)_\infty^4}\right)
\end{aligned} \tag{39.2.11}$$

and

$$\begin{aligned}
&\sum_{n\geq 0} c\phi_6(3n+2)q^n \\
&= \frac{(q^2;q^2)_\infty^5 (q^3;q^3)_\infty^6}{(q;q)_\infty^{22}(q^4;q^4)_\infty^2}\left(135a(q)^4\frac{(q^3;q^3)_\infty^6}{(q;q)_\infty^2}+2916qa(q)\frac{(q^3;q^3)_\infty^{15}}{(q;q)_\infty^5}\right) \\
&+18\frac{(q^3;q^3)_\infty^9(q^4;q^4)_\infty(q^6;q^6)_\infty^2}{(q;q)_\infty^{23}(q^2;q^2)_\infty(q^{12};q^{12})_\infty}\left(9a(q)^5\frac{(q^3;q^3)_\infty^3}{(q;q)_\infty}+486qa(q)^2\frac{(q^3;q^3)_\infty^{12}}{(q;q)_\infty^4}\right) \\
&-36q\frac{(q^3;q^3)_\infty^9(q^{12};q^{12})_\infty^2}{(q;q)_\infty^{23}(q^6;q^6)_\infty}\left(54a(q)^4\frac{(q^3;q^3)_\infty^6}{(q;q)_\infty^2}+729qa(q)\frac{(q^3;q^3)_\infty^{15}}{(q;q)_\infty^5}\right).
\end{aligned} \tag{39.2.12}$$

(39.1.1) and (39.1.2) follow from (39.2.11) and (39.2.12), respectively.

39.3 Proof of (39.1.3)

We have

$$\begin{aligned}
&\sum_{n\geq 0} c\phi_6(3n+1)q^n \\
&= \frac{(q^2;q^2)_\infty^5 (q^3;q^3)_\infty^6}{(q;q)_\infty^{22}(q^4;q^4)_\infty^2}\left(18a(q)^5\frac{(q^3;q^3)_\infty^3}{(q;q)_\infty}+1701qa(q)^2\frac{(q^3;q^3)_\infty^{12}}{(q;q)_\infty^4}\right) \\
&+18\frac{(q^3;q^3)_\infty^9(q^4;q^4)_\infty(q^6;q^6)_\infty^2}{(q;q)_\infty^{23}(q^2;q^2)_\infty(q^{12};q^{12})_\infty}\left(a(q)^6+189qa(q)^3\frac{(q^3;q^3)_\infty^9}{(q;q)_\infty^3}+729q^2\frac{(q^3;q^3)_\infty^{18}}{(q;q)_\infty^6}\right) \\
&-36q\frac{(q^3;q^3)_\infty^9(q^{12};q^{12})_\infty^2}{(q;q)_\infty^{23}(q^6;q^6)_\infty}\left(9a(q)^5\frac{(q^3;q^3)_\infty^3}{(q;q)_\infty}+486qa(q)^2\frac{(q^3;q^3)_\infty^{12}}{(q;q)_\infty^4}\right).
\end{aligned} \tag{39.3.1}$$

Modulo 27, this becomes

$$\begin{aligned}
&\sum_{n\geq 0} c\phi_6(3n+1)q^n \\
&\equiv 18\frac{(q^2;q^2)_\infty^5(q^3;q^3)_\infty^9}{(q;q)_\infty^{23}(q^4;q^4)_\infty^2}a(q)^5+18\frac{(q^3;q^3)_\infty^9(q^4;q^4)_\infty(q^6;q^6)_\infty^2}{(q;q)_\infty^{23}(q^2;q^2)_\infty(q^{12};q^{12})_\infty}a(q)^6 \\
&\equiv 18\frac{(q^2;q^2)_\infty^5(q^3;q^3)_\infty^9}{(q;q)_\infty^{23}(q^4;q^4)_\infty^2}+18\frac{(q^3;q^3)_\infty^9(q^4;q^4)_\infty(q^6;q^6)_\infty^2}{(q;q)_\infty^{23}(q^2;q^2)_\infty(q^{12};q^{12})_\infty} \\
&\equiv 18\phi(q)(q^3;q^3)_\infty^2\left(\frac{(q^3;q^3)_\infty}{(q;q)_\infty^3}\right)^7 \\
&\qquad +18\phi(q)(q^3;q^3)_\infty^2\left(\frac{(q^3;q^3)_\infty}{(q;q)_\infty^3}\right)^7\left(\frac{(q^6;q^6)_\infty}{(q^2;q^2)_\infty^3}\right)^2\left(\frac{(q^4;q^4)_\infty^3}{(q^{12};q^{12})_\infty}\right) \\
&\equiv 18\phi(q)(q^3;q^3)_\infty^2+18\phi(q)(q^3;q^3)_\infty^2 \\
&\equiv 9\phi(q)(q^3;q^3)_\infty^2 \\
&\equiv 9(q^3;q^3)_\infty^2\left(\phi(q^9)+2q\Omega(q^3)\right).
\end{aligned} \tag{39.3.2}$$

(39.1.3) follows from (39.3.2).

39.4 Proof of (39.1.4)

All congruences in this section are to the modulus 243 unless stated otherwise.

We have

$$\begin{aligned}
&\sum_{n\geq 0} c\phi_6(3n+1)q^n \\
&=\frac{(q^2;q^2)_\infty^5(q^3;q^3)_\infty^6}{(q;q)_\infty^{22}(q^4;q^4)_\infty^2}\left(18a(q)^5\frac{(q^3;q^3)_\infty^3}{(q;q)_\infty}+1701qa(q)^2\frac{(q^3;q^3)_\infty^{12}}{(q;q)_\infty^4}\right) \\
&+18\frac{(q^3;q^3)_\infty^9(q^4;q^4)_\infty(q^6;q^6)_\infty^2}{(q;q)_\infty^{23}(q^2;q^2)_\infty(q^{12};q^{12})_\infty}\left(a(q)^6+189qa(q)^3\frac{(q^3;q^3)_\infty^9}{(q;q)_\infty^3}+729q^2\frac{(q^3;q^3)_\infty^{18}}{(q;q)_\infty^6}\right) \\
&-36q\frac{(q^3;q^3)_\infty^9(q^{12};q^{12})_\infty^2}{(q;q)_\infty^{23}(q^6;q^6)_\infty}\left(9a(q)^5\frac{(q^3;q^3)_\infty^3}{(q;q)_\infty}+486qa(q)^2\frac{(q^3;q^3)_\infty^{12}}{(q;q)_\infty^4}\right)
\end{aligned} \tag{39.4.1}$$

$$\begin{aligned}
&\equiv 18\frac{(q^2;q^2)_\infty^5(q^3;q^3)_\infty^9}{(q;q)_\infty^{23}(q^4;q^4)_\infty^2}a(q)^5+18\frac{(q^3;q^3)_\infty^9(q^4;q^4)_\infty(q^6;q^6)_\infty^2}{(q;q)_\infty^{23}(q^2;q^2)_\infty(q^{12};q^{12})_\infty}a(q)^6 \\
&\qquad -81q\frac{(q^3;q^3)_\infty^{12}(q^{12};q^{12})_\infty^2}{(q;q)_\infty^{24}(q^6;q^6)_\infty}a(q)^5.
\end{aligned} \tag{39.4.2}$$

$$\equiv 18\phi(q)\left(\frac{(q^3;q^3)_\infty^9}{(q;q)_\infty^{27}}\right)(q;q)_\infty^6 a(q)^5 + 18\phi(-q^6)\psi(-q)\left(\frac{(q^3;q^3)_\infty^9}{(q;q)_\infty^{27}}\right)(q;q)_\infty^3 a(q)^6$$
$$-81q\psi(q^6)\left(\frac{(q^3;q^3)_\infty}{(q;q)_\infty^3}\right)^8 (q^3;q^3)_\infty^4 a(q)^5. \qquad (39.4.3)$$

We now make use of the fact that

$$(q;q)_\infty^{27} \equiv (q^3;q^3)_\infty^9 \pmod{27} \qquad (39.4.4)$$

and the 3-dissection (21.3.7),

$$\begin{aligned}(q;q)_\infty^3 &= (q^3;q^3)_\infty a(q^3) - 3q(q^9;q^9)_\infty^3 \\ &= (q^3;q^3)_\infty \left(a(q^3)-c(q^3)\right)\end{aligned} \qquad (39.4.5)$$

to find that

$$\sum_{n\geq 0} c\phi_6(3n+1)q^n$$
$$\equiv 18\left(\phi(q^9)+2q\Omega(q^3)\right)(q^3;q^3)_\infty^2\left(a(q^3)-c(q^3)\right)^2\left(a(q^3)+2c(q^3)\right)^5$$
$$+18\phi(-q^6)\left(\Pi(-q^3)-q\psi(-q^9)\right)(q^3;q^3)_\infty\left(a(q^3)-c(q^3)\right)\left(a(q^3)+2c(q^3)\right)^6$$
$$-81q\psi(q^6)(q^3;q^3)_\infty^4 \qquad (39.4.6)$$
$$\equiv 18\left(\phi(q^9)+2q\Omega(q^3)\right)(q^3;q^3)_\infty^2(a(q^3)^7+8a(q^3)^6c(q^3))$$
$$+18\phi(-q^6)\left(\Pi(-q^3)-q\psi(-q^9)\right)(q^3;q^3)_\infty(a(q^3)^7+2a(q^3)^6c(q^3))$$
$$-81q\psi(q^6)(q^3;q^3)_\infty^4. \qquad (39.4.7)$$

It follows that

$$\sum_{n\geq 0} c\phi_6(9n+7)q^n$$
$$\equiv 45q^{-\frac{1}{3}}\Omega(q)(q;q)_\infty^2 a(q)^6 c(q) - 36q^{-\frac{1}{3}}\phi(-q^2)\psi(-q^3)(q;q)_\infty a(q)^6 c(q) \qquad (39.4.8)$$
$$\equiv 27\frac{(q^2;q^2)_\infty^2(q^3;q^3)_\infty^4(q^{12};q^{12})_\infty}{(q^4;q^4)_\infty(q^6;q^6)_\infty}a(q)^6 \qquad (39.4.9)$$
$$\equiv 27\frac{(q^2;q^2)_\infty^2(q^3;q^3)_\infty^4(q^{12};q^{12})_\infty}{(q^4;q^4)_\infty(q^6;q^6)_\infty} \qquad (39.4.10)$$

$$\equiv 27\frac{(q^3;q^3)_\infty^4(q^{12};q^{12})_\infty}{(q^6;q^6)_\infty}\phi(-q^2) \tag{39.4.11}$$

$$\equiv 27\frac{(q^3;q^3)_\infty^4(q^{12};q^{12})_\infty}{(q^6;q^6)_\infty}(\phi(-q^{18})-2q^2\Omega(-q^6)). \tag{39.4.12}$$

(39.1.4) follows from (39.4.12).

Endnote.

For related literature, see Cu, Wang and Xia (2016) [54].

References

5. G.E. Andrews, *Generalized Frobenius Partitions*, vol. 301, Memoirs of the American Mathematical Society (AMS, Providence, 1984)
54. C. Gu, L. Wang, E.X.W. Xia, Congruences modulo 81 and 243 for generalized Frobenius partitions with 6 colors. Acta Arith. **175**(3), 291–300 (2016)
91. M.D. Hirschhorn, Some congruences for 6-coloured generalized Frobenius partitions. Ramanujan J. **40**, 463–471 (2016)

Chapter 40
Some Modular Equations of Ramanujan

40.1 Introduction

Let $r(q)$ be as defined in (8.6.5) and let $u = r(q)$, $v = r(q^2)$. Ramanujan ([111], p. 326) states that

$$\frac{v-u^2}{v+u^2} = uv^2 \tag{40.1.1}$$

and that

$$\text{if}\, u^5 = k\left(\frac{1-k}{1+k}\right)^2 \text{then}\, v^5 = k^2\left(\frac{1+k}{1-k}\right). \tag{40.1.2}$$

[We have taken the liberty of changing Ramanujan's n to k.]

We will examine his claim (40.1.2) at the end of §40.2.

Let us define

$$k = uv^2. \tag{40.1.3}$$

We will show that

$$\frac{v-u^2}{v+u^2} = k, \tag{40.1.4}$$

$$\frac{v}{u^2} = \frac{1+k}{1-k}, \tag{40.1.5}$$

$$u^5 = k\left(\frac{1-k}{1+k}\right)^2, \tag{40.1.6}$$

$$v^5 = k^2\left(\frac{1+k}{1-k}\right), \tag{40.1.7}$$

M.D. Hirschhorn, *The Power of q*, Developments in Mathematics 49,
DOI 10.1007/978-3-319-57762-3_40

as well as

$$(v+u^2)(1+uv^2) = 2v, \tag{40.1.8}$$
$$(v+u^2)(1-uv^2) = 2u^2, \tag{40.1.9}$$
$$(v-u^2)(1+uv^2) = 2uv^3, \tag{40.1.10}$$

and

$$(v-u^2)(1-uv^2) = 2u^3v^2. \tag{40.1.11}$$

Now let $u = r(q)$, $v = r(q^3)$. Then, in splendid isolation ([111], p. 321), Ramanujan writes

$$(v-u^3)(1+uv^3) = 3u^2v^2. \tag{40.1.12}$$

This we will prove in §40.3.

40.2 Proofs of (40.1.4)–(40.1.11)

We begin by proving

$$R(q)^2 = \left(\begin{matrix} q,q^4,q^6,q^9 \\ q^2,q^5,q^5,q^8 \end{matrix};q^{10}\right)_\infty - q\left(\begin{matrix} q,q,q^9,q^9 \\ q^3,q^5,q^5,q^7 \end{matrix};q^{10}\right)_\infty \tag{40.2.1}$$

and

$$R(q^2) = \left(\begin{matrix} q,q^4,q^6,q^9 \\ q^2,q^5,q^5,q^8 \end{matrix};q^{10}\right)_\infty + q\left(\begin{matrix} q,q,q^9,q^9 \\ q^3,q^5,q^5,q^7 \end{matrix};q^{10}\right)_\infty. \tag{40.2.2}$$

The right side of (40.2.1) is (see §41.2 for more detail)

$$\begin{aligned}
&\left(\begin{matrix} q,q^4,q^6,q^9 \\ q^2,q^5,q^5,q^8 \end{matrix};q^{10}\right)_\infty - q\left(\begin{matrix} q,q,q^9,q^9 \\ q^3,q^5,q^5,q^7 \end{matrix};q^{10}\right)_\infty \\
&= \left(\begin{matrix} q,q^9 \\ q^2,q^3,q^5,q^5,q^7,q^8,q^{10},q^{10} \end{matrix};q^{10}\right)_\infty \\
&\qquad \times ((q^3,q^4,q^6,q^7,q^{10},q^{10};q^{10})_\infty - q(q,q^2,q^8,q^9,q^{10},q^{10};q^{10})_\infty) \\
&= \left(\begin{matrix} q,q^9 \\ q^2,q^3,q^5,q^5,q^7,q^8,q^{10},q^{10} \end{matrix};q^{10}\right)_\infty (q,-q^2,-q^3,q^4,q^5,q^5;q^5)_\infty \\
&= \left(\begin{matrix} q,q,q^4,q^4,q^6,q^6,q^9,q^9 \\ q^2,q^2,q^3,q^3,q^7,q^7,q^8,q^8 \end{matrix};q^{10}\right)_\infty
\end{aligned}$$

$$= \left(\begin{matrix} q, q^4, q^6, q^9 \\ q^2, q^3, q^7, q^8 \end{matrix}; q^{10}\right)_\infty^2$$
$$= \left(\begin{matrix} q, q^4 \\ q^2, q^3 \end{matrix}; q^5\right)_\infty^2$$
$$= R(q)^2,$$

which is (40.2.1).

Similarly, the right side of (40.2.2) is

$$\left(\begin{matrix} q, q^4, q^6, q^9 \\ q^2, q^5, q^5, q^8 \end{matrix}; q^{10}\right)_\infty + q\left(\begin{matrix} q, q, q^9, q^9 \\ q^3, q^5, q^5, q^7 \end{matrix}; q^{10}\right)_\infty$$
$$= \left(\begin{matrix} q, q^9 \\ q^2, q^3, q^5, q^5, q^7, q^8, q^{10}, q^{10} \end{matrix}; q^{10}\right)_\infty$$
$$\times ((q^3, q^4, q^6, q^7, q^{10}, q^{10}; q^{10})_\infty + q(q, q^2, q^8, q^9, q^{10}, q^{10}; q^{10})_\infty)$$
$$= \left(\begin{matrix} q, q^9 \\ q^2, q^3, q^5, q^5, q^7, q^8, q^{10}, q^{10} \end{matrix}; q^{10}\right)_\infty (-q, q^2, q^3, -q^4, q^5, q^5; q^5)_\infty$$
$$= \left(\begin{matrix} q^2, q^8 \\ q^4, q^6 \end{matrix}; q^{10}\right)_\infty$$
$$= R(q^2),$$

which is (40.2.2).

From (40.2.1) and (40.2.2) we have

$$R(q^2) - R(q)^2 = 2q\left(\begin{matrix} q, q, q^9, q^9 \\ q^3, q^5, q^5, q^7 \end{matrix}; q^{10}\right)_\infty \tag{40.2.3}$$

and

$$R(q^2) + R(q)^2 = 2\left(\begin{matrix} q, q^4, q^6, q^9 \\ q^2, q^5, q^5, q^8 \end{matrix}; q^{10}\right)_\infty. \tag{40.2.4}$$

It follows from (40.2.3) and (40.2.4) that

$$\frac{R(q^2) - R(q)^2}{R(q^2) + R(q)^2} = q\left(\begin{matrix} q, q^2, q^8, q^9 \\ q^3, q^4, q^6, q^7 \end{matrix}; q^{10}\right)_\infty = q\left(\begin{matrix} q, q^4, q^6, q^9 \\ q^2, q^3, q^7, q^8 \end{matrix}; q^{10}\right)_\infty \left(\begin{matrix} q^2, q^8 \\ q^4, q^6 \end{matrix}; q^{10}\right)_\infty^2$$
$$= qR(q)R(q^2)^2 \tag{40.2.5}$$

and this translates to (40.1.4).

From (40.1.4) we deduce (40.1.5) by elementary manipulation.

If we divide (40.1.3) by the square of (40.1.5), we obtain (40.1.6), while if we multiply (40.1.5) by the square of (40.1.3), we obtain (40.1.7).

By (40.1.5), we have

$$v+u^2 = v + \frac{1-k}{1+k}v = \frac{2}{1+k}v,$$

or,

$$(v+u^2)(1+uv^2) = 2v,$$

which is (40.1.8). The proofs of (40.1.9)–(40.1.11) are similar, so left as exercises.

Now we address Ramanujan's claim. It is simply not true as it stands.

What is true is the following.

$$\text{If } 0 \leq q \leq 1 \text{ and if } u^5 = k\left(\frac{1-k}{1+k}\right)^2 \text{ with } k \leq \sqrt{5}-2 \text{ then } v^5 = k^2\left(\frac{1+k}{1-k}\right).$$

40.3 Proof of (40.1.12)

We start by proving

$$R(q^3) - R(q)^3 = 3q\left(\begin{matrix} q,q,q^4,q^4,q^6,q^9,q^{11},q^{11},q^{14},q^{14} \\ q^2,q^3,q^5,q^5,q^7,q^8,q^{10},q^{10},q^{12},q^{13} \end{matrix};q^{15}\right)_\infty \tag{40.3.1}$$

and

$$1+q^2R(q)R(q^3)^3 = \left(\begin{matrix} q^3,q^5,q^5,q^{10},q^{10},q^{12} \\ q^2,q^6,q^7,q^8,q^9,q^{13} \end{matrix};q^{15}\right)_\infty. \tag{40.3.2}$$

Our proofs of (40.3.1) and (40.3.2) were extracted from a recent paper of C. Gugg (2012) [56], to whom all credit is due.

In the quintuple product identity (10.1.2), replace q by q^5, then replace a by aq^{-2} and divide by $(q^5;q^5)_\infty$, and we obtain

$$\left(\begin{matrix} a^{-2}q^4,a^2q \\ a^{-1}q^2,aq^3 \end{matrix};q^5\right)_\infty = \left(\begin{matrix} a^3q^4,a^{-3}q^{11} \\ q^5,q^{10} \end{matrix};q^{15}\right)_\infty - a^2q\left(\begin{matrix} a^{-3}q,a^3q^{14} \\ q^5,q^{10} \end{matrix};q^{15}\right)_\infty. \tag{40.3.3}$$

If in (40.3.3) we set $a=1$, we obtain

$$R(q) = \left(\begin{matrix} q^4,q^{11} \\ q^5,q^{10} \end{matrix};q^{15}\right)_\infty - q\left(\begin{matrix} q,q^{14} \\ q^5,q^{10} \end{matrix};q^{15}\right)_\infty. \tag{40.3.4}$$

If in (40.3.3) we set $a = 1$, $a = \omega$ and $a = \omega^{-1}$ in turn, where ω is a cube root of unity other than 1, and multiply the three results, we obtain

$$R(q^3) = \left(\begin{matrix} q^4, q^{11} \\ q^5, q^{10} \end{matrix}; q^{15}\right)_\infty^3 - q^3 \left(\begin{matrix} q, q^{14} \\ q^5, q^{10} \end{matrix}; q^{15}\right)_\infty^3. \tag{40.3.5}$$

It follows from (40.3.4) and (40.3.5) that

$$\begin{aligned}
&R(q^3) - R(q)^3 \\
&= 3q \left(\begin{matrix} q^4, q^{11} \\ q^5, q^{10} \end{matrix}; q^{15}\right)_\infty \left(\begin{matrix} q, q^{14} \\ q^5, q^{10} \end{matrix}; q^{15}\right)_\infty \left(\left(\begin{matrix} q^4, q^{11} \\ q^5, q^{10} \end{matrix}; q^{15}\right)_\infty - q \left(\begin{matrix} q, q^{14} \\ q^5, q^{10} \end{matrix}; q^{15}\right)_\infty\right) \\
&= 3q \left(\begin{matrix} q^4, q^{11} \\ q^5, q^{10} \end{matrix}; q^{15}\right)_\infty \left(\begin{matrix} q, q^{14} \\ q^5, q^{10} \end{matrix}; q^{15}\right)_\infty \left(\begin{matrix} q, q^4 \\ q^2, q^3 \end{matrix}; q^5\right)_\infty \\
&= 3q \left(\begin{matrix} q, q, q^4, q^4, q^6, q^9, q^{11}, q^{11}, q^{14}, q^{14} \\ q^2, q^3, q^5, q^5, q^7, q^8, q^{10}, q^{10}, q^{12}, q^{13} \end{matrix}; q^{15}\right)_\infty,
\end{aligned}$$

which is (40.3.1).

Also,

$$\begin{aligned}
&1 + q^2 R(q) R(q^3)^3 \\
&= 1 + q^2 \left(\begin{matrix} q, q^4 \\ q^2, q^3 \end{matrix}; q^5\right) \left(\begin{matrix} q^3, q^{12} \\ q^6, q^9 \end{matrix}; q^{15}\right)^3 \\
&= \frac{1}{(q^2, q^3; q^5)_\infty (q^6, q^9; q^{15})_\infty^3} \\
&\qquad \times ((q^2, q^3; q^5)_\infty (q^6, q^9; q^{15})_\infty^3 + q^2 (q, q^4; q^5)_\infty (q^3, q^{12}; q^{15})_\infty^3) \\
&= \frac{1}{(q^2, q^3; q^5)_\infty (q^6, q^9; q^{15})_\infty^3} \\
&\qquad \times ((q^2, q^3, q^7, q^8, q^{12}, q^{13}; q^{15})_\infty (q^6, q^9; q^{15})_\infty^3 \\
&\qquad\qquad + q^2 (q, q^4, q^6, q^9, q^{11}, q^{14}; q^{15})_\infty (q^3, q^{12}; q^{15})_\infty^3) \\
&= \frac{1}{(q^2, q^3; q^5)_\infty (q^6, q^9; q^{15})_\infty^3} \cdot \frac{(q^3; q^3)_\infty}{(q^{15}; q^{15})_\infty} \\
&\qquad \times ((q^2, q^7, q^8, q^{13}; q^{15})_\infty (q^6, q^9; q^{15})_\infty^2 \\
&\qquad\qquad + q^2 (q, q^4, q^{11}, q^{14}; q^{15})_\infty (q^3, q^{12}; q^{15})_\infty^2) \\
&= \frac{(q^3; q^3)_\infty}{(q^2, q^3; q^5)_\infty (q^6, q^9; q^{15})_\infty^3 (q^{15}, q^{15})_\infty} \cdot \frac{1}{(q^{15}; q^{15})_\infty^4}
\end{aligned}$$

$$\times\big((q^2,q^6,q^9,q^{13},q^{15},q^{15};q^{15})_\infty(q^6,q^7,q^8,q^9,q^{15},q^{15};q^{15})_\infty$$
$$+q^2(q,q^3,q^{12},q^{14},q^{15},q^{15};q^{15})_\infty(q^3,q^4,q^{11},q^{12},q^{15},q^{15};q^{15})_\infty\big)$$
$$=\frac{(q^3;q^3)_\infty}{(q^2,q^3;q^5)_\infty(q^6,q^9;q^{15})^3_\infty(q^{15};q^{15})^5_\infty}$$
$$\times\big(\big((-q^8,-q^{11},-q^{19},-q^{22},q^{30},q^{30};q^{30})_\infty$$
$$-q^2(-q^4,-q^7,-q^{23},-q^{26},q^{30},q^{30};q^{30})_\infty\big)$$
$$\times\big((-q^{13},-q^{14},-q^{16},-q^{17},q^{30},q^{30};q^{30})_\infty$$
$$-q^6(-q,-q^2,-q^{28},-q^{29},q^{30},q^{30};q^{30})_\infty\big)$$
$$+q^2\big((-q^4,-q^{13},-q^{17},-q^{26},q^{30},q^{30};q^{30})_\infty$$
$$-q(-q^2,-q^{11},-q^{19},-q^{28},q^{30},q^{30};q^{30})_\infty\big)$$
$$\times\big((-q^7,-q^{14},-q^{16},-q^{23},q^{30},q^{30};q^{30})_\infty$$
$$-q^3(-q,-q^8,-q^{22},-q^{29},q^{30},q^{30};q^{30})_\infty\big)\big)$$
$$=\frac{(q^3;q^3)_\infty}{(q^2,q^3;q^5)_\infty(q^6,q^9;q^{15})^3_\infty(q^{15};q^{15})^5_\infty}$$
$$\times\big((-q^8,-q^{11},-q^{13},-q^{14},-q^{16},-q^{17},-q^{19},-q^{22},q^{30},q^{30},q^{30},q^{30};q^{30})_\infty$$
$$-q^3(-q^2,-q^7,-q^{11},-q^{14},-q^{16},-q^{19},-q^{23},-q^{28},q^{30},q^{30},q^{30},q^{30};q^{30})_\infty$$
$$-q^5(-q,-q^4,-q^8,-q^{13},-q^{17},-q^{22},-q^{26},-q^{29},q^{30},q^{30},q^{30},q^{30};q^{30})_\infty$$
$$+q^8(-q,-q^2,-q^4,-q^7,-q^{23},-q^{26},-q^{28},-q^{29},q^{30},q^{30},q^{30},q^{30};q^{30})_\infty\big)$$
$$=\frac{(q^3;q^3)_\infty}{(q^2,q^3;q^5)_\infty(q^6,q^9;q^{15})^3_\infty(q^{15};q^{15})^5_\infty}$$
$$\times\big((-q^8,-q^{13},-q^{17},-q^{22},q^{30},q^{30};q^{30})_\infty$$
$$-q^3(-q^2,-q^7,-q^{23},-q^{28},q^{30},q^{30};q^{30})_\infty\big)$$
$$\times\big((-q^{11},-q^{14},-q^{16},-q^{19},q^{30},q^{30};q^{30})_\infty$$
$$-q^5(-q,-q^4,-q^{26},-q^{29},q^{30},q^{30};q^{30})_\infty\big)$$
$$=\frac{(q^3;q^3)_\infty}{(q^2,q^3;q^5)_\infty(q^6,q^9;q^{15})^3_\infty(q^{15};q^{15})^5_\infty}$$
$$\times(q^3,q^5,q^{10},q^{12},q^{15},q^{15};q^{15})_\infty(q^5,q^6,q^9,q^{10},q^{15},q^{15};q^{15})_\infty$$
$$=\left(\begin{matrix}q^3,q^5,q^5,q^{10},q^{10},q^{12}\\ q^2,q^6,q^7,q^8,q^9,q^{13}\end{matrix};q^{15}\right)_\infty,$$

which is (40.3.2).
If we multiply (40.3.1) by (40.3.2), we obtain

$$\begin{aligned}
\left(R(q^3)-R(q)^3\right)\left(1+q^2R(q)R(q^3)^3\right) &= 3q\left(\begin{matrix}q,q^4,q^{11},q^{14}\\ q^2,q^7,q^8,q^{13}\end{matrix};q^{15}\right)_\infty^2\\
&= 3q\left(\left(\begin{matrix}q,q^4,q^6,q^9,q^{11},q^{14}\\ q^2,q^3,q^7,q^8,q^{12},q^{13}\end{matrix};q^{15}\right)_\infty\left(\begin{matrix}q^3,q^{12}\\ q^6,q^9\end{matrix};q^{15}\right)_\infty\right)^2\\
&= 3q\left(R(q)R(q^3)\right)^2, \qquad (40.3.6)
\end{aligned}$$

and this translates to (40.1.12).

Exercise: Show that

$$H_5\left(\left(\frac{\eta(\tau)\eta(3\tau)}{\eta(25\tau)\eta(75\tau)}\right)^2\right) = -5.$$

References

56. C. Gugg, Modular equations for cubes of the Rogers-Ramanujan and Ramanujan-Göllnitz functions and their associated continued fractions. J. Number Theory **132**, 1519–1553 (2012)
111. S. Ramanujan, Notebooks, II, Tata Institute of Fundamental Research, Mumbai (2012)

Chapter 41
Identities Involving $k = r(q)r(q^2)^2$

41.1 Introduction

In this chapter we investigate a number of identities involving

$$k(q) = r(q)r(q^2)^2 = q\begin{pmatrix} q, q^2, q^8, q^9 \\ q^3, q^4, q^6, q^7 \end{pmatrix}; q^{10}\Big)_\infty. \tag{41.1.1}$$

We will show, among other things, that

$$1+k, \quad 1-k, \quad \frac{1}{k}-k, \quad \frac{1}{k}+1-k \quad \text{and} \quad \frac{1}{k}-4-k$$

are all products.

Indeed, we show that

$$1+k = \left(\begin{matrix} q^2, q^2, q^5, q^5, q^8, q^8 \\ q, q^4, q^4, q^6, q^6, q^9 \end{matrix}; q^{10}\right)_\infty, \tag{41.1.2}$$

$$1-k = \left(\begin{matrix} q, q^4, q^5, q^5, q^6, q^9 \\ q^2, q^3, q^3, q^7, q^7, q^8 \end{matrix}; q^{10}\right)_\infty, \tag{41.1.3}$$

$$\frac{1}{k}-k = \frac{1}{q}\left(\begin{matrix} q^5, q^5, q^5, q^5 \\ q, q^3, q^7, q^9 \end{matrix}; q^{10}\right)_\infty = \frac{(q^2;q^2)_\infty (q^5;q^5)_\infty^5}{q(q;q)_\infty (q^{10};q^{10})_\infty^5}, \tag{41.1.4}$$

$$\frac{1}{k}+1-k = \frac{1}{q}\left(\begin{matrix} q^2, q^4, q^6, q^8 \\ q, q^3, q^7, q^9 \end{matrix}; q^{10}\right)_\infty^2 = \frac{(q^2;q^2)_\infty^4 (q^5;q^5)_\infty^2}{q(q;q)_\infty^2 (q^{10};q^{10})_\infty^4} \tag{41.1.5}$$

M.D. Hirschhorn, *The Power of q*, Developments in Mathematics 49,
DOI 10.1007/978-3-319-57762-3_41

and

$$\begin{aligned}\frac{1}{k} - 4 - k &= \frac{1}{q}(q,q^3,q^7,q^9;q^{10})_\infty^3 (q^2,q^4,q^5,q^5,q^6,q^8;q^{10})_\infty^2 \\ &= \frac{(q;q)_\infty^3 (q^5;q^5)_\infty}{q(q^2;q^2)_\infty (q^{10};q^{10})_\infty^3}. \end{aligned} \tag{41.1.6}$$

We also give the following factorisations of (41.1.4), (41.1.5) and (41.1.6).

$$\frac{1}{\sqrt{k}} + \sqrt{k} = \frac{1}{q^{\frac{1}{5}} \sqrt{r(q)^3}} \begin{pmatrix} q^5, q^5 \\ q^3, q^7 \end{pmatrix} ; q^{10} \Bigg)_\infty, \tag{41.1.7}$$

$$\frac{1}{\sqrt{k}} - \sqrt{k} = \frac{\sqrt{r(q)^3}}{q^{\frac{4}{5}}} \begin{pmatrix} q^5, q^5 \\ q, q^9 \end{pmatrix} ; q^{10} \Bigg)_\infty, \tag{41.1.8}$$

$$\frac{1}{\sqrt{k}} + \alpha\sqrt{k} = \sqrt{\frac{(q^2;q^2)_\infty^4 (q^5;q^5)_\infty}{q(q;q)_\infty (q^{10};q^{10})_\infty^4}} \prod_{k\geq 1} (1 + \alpha q^{2k-1} + q^{4k-2})(1 + \beta q^{2k} + q^{4k}), \tag{41.1.9}$$

$$\frac{1}{\sqrt{k}} + \beta\sqrt{k} = \sqrt{\frac{(q^2;q^2)_\infty^4 (q^5;q^5)_\infty}{q(q;q)_\infty (q^{10};q^{10})_\infty^4}} \prod_{k\geq 1} (1 + \beta q^{2k-1} + q^{4k-2})(1 + \alpha q^{2k} + q^{4k}) \tag{41.1.10}$$

and

$$\frac{1}{\sqrt{k}} - (2\alpha + 1)\sqrt{k} = \sqrt{\frac{(q;q)_\infty^3 (q^5;q^5)_\infty}{q(q^{10};q^{10})_\infty^4}} \prod_{k\geq 1} (1 - \alpha q^k + q^{2k})(1 + \beta q^k + q^{2k}), \tag{41.1.11}$$

$$\frac{1}{\sqrt{k}} - (2\beta + 1)\sqrt{k} = \sqrt{\frac{(q;q)_\infty^3 (q^5;q^5)_\infty}{q(q^{10};q^{10})_\infty^4}} \prod_{k\geq 1} (1 + \alpha q^k + q^{2k})(1 - \beta q^k + q^{2k}). \tag{41.1.12}$$

It appears that all of (41.1.2)–(41.1.12) are due to Ramanujan [112], p. 56, Cooper (2009) [34] or to Cooper and Hirschhorn (2011) [38].

41.2 Proofs of (41.1.2) and (41.1.3)

We start by noting that (41.1.2) and (41.1.3) are equivalent to

$$(q^3,q^4,q^6,q^7,q^{10},q^{10};q^{10})_\infty + q(q,q^2,q^8,q^9,q^{10},q^{10};q^{10})_\infty \\ = (-q,q^2,q^3,-q^4,q^5,q^5;q^5)_\infty \tag{41.2.1}$$

and

$$(q^3,q^4,q^6,q^7,q^{10},q^{10};q^{10})_\infty - q(q,q^2,q^8,q^9,q^{10},q^{10};q^{10})_\infty \\ = (q,-q^2,-q^3,q^4,q^5,q^5;q^5)_\infty. \tag{41.2.2}$$

We have

$$\begin{aligned}
&(q^3,q^4,q^6,q^7,q^{10},q^{10};q^{10})_\infty + q(q,q^2,q^8,q^9,q^{10},q^{10};q^{10})_\infty \\
&= \sum_{m,n=-\infty}^{\infty} (-1)^{m+n} q^{5m^2-2m+5n^2-n} + q \sum_{m,n=-\infty}^{\infty} (-1)^{m+n} q^{5m^2-4m+5n^2-3n} \\
&= \sum_{r,s=-\infty}^{\infty} q^{5(r+s)^2-2(r+s)+5(r-s)^2-(r-s)} - \sum_{r,s=-\infty}^{\infty} q^{5(r+s+1)^2-2(r+s+1)+5(r-s)^2-(r-s)} \\
&+ q \sum_{r,s=-\infty}^{\infty} q^{5(r+s)^2-4(r+s)+5(r-s)^2-3(r-s)} - q \sum_{r,s=-\infty}^{\infty} q^{5(r+s+1)^2-4(r+s+1)+5(r-s)^2-3(r-s)} \\
&= \sum_{r,s=-\infty}^{\infty} q^{10r^2-3r+10s^2-s} - q^3 \sum_{r,s=-\infty}^{\infty} q^{10r^2+7r+10s^2+9s} \\
&+ q \sum_{r,s=-\infty}^{\infty} q^{10r^2-7r+10s^2-s} - q^2 \sum_{r,s=-\infty}^{\infty} q^{10r^2+3r+10s^2+9s} \\
&= \left(\sum_{r,s=-\infty}^{\infty} q^{10r^2+3r} + q \sum_{r,s=-\infty}^{\infty} q^{10r^2-7r} \right) \left(\sum_{r,s=-\infty}^{\infty} q^{10s^2-s} - q^2 \sum_{r,s=-\infty}^{\infty} q^{10s^2+9s} \right) \\
&= \sum_{-\infty}^{\infty} q^{(5r^2-3r)/2} \cdot \sum_{-\infty}^{\infty} (-1)^s q^{(5s^2-s)/2} \\
&= (-q,q^2,q^3,-q^4,q^5,q^5;q^5)_\infty,
\end{aligned}$$

which is (41.2.1).

The proof of (41.2.2) is similar, and left as an exercise.

41.3 Proof of (41.1.4)

If we multiply (41.1.2) by (41.1.3) and divide by (41.1.1), we obtain (41.1.4).

41.4 Proof of (41.1.5)

We have, by (41.1.4) and (34.1.21),

$$\begin{aligned}
\frac{1}{k}+1-k &= \frac{(q^2;q^2)_\infty(q^5;q^5)_\infty^5}{q(q;q)_\infty(q^{10};q^{10})_\infty^5}+1 \\
&= \frac{(q^5;q^5)_\infty^2}{q(q^{10};q^{10})_\infty^4}\left(\frac{(q^2;q^2)_\infty(q^5;q^5)_\infty^3}{(q;q)_\infty(q^{10};q^{10})_\infty}+q\frac{(q^{10};q^{10})_\infty^4}{(q^5;q^5)_\infty^2}\right) \\
&= \frac{(q^5;q^5)_\infty^2}{q(q^{10};q^{10})_\infty^4}((\psi(q)^2-q\psi(q^5)^2)+q\psi(q^5)^2) \\
&= \frac{(q^5;q^5)_\infty^2}{q(q^{10};q^{10})_\infty^4}\psi(q)^2 \\
&= \frac{(q^2;q^2)_\infty^4(q^5;q^5)_\infty^2}{q(q;q)_\infty^2(q^{10};q^{10})_\infty^4},
\end{aligned}$$

which is (41.1.5).

41.5 Proof of (41.1.6)

We have, by (41.1.4), (34.1.21) and (34.1.23),

$$\begin{aligned}
\frac{1}{k}-4-k &= \frac{(q^2;q^2)_\infty(q^5;q^5)_\infty^5}{q(q;q)_\infty(q^{10};q^{10})_\infty^5}-4 \\
&= \frac{(q^5;q^5)_\infty^2}{q(q^{10};q^{10})_\infty^4}\left(\frac{(q^2;q^2)_\infty(q^5;q^5)_\infty^3}{(q;q)_\infty(q^{10};q^{10})_\infty}-4q\frac{(q^{10};q^{10})_\infty^4}{(q^5;q^5)_\infty^2}\right) \\
&= \frac{(q^5;q^5)_\infty^2}{q(q^{10};q^{10})_\infty^4}((\psi(q)^2-q\psi(q^5)^2)-4q\psi(q^5)^2) \\
&= \frac{(q^5;q^5)_\infty^2}{q(q^{10};q^{10})_\infty^4}(\psi(q)^2-5q\psi(q^5)^2)
\end{aligned}$$

$$= \frac{(q^5;q^5)_\infty^2}{q(q^{10};q^{10})_\infty^4} \cdot \frac{(q;q)_\infty^3(q^{10};q^{10})_\infty}{(q^2;q^2)_\infty(q^5;q^5)_\infty}$$
$$= \frac{(q;q)_\infty^3(q^5;q^5)_\infty}{q(q^2;q^2)_\infty(q^{10};q^{10})_\infty^3},$$

which is (41.1.6).

41.6 Proofs of (41.1.7) and (41.1.8)

If we square (41.1.2) and divide by (41.1.1), we obtain

$$\left(\frac{1}{\sqrt{k}}+\sqrt{k}\right)^2 = \frac{1}{q}\left(\begin{matrix}q^2,q^3\\ q,q^4\end{matrix};q^5\right)_\infty^3 \left(\begin{matrix}q^5,q^5\\ q^3,q^7\end{matrix};q^{10}\right)_\infty^2 = \frac{1}{q^{\frac{2}{5}}r(q)^3}\left(\begin{matrix}q^5,q^5\\ q^3,q^7\end{matrix};q^{10}\right)_\infty^2,$$

from which (41.1.7) follows.

The proof of (41.1.8) is similar, and left as an exercise.

41.7 Proofs of (41.1.9) and (41.1.10)

We start by noting that (41.1.9) and (41.1.10) are equivalent to

$$(q^3,q^4,q^6,q^7,q^{10},q^{10};q^{10})_\infty + \alpha q(q,q^2,q^8,q^9,q^{10},q^{10};q^{10})_\infty$$
$$= (\eta^{-2}q,\eta^2 q,q^2;q^2)_\infty(\eta^{-1}q^2,\eta q^2,q^2;q^2)_\infty) \tag{41.7.1}$$

and

$$(q^3,q^4,q^6,q^7,q^{10},q^{10};q^{10})_\infty + \beta q(q,q^2,q^8,q^9,q^{10},q^{10};q^{10})_\infty$$
$$= (\eta^{-1}q,\eta q,q^2;q^2)_\infty(\eta^{-2}q^2,\eta^2 q^2,q^2;q^2)_\infty) \tag{41.7.2}$$

where $\eta = e^{\frac{2\pi i}{5}}$.

In order to prove (41.7.1) and (41.7.2), we make use of JTP (1.1.1), (8.3.6) and (8.3.7) to find that

$$\begin{aligned}&(\eta^{-1}q,\eta q,q^2;q^2)_\infty\\&=(q^{25},q^{25},q^{50};q^{50})_\infty+\beta q(q^{15},q^{35},q^{50};q^{50})_\infty-\alpha q^4(q^5,q^{45},q^{50};q^{50})_\infty\\&=A_0+\beta qA_2-\alpha q^4A_4, \qquad (41.7.3)\\&(\eta^{-2}q,\eta^2q,q^2;q^2)_\infty\\&=(q^{25},q^{25},q^{50};q^{50})_\infty+\alpha q(q^{15},q^{35},q^{50};q^{50})_\infty-\beta q^4(q^5,q^{45},q^{50};q^{50})_\infty\\&=A_0+\alpha qA_2-\beta q^4A_4, \qquad (41.7.4)\end{aligned}$$

$$\begin{aligned}(\eta^{-1}q^2,\eta q^2,q^2;q^2)_\infty&=(q^{20},q^{30},q^{50};q^{50})_\infty-\alpha q^2(q^{10},q^{40},q^{50};q^{50})_\infty\\&=A_1-\alpha q^2A_3 \qquad (41.7.5)\end{aligned}$$

and

$$\begin{aligned}(\eta^{-2}q^2,\eta^2q^2,q^2;q^2)_\infty&=(q^{20},q^{30},q^{50};q^{50})_\infty-\beta q^2(q^{10},q^{40},q^{50})_\infty\\&=A_1-\beta q^2A_3, \qquad (41.7.6)\end{aligned}$$

where

$$A_k=(q^{25-5k},q^{25+5k},q^{50};q^{50})_\infty. \qquad (41.7.7)$$

If we multiply (41.7.4) and (41.7.5), we see that the right side of (41.7.1) is

$$\begin{aligned}&(\eta^{-2}q,\eta^2q,q^2;q^2)_\infty(\eta^{-1}q^2,\eta q^2,q^2;q^2)_\infty=\left(A_0+\alpha qA_2-\beta q^4A_4\right)\left(A_1-\alpha q^2A_3\right)\\&=A_0A_1+\alpha qA_1A_2-\alpha q^2A_0A_3-\alpha^2q^3A_2A_3-\beta q^4A_1A_4-q^6A_3A_4. \qquad (41.7.8)\end{aligned}$$

Note that (41.7.8) is essentially a 5-dissection. We now calculate the 5-dissection of the left side of (41.7.1).

We have

$$(q^3,q^4,q^6,q^7,q^{10},q^{10};q^{10})_\infty=\sum_{m,n=-\infty}^{\infty}(-1)^{m+n}q^{5m^2-2m+5n^2-n}. \qquad (41.7.9)$$

We split the sum according to the residue modulo 5 of $2m+n$.

If $2m+n\equiv 0$ then $m-2n\equiv 0$, $2m+n=5k$, $m-2n=5l$, $m=2k+l$, $n=k-2l$,

if $2m+n\equiv 1$, $m-2n\equiv -2$, $2m+n=5k+1$, $m-2n=5l-2$, $m=2k+l$, $n=k-2l+1$,

if $2m+n\equiv -1$, $m-2n\equiv 2$, $2m+n=5k-1$, $m-2n=5l+2$, $m=2k+l$, $n=k-2l-1$,

if $2m+n\equiv 2$, $m-2n\equiv 1$, $2m+n=5k+2$, $m-2n=5l+1$, $m=2k+l+1$, $n=k-2l$,

and if $2m+n \equiv -2$, $m-2n \equiv -1$, $2m+n = 5k-2$, $m-2n = 5l-1$, $m = 2k+l-1$, $n = k-2l$,
and (41.7.9) becomes

$$\begin{aligned}
&(q^3,q^4,q^6,q^7,q^{10},q^{10};q^{10})_\infty \\
&= \sum_{k,l=-\infty}^{\infty} (-1)^{k+l} q^{5(2k+l)^2-2(2k+l)+5(k-2l)^2-(k-2l)} \\
&+ \sum_{k,l=-\infty}^{\infty} (-1)^{k+l+1} q^{5(2k+l)^2-2(2k+l)+5(k-2l+1)^2-(k-2l+1)} \\
&+ \sum_{k,l=-\infty}^{\infty} (-1)^{k+l+1} q^{5(2k+l)^2-2(2k+l)+5(k-2l-1)^2-(k-2l-1)} \\
&+ \sum_{k,l=-\infty}^{\infty} (-1)^{k+l+1} q^{5(2k+l+1)^2-2(2k+l+1)+5(k-2l)^2-(k-2l)} \\
&+ \sum_{k,l=-\infty}^{\infty} (-1)^{k+l+1} q^{5(2k+l-1)^2-2(2k+l-1)+5(k-2l)^2-(k-2l)} \\
&= \sum_{k,l=-\infty}^{\infty} (-1)^{k+l} q^{25k^2-5k+25l^2} - q^4 \sum_{k,=-\infty}^{\infty} (-1)^{k+l} q^{25k^2+5k+25l^2-20l} \\
&- q^6 \sum_{k,l=-\infty}^{\infty} (-1)^{k+l} q^{25k^2-15k+25l^2+20l} - q^3 \sum_{k,l=-\infty}^{\infty} (-1)^{k+l} q^{25k^2+15k+25l^2+10l} \\
&- q^7 \sum_{k,l=-\infty}^{\infty} (-1)^{k+l} q^{25k^2-25k+25l^2-10l} \\
&= (q^{20},q^{25},q^{25},q^{30},q^{50},q^{50};q^{50})_\infty - q^6(q^5,q^{10},q^{40},q^{45},q^{50},q^{50};q^{50})_\infty \\
&- q^3(q^{10},q^{15},q^{35},q^{40},q^{50},q^{50};q^{50})_\infty - q^4(q^5,q^{20},q^{30},q^{45},q^{50},q^{50};q^{50})_\infty \\
&= A_0A_1 - q^6A_3A_4 - q^3A_2A_3 - q^4A_1A_4. \qquad (41.7.10)
\end{aligned}$$

Similarly,

$$(q,q^2,q^8,q^9,q^{10},q^{10};q^{10})_\infty = A_1A_2 - qA_0A_3 - q^2A_2A_3 + q^3A_1A_4. \qquad (41.7.11)$$

From (41.7.10) and (41.7.11), we see that the left side of (41.7.1) is

$$
\begin{aligned}
&(q^3,q^4,q^6,q^7,q^{10},q^{10};q^{10})_\infty + \alpha q(q,q^2,q^8,q^9,q^{10},q^{10};q^{10})_\infty \\
&= (A_0A_1 - q^6A_3A_4 - q^3A_2A_3 - q^4A_1A_4) \\
&\qquad\qquad\qquad + \alpha q\left(A_1A_2 - qA_0A_3 - q^2A_2A_3 + q^3A_1A_4\right) \\
&= A_0A_1 + \alpha qA_1A_2 - \alpha q^2A_0A_3 - (1+\alpha)q^3A_2A_3 + (\alpha-1)q^4A_1A_4 - q^6A_3A_4.
\end{aligned}
\tag{41.7.12}
$$

Since (41.7.12) and (41.7.8) are in agreement, (41.7.1) is proved.
The proof of (41.7.2) is similar, so omitted.

41.8 Proofs of (41.1.11) and (41.1.12)

We start by noting that (41.1.11) and (41.1.12) are equivalent to

$$
\begin{aligned}
&(q^3,q^4,q^6,q^7,q^{10},q^{10};q^{10})_\infty - (2\alpha+1)q(q,q^2,q^8,q^9,q^{10},q^{10};q^{10})_\infty \\
&\qquad = (\eta^{-1}q,\eta q,q;q)_\infty(-\eta^{-2}q,-\eta^2q,q;q)_\infty
\end{aligned}
\tag{41.8.1}
$$

and

$$
\begin{aligned}
&(q^3,q^4,q^6,q^7,q^{10},q^{10};q^{10})_\infty - (2\beta+1)q(q,q^2,q^8,q^9,q^{10},q^{10};q^{10})_\infty \\
&\qquad = (\eta^{-2}q,\eta^2q,q;q)_\infty(-\eta^{-1}q,-\eta q,q;q)_\infty
\end{aligned}
\tag{41.8.2}
$$

where $\eta = e^{\frac{2\pi i}{5}}$.

In order to prove (41.8.1) and (41.8.2), we require (8.3.6) and (8.3.7),

$$
(\eta^{-1}q,\eta q,q;q)_\infty = (q^{10},q^{15},q^{25};q^{25})_\infty - \alpha q(q^5,q^{20},q^{25};q^{25})_\infty \tag{41.8.3}
$$

and

$$
(\eta^{-2}q,\eta^2q,q;q)_\infty = (q^{10},q^{15},q^{25};q^{25})_\infty - \beta q(q^5,q^{20},q^{25};q^{25})_\infty \tag{41.8.4}
$$

as well as the special cases $a = -\eta^2$ and $a = -\eta$ of (8.3.4), namely

$$
\begin{aligned}
(-\eta^{-2}q,-\eta^2q,q;q)_\infty &= (-q^{10},-q^{15},q^{25};q^{25})_\infty - (\alpha+1)q(-q^5,-q^{20},q^{25};q^{25})_\infty \\
&\quad + 2\alpha q^3(-q^{25},-q^{25},q^{25};q^{25})_\infty
\end{aligned}
\tag{41.8.5}
$$

and

$$(-\eta^{-1}q,-\eta q,q;q)_\infty = (-q^{10},-q^{15},q^{25};q^{25})_\infty - (\beta+1)q(-q^5,-q^{20},q^{25};q^{25})_\infty \\ +2\beta q^3(-q^{25},-q^{25},q^{25};q^{25})_\infty. \tag{41.8.6}$$

If we multiply (41.8.3) and (41.8.5) and make use of (41.2.1) and (41.2.2) with q^5 for q, we see that the right side of (41.8.1) is

$$\begin{aligned}
&(\eta^{-1}q,\eta q,q;q)_\infty(-\eta^{-2}q,-\eta^2q,q;q)_\infty \\
&= (q^{20},q^{25},q^{25},q^{30},q^{50},q^{50};q^{50})_\infty - \alpha q(q^5,-q^{10},-q^{15},q^{20},q^{25},q^{25};q^{25})_\infty \\
&- (\alpha+1)q(-q^5,q^{10},q^{15},-q^{20},q^{25},q^{25};q^{25})_\infty \\
&\qquad + (\alpha^2+\alpha)q^2(q^{10},q^{25},q^{25},q^{40},q^{50},q^{50};q^{50})_\infty \\
&+2\alpha q^3(q^{10},q^{15},q^{35},q^{40},q^{50},q^{50};q^{50})_\infty - 2\alpha^2q^4(q^5,q^{20},q^{30},q^{45},q^{50},q^{50};q^{50})_\infty \\
&= A_0A_1 - \alpha q(A_1A_2 - q^5A_3A_4) - (\alpha+1)q(A_1A_2+q^5A_3A_4) \\
&\qquad + (2\alpha+1)q^2A_0A_3 + 2\alpha q^3A_2A_3 - (2\alpha+2)A_1A_4 \\
&= A_0A_1 - (2\alpha+1)qA_1A_2 + (2\alpha+1)q^2A_0A_3 + 2\alpha q^3A_2A_3 \\
&\qquad - (2\alpha+2)q^4A_1A_4 - q^6A_3A_4.
\end{aligned} \tag{41.8.7}$$

The left side of (41.8.1) is, by (41.7.10) and (41.7.11),

$$\begin{aligned}
&(q^3,q^4,q^6,q^7,q^{10},q^{10};q^{10})_\infty - (2\alpha+1)q(q,q^2,q^8,q^9,q^{10},q^{10};q^{10})_\infty \\
&= (A_0A_1 - q^6A_3A_4 - q^3A_2A_3 - q^4A_1A_4) \\
&\qquad - (2\alpha+1)q\left(A_1A_2 - qA_0A_3 - q^2A_2A_3 + q^3A_1A_4\right) \\
&= A_0A_1 - (2\alpha+1)qA_1A_2 + (2\alpha+1)q^2A_0A_3 + 2\alpha q^3A_2A_3 \\
&\qquad - (2\alpha+2)q^4A_1A_4 - q^6A_3A_4.
\end{aligned} \tag{41.8.8}$$

Since (41.8.7) and (41.8.8) are in agreement, (41.8.1) is proved.
The proof of (41.8.2) is similar, so omitted.

References

34. S. Cooper, On Ramanujan's function $k(q) = r(q)r(q^2)^2$. Ramanujan J. **20**, 311–328 (2009)
38. S. Cooper, M.D. Hirschhorn, Factorizations that involve Ramanujan's function $k(q) = r(q)r(q^2)^2$. Acta Math. Sin. Engl. Ser. **27**(12), 2301–2308 (2011)
112. S. Ramanujan, *The Lost Notebook and Other Unpublished Papers* (Springer, Berlin, 1988)

Chapter 42
Identities Involving $v = q^{\frac{1}{2}}(q,q^7;q^8)_\infty/(q^3,q^5;q^8)_\infty$

42.1 Introduction

In this chapter we investigate a number of identities involving

$$v = q^{\frac{1}{2}} \left(\begin{matrix} q,q^7 \\ q^3,q^5 \end{matrix} ; q^8 \right)_\infty . \tag{42.1.1}$$

We will show, among other things, that

$$\frac{1}{v} - v,\ \frac{1}{v} + v,\ \frac{1}{v^2} - 6 + v^2,\ \frac{1}{v} + 2 - v,\ \frac{1}{v} - 2 - v,\ \frac{1}{v} - 2\sqrt{2} + v,\ \frac{1}{v} + 2\sqrt{2} + v$$

are all products.

Indeed, we show that

$$\frac{1}{v} - v = \frac{\phi(q^2)}{q^{\frac{1}{2}}\psi(q^4)}, \tag{42.1.2}$$

$$\frac{1}{v} + v = \frac{\phi(q)}{q^{\frac{1}{2}}\psi(q^4)}, \tag{42.1.3}$$

$$\frac{1}{v^2} - 6 + v^2 = \frac{\phi(-q)^2}{q\psi(q^4)^2}, \tag{42.1.4}$$

$$\frac{1}{v} + 2 - v = \frac{\phi(q^{\frac{1}{2}})}{q^{\frac{1}{2}}\psi(q^4)}, \tag{42.1.5}$$

$$\frac{1}{v} - 2 - v = \frac{\phi(-q^{\frac{1}{2}})}{q^{\frac{1}{2}}\psi(q^4)}, \tag{42.1.6}$$

M.D. Hirschhorn, *The Power of q*, Developments in Mathematics 49,
DOI 10.1007/978-3-319-57762-3_42

and

$$\frac{1}{v} - 2\sqrt{2} + v = \frac{\phi(-q)}{q^{\frac{1}{2}}\psi(q^4)} \prod_{n \text{ odd}} \frac{1-\sqrt{2}q^{\frac{n}{2}}+q^n}{1+\sqrt{2}q^{\frac{n}{2}}+q^n}, \tag{42.1.7}$$

$$\frac{1}{v} + 2\sqrt{2} + v = \frac{\phi(-q)}{q^{\frac{1}{2}}\psi(q^4)} \prod_{n \text{ odd}} \frac{1+\sqrt{2}q^{\frac{n}{2}}+q^n}{1-\sqrt{2}q^{\frac{n}{2}}+q^n}. \tag{42.1.8}$$

We will also give the following factorisations of (42.1.2), (42.1.3) and (42.1.5)–(42.1.8).

$$\frac{1}{\sqrt{v}} + \sqrt{v} = \sqrt{\frac{\phi(q^2)}{q^{\frac{1}{2}}\psi(q^4)} \left(\begin{matrix} -q^{\frac{1}{2}}, q^{\frac{3}{2}}, q^{\frac{5}{2}}, -q^{\frac{7}{2}} \\ q^{\frac{1}{2}}, -q^{\frac{3}{2}}, -q^{\frac{5}{2}}, q^{\frac{7}{2}} \end{matrix} ; q^4 \right)_\infty}, \tag{42.1.9}$$

$$\frac{1}{\sqrt{v}} - \sqrt{v} = \sqrt{\frac{\phi(q^2)}{q^{\frac{1}{2}}\psi(q^4)} \left(\begin{matrix} q^{\frac{1}{2}}, -q^{\frac{3}{2}}, -q^{\frac{5}{2}}, q^{\frac{7}{2}} \\ -q^{-\frac{1}{2}}, q^{\frac{3}{2}}, q^{\frac{5}{2}}, -q^{\frac{7}{2}} \end{matrix} ; q^4 \right)_\infty}, \tag{42.1.10}$$

$$\frac{1}{\sqrt{v}} + i\sqrt{v} = \sqrt{\frac{\phi(q)}{q^{\frac{1}{2}}\psi(q^4)} \left(\begin{matrix} -iq^{\frac{1}{2}}, iq^{\frac{3}{2}} \\ iq^{\frac{1}{2}}, -iq^{\frac{3}{2}} \end{matrix} ; q^2 \right)_\infty}, \tag{42.1.11}$$

$$\frac{1}{\sqrt{v}} - i\sqrt{v} = \sqrt{\frac{\phi(q)}{q^{\frac{1}{2}}\psi(q^4)} \left(\begin{matrix} iq^{\frac{1}{2}}, -iq^{\frac{3}{2}} \\ -iq^{\frac{1}{2}}, iq^{\frac{3}{2}} \end{matrix} ; q^2 \right)_\infty}, \tag{42.1.12}$$

$$\frac{1}{\sqrt{v}} - (\sqrt{2}+1)\sqrt{v} = \sqrt{\frac{\phi(-q^{\frac{1}{2}})}{q^{\frac{1}{2}}\psi(q^4)} \prod_{n\geq 1} \frac{1-\sqrt{2}q^{\frac{n}{2}}+q^n}{1+\sqrt{2}q^{\frac{n}{2}}+q^n}}, \tag{42.1.13}$$

$$\frac{1}{\sqrt{v}} + (\sqrt{2}-1)\sqrt{v} = \sqrt{\frac{\phi(-q^{\frac{1}{2}})}{q^{\frac{1}{2}}\psi(q^4)} \prod_{n\geq 1} \frac{1+\sqrt{2}q^{\frac{n}{2}}+q^n}{1-\sqrt{2}q^{\frac{n}{2}}+q^n}}, \tag{42.1.14}$$

and

$$\frac{1}{\sqrt{v}} + (\sqrt{2}+1)\sqrt{v} = \sqrt{\frac{\phi(q^{\frac{1}{2}})}{q^{\frac{1}{2}}\psi(q^4)} \prod_{n \text{ odd}} \frac{1+\sqrt{2}q^{\frac{n}{2}}+q^n}{1-\sqrt{2}q^{\frac{n}{2}}+q^n} \prod_{n \text{ even}} \frac{1-\sqrt{2}q^{\frac{n}{2}}+q^n}{1+\sqrt{2}q^{\frac{n}{2}}+q^n}}, \tag{42.1.15}$$

$$\frac{1}{\sqrt{v}} - (\sqrt{2}-1)\sqrt{v} = \sqrt{\frac{\phi(q^{\frac{1}{2}})}{q^{\frac{1}{2}}\psi(q^4)} \prod_{n \text{ odd}} \frac{1-\sqrt{2}q^{\frac{n}{2}}+q^n}{1+\sqrt{2}q^{\frac{n}{2}}+q^n} \prod_{n \text{ even}} \frac{1+\sqrt{2}q^{\frac{n}{2}}+q^n}{1-\sqrt{2}q^{\frac{n}{2}}+q^n}}. \tag{42.1.16}$$

All the factorisations are clear, with the possible exceptions that (42.1.7) is the product of (42.1.13) and (42.1.16), and (42.1.8) is the product of (42.1.14) and (42.1.15).

In our proofs, we will make use of the elementary fact that

$$\text{If } xy=P \text{ and } x/y=Q, \text{ then } x=\sqrt{PQ} \text{ and } y=\sqrt{P/Q}. \tag{42.1.17}$$

42.2 Proofs of (42.1.2), (42.1.9) and (42.1.10)

We start with the following 2-dissections:

$$(-q,q^3,q^5,-q^7,q^8,q^8;q^8)_\infty = \frac{(q^4;q^4)_\infty(q^{16};q^{16})_\infty^2}{(q^8;q^8)_\infty} \left((q^6,q^{10};q^{16})_\infty + q(q^2,q^{14};q^{16})_\infty\right) \tag{42.2.1}$$

and

$$(q,-q^3,-q^5,q^7,q^8,q^8;q^8)_\infty = \frac{(q^4;q^4)_\infty(q^{16};q^{16})_\infty^2}{(q^8;q^8)_\infty} \left((q^6,q^{10};q^{16})_\infty - q(q^2,q^{14};q^{16})_\infty\right). \tag{42.2.2}$$

We have

$$\begin{aligned}
(-q,q^3,q^5,-q^7,q^8,q^8;q^8)_\infty &= \sum_{m,n=-\infty}^{\infty} (-1)^n q^{4m^2-3m+4n^2-n} \\
&= \sum_{r,s=-\infty}^{\infty} (-1)^{r-s} q^{4(r+s)^2-3(r+s)+4(r-s)^2-(r-s)} \\
&\quad + \sum_{r.s=-\infty}^{\infty} (-1)^{r-s} q^{4(r+s+1)^2-3(r+s+1)+4(r-s)^2-(r-s)} \\
&= \sum_{r,s=-\infty}^{\infty} (-1)^{r+s} q^{8r^2-4r+8s^2-2s} + \sum_{r,s=-\infty}^{\infty} (-1)^{r+s} q^{8r^2+4r+8s^2+6s+1} \\
&= (q^4,q^6,q^{10},q^{12},q^{16},q^{16};q^{16})_\infty + q(q^2,q^4,q^{12},q^{14},q^{16},q^{16};q^{16})_\infty \\
&= (q^4,q^{12},q^{16};q^{16})_\infty\left((q^6,q^{10},q^{16};q^{16})_\infty + q(q^2,q^{14},q^{16};q^{16})_\infty\right) \\
&= \frac{(q^4;q^4)_\infty(q^{16};q^{16})_\infty^2}{(q^8;q^8)_\infty}\left((q^6,q^{10};q^{16})_\infty + q(q^2,q^{14};q^{16})_\infty\right),
\end{aligned}$$

which is (42.2.1).

If we put $-q$ for q in (42.2.1), we obtain (42.2.2).

From (42.2.1) and (42.2.2) we deduce

$$(q^3,q^5;q^8)_\infty + q^{\frac{1}{2}}(q,q^7;q^8)_\infty = \frac{(q^4;q^4)_\infty^3}{(q^2;q^2)_\infty (q^8;q^8)_\infty^2}(-q^{\frac{1}{2}},q^{\frac{3}{2}},q^{\frac{5}{2}},-q^{\frac{7}{2}};q^4)_\infty \tag{42.2.3}$$

and

$$(q^3,q^5;q^8)_\infty - q^{\frac{1}{2}}(q,q^7;q^8)_\infty = \frac{(q^4;q^4)_\infty^3}{(q^2;q^2)_\infty (q^8;q^8)_\infty^2}(q^{\frac{1}{2}},-q^{\frac{3}{2}},-q^{\frac{5}{2}},q^{\frac{7}{2}};q^4)_\infty. \tag{42.2.4}$$

If we multiply (42.2.3) and (42.2.4), we find

$$(q^3,q^5;q^8)_\infty^2 - q(q,q^7;q^8)_\infty^2 = \frac{(q^4;q^4)_\infty^6}{(q^2;q^2)_\infty^2 (q^8;q^8)_\infty^4}(q,q^3,q^5,q^7;q^8)_\infty. \tag{42.2.5}$$

If we divide (42.2.5) by $q^{\frac{1}{2}}(q,q^3,q^5,q^7;q^8)_\infty$, we obtain

$$\frac{1}{q^{\frac{1}{2}}}\begin{pmatrix} q^3,q^5 \\ q,q^7 \end{pmatrix}_\infty - q^{\frac{1}{2}}\begin{pmatrix} q,q^7 \\ q^3,q^5 \end{pmatrix}_\infty = \frac{(q^4;q^4)_\infty^6}{q^{\frac{1}{2}}(q^2;q^2)_\infty^2 (q^8;q^8)_\infty^4} = \frac{\phi(q^2)}{q^{\frac{1}{2}}\psi(q^4)}, \tag{42.2.6}$$

where the parenthesized arrays carry $;q^8$ as printed:
$\left(\begin{smallmatrix} q^3,q^5 \\ q,q^7 \end{smallmatrix};q^8\right)_\infty$ and $\left(\begin{smallmatrix} q,q^7 \\ q^3,q^5 \end{smallmatrix};q^8\right)_\infty$

or,

$$\frac{1}{v} - v = \frac{\phi(q^2)}{q^{\frac{1}{2}}\psi(q^4)},$$

which is (42.1.2).

We have

$$\left(\frac{1}{\sqrt{v}} + \sqrt{v}\right)\left(\frac{1}{\sqrt{v}} - \sqrt{v}\right) = \frac{1}{v} - v = \frac{\phi(q^2)}{q^{\frac{1}{2}}\psi(q^4)} \tag{42.2.7}$$

and, by (42.2.3) and (42.2.4),

$$\begin{aligned}\left(\frac{1}{\sqrt{v}} + \sqrt{v}\right)\Big/\left(\frac{1}{\sqrt{v}} - \sqrt{v}\right) &= \frac{1+v}{1-v} \\ &= \frac{(q^3,q^5;q^8)_\infty + q^{\frac{1}{2}}(q,q^7;q^8)_\infty}{(q^3,q^5;q^8)_\infty - q^{\frac{1}{2}}(q,q^7;q^8)_\infty} \\ &= \left(\begin{matrix} -q^{\frac{1}{2}},q^{\frac{3}{2}},q^{\frac{5}{2}},-q^{\frac{7}{2}} \\ q^{\frac{1}{2}},-q^{\frac{3}{2}},-q^{\frac{5}{2}},q^{\frac{7}{2}} \end{matrix};q^4\right)_\infty.\end{aligned} \tag{42.2.8}$$

It follows from (42.2.7) and (42.2.8) together with (42.1.17) that

$$\frac{1}{\sqrt{v}}+\sqrt{v}=\sqrt{\frac{\phi(q^2)}{q^{\frac{1}{2}}\psi(q^4)}\begin{pmatrix}-q^{\frac{1}{2}},q^{\frac{3}{2}},q^{\frac{5}{2}},-q^{\frac{7}{2}} \\ q^{\frac{1}{2}},-q^{\frac{3}{2}},-q^{\frac{5}{2}},q^{\frac{7}{2}};q^4\end{pmatrix}_\infty}$$

and

$$\frac{1}{\sqrt{v}}-\sqrt{v}=\sqrt{\frac{\phi(q^2)}{q^{\frac{1}{2}}\psi(q^4)}\begin{pmatrix}q^{\frac{1}{2}},-q^{\frac{3}{2}},-q^{\frac{5}{2}},q^{\frac{7}{2}} \\ -q^{\frac{1}{2}},q^{\frac{3}{2}},q^{\frac{5}{2}},-q^{\frac{7}{2}};q^4\end{pmatrix}_\infty}$$

which are (42.1.9) and (42.1.10), respectively.

42.3 Proofs of (42.1.3), (42.1.11) and (42.1.12)

We start with the following 2-dissections.

$$(-iq,iq^3,q^4;q^4)_\infty=(q^6,q^{10},q^{16};q^{16})_\infty+iq(q^2,q^{14},q^{16};q^{16})_\infty \tag{42.3.1}$$

and

$$(iq,-iq^3,q^4;q^4)_\infty=(q^6,q^{10},q^{16};q^{16})_\infty-iq(q^2,q^{14},q^{16};q^{16})_\infty. \tag{42.3.2}$$

We have JTP (1.1.1),

$$(a^{-1}q,aq,q^2;q^2)_\infty=\sum_{-\infty}^{\infty}(-1)^n a^n q^{n^2}.$$

If we replace q by q^2, and set $a=iq$, we find that

$$\begin{aligned}(-iq,iq^3,q^4;q^4)_\infty&=\sum_{-\infty}^{\infty}(-1)^n i^n q^{2n^2+n}\\&=\sum_{-\infty}^{\infty}(-1)^n q^{8n^2+2n}+i\sum_{-\infty}^{\infty}(-1)^n q^{8n^2-6n+1}\\&=(q^6,q^{10},q^{16};q^{16})_\infty+iq(q^2,q^{14},q^{16};q^{16})_\infty,\end{aligned}$$

which is (42.3.1).

If we replace i by $-i$ (or q by $-q$) in (42.3.1), we obtain (42.3.2).

We deduce that

$$(q^3,q^5;q^8)_\infty + iq^{\frac{1}{2}}(q,q^7;q^8)_\infty = \frac{(q^2;q^2)_\infty}{(q^8;q^8)_\infty}(-iq^{\frac{1}{2}},iq^{\frac{3}{2}};q^2)_\infty \tag{42.3.3}$$

and

$$(q^3,q^5;q^8)_\infty - iq^{\frac{1}{2}}(q,q^7;q^8)_\infty = \frac{(q^2;q^2)_\infty}{(q^8;q^8)_\infty}(iq^{\frac{1}{2}},-iq^{\frac{3}{2}};q^2)_\infty. \tag{42.3.4}$$

If we multiply (42.3.3) and (42.3.4), we find that

$$\begin{aligned}(q^3,q^5;q^8)_\infty^2 + q(q,q^7;q^8)_\infty^2 &= \frac{(q^2;q^2)_\infty^2}{(q^8;q^8)_\infty^2}(-q,-q^3;q^4)_\infty \\ &= \frac{(q^2;q^2)_\infty^4}{(q;q)_\infty(q^4;q^4)_\infty(q^8;q^8)_\infty^2}.\end{aligned} \tag{42.3.5}$$

If we divide (42.3.5) by $q^{\frac{1}{2}}(q,q^3,q^5,q^7;q^8)_\infty = q^{\frac{1}{2}}\dfrac{(q;q)_\infty}{(q^2;q^2)_\infty}$, we obtain

$$\begin{aligned}\frac{1}{q^{\frac{1}{2}}}\begin{pmatrix} q^3,q^5 \\ q,q^7 \end{pmatrix};q^8\Bigg)_\infty + q^{\frac{1}{2}}\begin{pmatrix} q,q^7 \\ q^3,q^5 \end{pmatrix};q^8\Bigg)_\infty &= \frac{(q^2;q^2)_\infty^5}{(q;q)_\infty^2(q^4;q^4)_\infty(q^8;q^8)_\infty^2} \\ &= \frac{\phi(q)}{q^{\frac{1}{2}}\psi(q^4)},\end{aligned} \tag{42.3.6}$$

or,

$$\frac{1}{v} + v = \frac{\phi(q)}{q^{\frac{1}{2}}\psi(q^4)},$$

which is (42.1.3).

We have

$$\left(\frac{1}{\sqrt{v}} + i\sqrt{v}\right)\left(\frac{1}{\sqrt{v}} - i\sqrt{v}\right) = \frac{1}{v} + v = \frac{\phi(q)}{q^{\frac{1}{2}}\psi(q^4)} \tag{42.3.7}$$

and

$$\begin{aligned}\left(\frac{1}{\sqrt{v}} + i\sqrt{v}\right)\Big/\left(\frac{1}{\sqrt{v}} - i\sqrt{v}\right) &= \frac{1+iv}{1-iv} \\ &= \frac{(q^3,q^5;q^8)_\infty + iq^{\frac{1}{2}}(q,q^7;q^8)_\infty}{(q^3,q^5;q^8)_\infty - iq^{\frac{1}{2}}(q,q^7;q^8)_\infty} \\ &= \begin{pmatrix} -iq^{\frac{1}{2}},iq^{\frac{3}{2}} \\ iq^{\frac{1}{2}},-iq^{\frac{3}{2}} \end{pmatrix};q^2\Bigg)_\infty.\end{aligned} \tag{42.3.8}$$

It follows from (42.3.7) and (42.3.8) together with (42.1.17) that

$$\frac{1}{\sqrt{v}}+i\sqrt{v}=\sqrt{\frac{\phi(q)}{q^{\frac{1}{2}}\psi(q^4)}\left(\begin{matrix}-iq^{\frac{1}{2}},iq^{\frac{3}{2}}\\ iq^{\frac{1}{2}},-iq^{\frac{3}{2}}\end{matrix};q^2\right)_\infty}$$

and

$$\frac{1}{\sqrt{v}}-i\sqrt{v}=\sqrt{\frac{\phi(q)}{q^{\frac{1}{2}}\psi(q^4)}\left(\begin{matrix}iq^{\frac{1}{2}},-iq^{\frac{3}{2}}\\ -iq^{\frac{1}{2}},iq^{\frac{3}{2}}\end{matrix};q^2\right)_\infty},$$

which are (42.1.11) and (42.1.12) respectively.

42.4 Proofs of (42.1.4), (42.1.5) and (42.1.6)

We have, by (42.1.2), (42.1.3), (19.2.4) and (19.2.5),

$$\begin{aligned}\frac{1}{v^2}-6-v^2&=2\left(\frac{1}{v}-v\right)^2-\left(\frac{1}{v}+v\right)^2\\&=\frac{2\phi(q^2)^2-\phi(q)^2}{q\psi(q^4)^2}\\&=\frac{\phi(-q)^2}{q\psi(q^4)^2},\end{aligned}$$

which is (42.1.4).

We have, from (42.1.2) and (19.2.1) with $q^{\frac{1}{2}}$ for q,

$$\frac{1}{v}+2-v=\frac{\phi(q^2)+2q^{\frac{1}{2}}\psi(q^4)}{q^{\frac{1}{2}}\psi(q^4)}=\frac{\phi(q^{\frac{1}{2}})}{q^{\frac{1}{2}}\psi(q^4)},$$

which is (42.1.5), and from (42.1.2) and (19.2.2) with $q^{\frac{1}{2}}$ for q,

$$\frac{1}{v}-2-v=\frac{\phi(q^2)-2q^{\frac{1}{2}}\psi(q^4)}{q^{\frac{1}{2}}\psi(q^4)}=\frac{\phi(-q^{\frac{1}{2}})}{q^{\frac{1}{2}}\psi(q^4)},$$

which is (42.1.6).

Note that the pair (42.1.5), (42.1.6) constitute a factorisation of (42.1.4), by virtue of (1.5.14).

42.5 Proofs of (42.1.13) and (42.1.14)

We start with the 2-dissections

$$\prod_{n\geq 1}(1-\sqrt{2}q^n+q^{2n})(1-q^n) = (q^6,q^{10},q^{16};q^{16})_\infty - (\sqrt{2}+1)q(q^2,q^{14},q^{16};q^{16})_\infty \tag{42.5.1}$$

and

$$\prod_{n\geq 1}(1+\sqrt{2}q^n+q^{2n})(1-q^n) = (q^6,q^{10},q^{16};q^{16})_\infty + (\sqrt{2}-1)q(q^2,q^{14},q^{16};q^{16})_\infty. \tag{42.5.2}$$

We have (1.7.5)

$$(a^{-1}q,aq,q;q)_\infty = 1+\sum_{n\geq 1}(-1)^n\left(\frac{a^{n+\frac{1}{2}}-a^{-n-\frac{1}{2}}}{a^{\frac{1}{2}}-a^{-\frac{1}{2}}}\right)q^{(n^2+n)/2}. \tag{42.5.3}$$

If in (42.5.3) we set $a = \exp\{2i\theta\}$, where $\theta \neq k\pi$, k an integer, we obtain

$$\prod_{n\geq 1}(1-2\cos 2\theta q^n+q^{2n})(1-q^n) = \sum_{n\geq 0}(-1)^n\frac{\sin{(2n+1)\theta}}{\sin\theta}q^{(n^2+n)/2}. \tag{42.5.4}$$

If in (42.5.4) we set $\theta = \dfrac{\pi}{8}$, we obtain

$$\begin{aligned}\prod_{n\geq 1}&(1-\sqrt{2}q^n+q^{2n})(1-q^n)\\ &= 1-(\sqrt{2}+1)q+(\sqrt{2}+1)q^3-q^6-q^{10}+\cdots\\ &= (q^6,q^{10},q^{16};q^{16})_\infty-(\sqrt{2}+1)q(q^2,q^{14},q^{16};q^{16})_\infty,\end{aligned}$$

which is (42.5.1), while if in (42.5.4) we set $\theta = \dfrac{3\pi}{8}$, or if in (42.5.1) we change $\sqrt{2}$ to $-\sqrt{2}$, we obtain (42.5.2).

From (42.5.1) and (42.5.2) it follows that

$$(q^3,q^5;q^8)_\infty-(\sqrt{2}+1)q^{\frac{1}{2}}(q,q^7;q^8)_\infty = \frac{(q^{\frac{1}{2}};q^{\frac{1}{2}})_\infty}{(q^8;q^8)_\infty}\prod_{n\geq 1}(1-\sqrt{2}q^{\frac{n}{2}}+q^n) \tag{42.5.5}$$

and

$$(q^3,q^5;q^8)_\infty+(\sqrt{2}-1)q^{\frac{1}{2}}(q,q^7;q^8)_\infty=\frac{(q^{\frac{1}{2}};q^{\frac{1}{2}})_\infty}{(q^8;q^8)_\infty}\prod_{n\geq 1}(1+\sqrt{2}q^{\frac{n}{2}}+q^n). \quad (42.5.6)$$

If we multiply (42.5.5) and (42.5.6), we obtain

$$\begin{aligned}(q^3,q^5;q^8)_\infty^2 &-2q^{\frac{1}{2}}(q,q^3,q^5,q^7;q^8)_\infty-q(q,q^7;q^8)_\infty^2\\ &=\frac{(q^{\frac{1}{2}};q^{\frac{1}{2}})_\infty^2}{(q^8;q^8)_\infty^2}\prod_{n\geq 1}(1+q^{2n})\\ &=\frac{(q^{\frac{1}{2}};q^{\frac{1}{2}})_\infty^2(q^4;q^4)_\infty}{(q^2;q^2)_\infty(q^8;q^8)_\infty^2}. \quad (42.5.7)\end{aligned}$$

If we divide (42.5.7) by $q^{\frac{1}{2}}(q;q^2)_\infty$, we obtain

$$\frac{1}{q^{\frac{1}{2}}}\begin{pmatrix}q^3,q^5\\q,q^7;q^8\end{pmatrix}_\infty-2-q^{\frac{1}{2}}\begin{pmatrix}q,q^7\\q^3,q^5;q^8\end{pmatrix}_\infty=\frac{(q^{\frac{1}{2}};q^{\frac{1}{2}})_\infty^2(q^4;q^4)_\infty}{(q^2;q^2)_\infty(q^8;q^8)_\infty^2}=\frac{\phi(-q^{\frac{1}{2}})}{q^{\frac{1}{2}}\psi(q^4)}, \quad (42.5.8)$$

or,

$$\frac{1}{v}-2-v=\frac{\phi(-q^{\frac{1}{2}})}{q^{\frac{1}{2}}\psi(q^4)},$$

which is (42.1.6) (again).

We have

$$\left(\frac{1}{\sqrt{v}}-(\sqrt{2}+1)\sqrt{v}\right)\left(\frac{1}{\sqrt{v}}+(\sqrt{2}-1)\sqrt{v}\right)=\frac{1}{v}-2-v=\frac{\phi(-q^{\frac{1}{2}})}{q^{\frac{1}{2}}\psi(q^4)} \quad (42.5.9)$$

and, by (42.5.5) and (42.5.6),

$$\begin{aligned}\left(\frac{1}{\sqrt{v}}-(\sqrt{2}+1)\sqrt{v}\right)\Big/\left(\frac{1}{\sqrt{v}}+(\sqrt{2}-1)\sqrt{v}\right)&=\frac{1-(\sqrt{2}+1)v}{1+(\sqrt{2}-1)v}\\ &=\frac{(q^3,q^5;q^8)_\infty-(\sqrt{2}+1)q^{\frac{1}{2}}(q,q^7;q^8)_\infty}{(q^3,q^5;q^8)_\infty+(\sqrt{2}-1)q^{\frac{1}{2}}(q,q^7;q^8)_\infty}\\ &=\prod_{n\geq 1}\frac{1-\sqrt{2}q^{\frac{n}{2}}+q^n}{1+\sqrt{2}q^{\frac{n}{2}}+q^n}. \quad (42.5.10)\end{aligned}$$

From (42.5.9), (42.5.10) and (42.1.17), we deduce (42.1.13) and (42.1.14).

42.6 Proofs of (42.1.15) and (42.1.16), (42.1.7) and (42.1.8)

If in (42.5.1) and (42.5.2) we set $-q$ for q, we find

$$\prod_{n \text{ odd}} (1+\sqrt{2}q^n+q^{2n})(1+q^n) \prod_{n \text{ even}} (1+\sqrt{2}q^n+q^{2n})(1-q^n)$$
$$= (q^6,q^{10},q^{16};q^{16})_\infty + (\sqrt{2}+1)q(q^2,q^{14},q^{16};q^{16})_\infty \tag{42.6.1}$$

and

$$\prod_{n \text{ odd}} (1-\sqrt{2}q^n+q^{2n})(1+q^n) \prod_{n \text{ even}} (1+\sqrt{2}q^n+q^{2n})(1-q^n)$$
$$= (q^6,q^{10},q^{16};q^{16})_\infty - (\sqrt{2}-1)q(q^2,q^{14},q^{16};q^{16})_\infty. \tag{42.6.2}$$

Identities (42.1.15) and (42.1.16) follow from (42.6.1) and (42.6.2) as did (42.1.14) and (42.1.15) from (42.5.9) and (42.5.10). This is left as an exercise.

The identities (42.1.7) and (42.1.8) follow from (42.1.13) and (42.1.16) and from (42.1.14) and (42.1.15), respectively.

Endnote.

My interest in this work was sparked by B. Yuttanan (2012) [128].

References

88. M.D. Hirschhorn, Factorizations of certain q-series identities of Ramanujan and others. Ramanujan J. **31**, 15–22 (2013)
111. S. Ramanujan, Notebooks, II, Tata Institute of Fundamental Research, Mumbai (2012)
128. B. Yuttanan, New properties for the Ramanujan-Göllnitz-Gordon continued fraction. Acta Arith. **151**, 293–310 (2012)

Chapter 43
Ramanujan's Tau Function

43.1 Introduction

Ramanujan's tau function is defined by

$$\sum_{n\geq 1} \tau(n)q^n = qE(q)^{24}. \tag{43.1.1}$$

The tau function has many fascinating properties. One of these is that for prime p,

$$\sum_{n\geq 1} \tau(pn)q^n = \tau(p)\sum_{n\geq 1} \tau(n)q^n - p^{11}\sum_{n\geq 1} \tau(n)q^{pn}, \tag{43.1.2}$$

or,

$$\tau(pn) = \tau(p)\tau(n) - p^{11}\tau\left(\frac{n}{p}\right) \tag{43.1.3}$$

where it is understood that $\tau\left(\frac{n}{p}\right) = 0$ if $p \nmid n$.

We can use the identities in this book and give elementary proofs of (43.1.3) for $p = 2,\ 3,\ 5$ and 7.

43.2 $p = 2$

We can write (19.4.5)

$$\frac{E(q)^8}{E(q^4)^8} = F(q^2) - 8q, \tag{43.2.1}$$

for some even function $F(q^2)$.

M.D. Hirschhorn, *The Power of q*, Developments in Mathematics 49,
DOI 10.1007/978-3-319-57762-3_43

If we define

$$\zeta = \frac{E(q)^8}{qE(q^4)^8}, \tag{43.2.2}$$

then (43.2.1) yields

$$\zeta = \frac{F(q^2)}{q} - 8. \tag{43.2.3}$$

If we square (43.2.3), we find

$$\zeta^2 = \frac{F(q^2)^2}{q^2} - 16\frac{F(q^2)}{q} + 64. \tag{43.2.4}$$

From (43.2.3) and (43.2.4) it follows that

$$\zeta^2 + 16\zeta = \frac{F(q^2)^2}{q^2} - 64 = -\zeta(q)\zeta(-q). \tag{43.2.5}$$

Now,

$$\zeta(q) = \frac{E(q^2)^8}{qE(q^4)^8}(q;q^2)_\infty^8, \tag{43.2.6}$$

so

$$\begin{aligned} -\zeta(q)\zeta(-q) &= \frac{E(q^2)^{16}}{q^2E(q^4)^{16}}(q;q^2)_\infty^8(-q;q^2)_\infty^8 \\ &= \frac{E(q^2)^{16}}{q^2E(q^4)^{16}}(q^2;q^4)_\infty^8 \\ &= \frac{E(q^2)^{16}}{q^2E(q^4)^{16}}\frac{E(q^2)^8}{E(q^4)^8} \\ &= \frac{E(q^2)^{24}}{q^2E(q^4)^{24}}. \end{aligned} \tag{43.2.7}$$

and (43.2.5) becomes

$$\zeta^2 + 16\zeta = \frac{E(q^2)^{24}}{q^2E(q^4)^{24}}. \tag{43.2.8}$$

If we define

$$T = \frac{E(q^2)^{24}}{q^2E(q^4)^{24}} \tag{43.2.9}$$

then (43.2.8) becomes

$$\zeta^2 + 16\zeta = T. \tag{43.2.10}$$

It follows from (43.2.10) that for $i \geq 2$,

$$\zeta^i + 16\zeta^{i-1} - T\zeta^{i-2} = 0. \tag{43.2.11}$$

Let H_2 be the huffing operator modulo 2, and for $i \geq 0$ let

$$u_i = H_2(\zeta^i). \tag{43.2.12}$$

Then

$$u_0 = 1, \quad u_1 = -8, \tag{43.2.13}$$

and for $i \geq 2$,

$$u_i + 16u_{i-1} - Tu_{i-2} = 0. \tag{43.2.14}$$

Let

$$U = \sum_{i \geq 0} u_i z^i. \tag{43.2.15}$$

It follows from (43.2.13) and (43.2.14) that

$$(1 + 16z - Tz^2)U = u_0 + (u_1 + 16u_0)z = 1 + 8z, \tag{43.2.16}$$

and

$$U = \frac{1 + 8z}{1 + 16z - Tz^2}. \tag{43.2.17}$$

If we expand (43.2.17) to the term in z^3, we find

$$u_3 = -24T - 2048. \tag{43.2.18}$$

That is,

$$H_2\left(\frac{E(q)^{24}}{q^3 E(q^4)^{24}}\right) = -24\frac{E(q^2)^{24}}{q^2 E(q^4)^{24}} - 2^{11}. \tag{43.2.19}$$

If we multiply (43.2.19) by $q^4 E(q^4)^{24}$, we obtain

$$H_2(qE(q)^{24}) = -24q^2 E(q^2)^{24} - 2^{11} q^4 E(q^4)^{24}, \tag{43.2.20}$$

or,

$$\sum_{n \geq 1} \tau(2n) q^{2n} = -24 \sum_{n \geq 1} \tau(n) q^{2n} - 2^{11} \sum_{n \geq 1} \tau(n) q^{4n}. \tag{43.2.21}$$

If we replace q^2 by q, we obtain

$$\sum_{n \geq 1} \tau(2n) q^n = -24 \sum_{n \geq 1} \tau(n) q^n - 2^{11} \sum_{n \geq 1} \tau(n) q^{2n}. \tag{43.2.22}$$

The term $n=1$ gives $\tau(2) = -24\tau(1) = -24$, so (43.2.22) becomes

$$\sum_{n\geq 1} \tau(2n)q^n = \tau(2)\sum_{n\geq 1}\tau(n)q^n - 2^{11}\sum_{n\geq 1}\tau(n)q^{2n}, \tag{43.2.23}$$

as claimed.

43.3 $p=3$

We can write (1.7.1)

$$E(q)^3 = G(q^3) - 3qE(q^9)^3 \tag{43.3.1}$$

where $G(q^3)$ is some function of q^3.

If we define

$$\zeta = \frac{E(q)^3}{qE(q^9)^3} \tag{43.3.2}$$

then (43.3.1) yields

$$\zeta = \frac{G(q^3)}{q} - 3. \tag{43.3.3}$$

It follows from (43.3.3) that

$$\zeta^2 = \frac{G(q^3)^2}{q^2} - 6\frac{G(q^3)}{q} + 9 \tag{43.3.4}$$

and

$$\zeta^3 = \frac{G(q^3)^3}{q^3} - 9\frac{G(q^3)^2}{q^2} + 27\frac{G(q^3)}{q} - 27. \tag{43.3.5}$$

It follows from (43.3.3)–(43.3.5) that

$$\begin{aligned}\zeta^3 + 9\zeta^2 + 27\zeta &= \frac{G(q^3)^3}{q^3} - 27\\ &= \zeta(q)\zeta(\omega q)\zeta(\omega^2 q),\end{aligned} \tag{43.3.6}$$

where ω is a cube root of unity other than 1.

Now,

$$\zeta(q) = \frac{E(q^3)^3}{qE(q^9)^3}(q;q^3)^3_\infty(q^2;q^3)^3_\infty, \tag{43.3.7}$$

so

$$\zeta(q)\zeta(\omega q)\zeta(\omega^2 q) = \frac{E(q^3)^9}{q^3E(q^9)^9}(q^3;q^9)_\infty^3(q^6;q^9)_\infty^3$$
$$= \frac{E(q^3)^9}{E(q^9)^9}\frac{E(q^3)^3}{E(q^9)^3}$$
$$= \frac{E(q^3)^{12}}{q^3E(q^9)^{12}}. \qquad (43.3.8)$$

So (43.3.6) becomes

$$\zeta^3 + 9\zeta^2 + 27\zeta = \frac{E(q^3)^{12}}{q^3E(q^9)^{12}}. \qquad (43.3.9)$$

If we define

$$T = \frac{E(q^3)^{12}}{q^3E(q^9)^{12}} \qquad (43.3.10)$$

then (43.3.9) becomes

$$\zeta^3 + 9\zeta^2 + 27\zeta = T. \qquad (43.3.11)$$

It follows that for $i \geq 3$,

$$\zeta^i + 9\zeta^{i-1} + 27\zeta^{i-2} - T\zeta^{i-3} = 0. \qquad (43.3.12)$$

Let H_3 be the huffing operator modulo 3, and for $i \geq 0$ let

$$u_i = H_3(\zeta^i). \qquad (43.3.13)$$

Then

$$u_0 = 1, \;\; u_1 = -3, \;\; u_2 = 9 \qquad (43.3.14)$$

and for $i \geq 3$,

$$u_i + 9u_{i-1} + 27u_{i-2} - Tu_{i-3} = 0. \qquad (43.3.15)$$

Let

$$U = \sum_{i\geq 0} u_i z^i. \qquad (43.3.16)$$

Then

$$(1 + 9z + 27z^2 - Tz^3)U = u_0 + (u_1 + 9u_0)z + (u_2 + 9u_1 + 27u_0)z^2$$
$$= 1 + 6z + 9z^2, \qquad (43.3.17)$$

and

$$U=\frac{1+6z+9z^2}{1+9z+27z^2-Tz^3}. \tag{43.3.18}$$

If we expand (43.3.18) to the term in z^8, we find

$$u_8=252T^2-3^{11}. \tag{43.3.19}$$

That is,

$$H_3\left(\frac{E(q)^{24}}{q^8E(q^9)^{24}}\right)=252\frac{E(q^3)^{24}}{q^6E(q^9)^{24}}-3^{11}. \tag{43.3.20}$$

This yields (43.1.3) for $p=3$.

43.4 $p=5$

As in Chapter 6, let

$$\zeta=\frac{E(q)}{qE(q^{25})},\quad T=\frac{E(q^5)^6}{q^5E(q^{25})^6}. \tag{43.4.1}$$

We have (6.2.14)

$$\zeta^5+5\zeta^4+15\zeta^3+25\zeta^2+25\zeta-T=0, \tag{43.4.2}$$

so for $i\geq 5$,

$$\zeta^i+5\zeta^{i-1}+15\zeta^{i-2}+25\zeta^{i-3}+25\zeta^{i-4}-T\zeta^{i-5}=0. \tag{43.4.3}$$

If we write

$$u_i=H_5(\zeta^i) \tag{43.4.4}$$

then

$$u_0=1,\ u_1=-1,\ u_2=-1,\ u_3=5,\ u_4=-5 \tag{43.4.5}$$

and for $i\geq 5$,

$$u_i+5u_{i-1}+15u_{i-2}+25u_{i-3}+25u_{i-4}-Tu_{i-5}=0. \tag{43.4.6}$$

Now write

$$U=\sum_{i\geq 0}u_iz^i. \tag{43.4.7}$$

It follows from (43.4.5) and (43.4.6) that

$$\left(1+5z+15z^2+25z^3+25z^4-Tz^5\right)U = 1+4z+9z^2+10z^3+5z^4 \tag{43.4.8}$$

so

$$U = \frac{1+4z+9z^2+10z^3+5z^4}{1+5z+15z^2+25z^3+25z^4-Tz^5}. \tag{43.4.9}$$

If we expand U to the term z^{24} we find

$$u_{24} = 4830T^4 - 5^{11}. \tag{43.4.10}$$

That is,

$$H_5\left(\frac{E(q)^{24}}{q^{24}E(q^{25})^{24}}\right) = 4830\frac{E(q^5)^{24}}{q^{20}E(q^{25})^{24}} - 5^{11}. \tag{43.4.11}$$

This yields (43.1.3) for $p = 5$.

43.5 $p = 7$

As in Chapter 7, let

$$\zeta = \frac{E(q)}{q^2E(q^{49})}, \quad T = \frac{E(q^7)^4}{q^7E(q^{49})^4}. \tag{43.5.1}$$

We have (7.4.2)

$$\zeta^7+7\zeta^6+21\zeta^5+49\zeta^4+147\zeta^3+343\zeta^2+343\zeta+(7\zeta^3+35\zeta^2+49\zeta)T-T^2=0. \tag{43.5.2}$$

If we write

$$u_i = H_7(\zeta^i) \tag{43.5.3}$$

then

$$u_0 = 1,\ u_1 = -1,\ u_2 = 1,\ u_3 = -7,\ u_4 = -4T-7,\ u_5 = 10T+49,\ u_6 = 49 \tag{43.5.4}$$

and for $i \geq 7$,

$$\begin{aligned} &u_i+7u_{i-1}+49u_{i-2}+147u_{i-4}+343u_{i-5}+343u_{i-6} \\ &\qquad +(7u_{i-4}+35u_{i-5}+49u_{i-6})T-u_{i-7}T^2=0. \end{aligned} \tag{43.5.5}$$

If we write

$$U = \sum_{i \geq 0} u_i z^i \tag{43.5.6}$$

then

$$\left(1+7z+21z^2+49z^3+(147+7T)z^4+(343+35T)z^5+(343+49T)z^6-T^2z^7\right)U$$
$$= 1+6z+15z^2+28z^3+(63+3T)z^4+(98+10T)z^5+(49+7T)z^6. \tag{43.5.7}$$

so

$$U = \frac{1+6z+15z^2+28z^3+(63+3T)z^4+(98+10T)z^5+(49+7T)z^6}{1+7z+21z^2+49z^3+(147+7T)z^4+(343+35T)z^5+(343+49T)z^6-T^2z^7}. \tag{43.5.8}$$

If we expand U to the term z^{24} we find

$$u_{24} = -16744T^6 - 7^{11}. \tag{43.5.9}$$

That is,

$$H_7\left(\frac{E(q)^{24}}{q^{48}E(q^{49})^{24}}\right) = -16744\frac{E(q^7)^{24}}{q^{42}E(q^{49})^{24}} - 7^{11}. \tag{43.5.10}$$

This yields (43.1.3) for $p = 7$.

Erratum to: The Power of q

Michael D. Hirschhorn

Erratum to: M.D. Hirschhorn, *The Power of q*, Developments in Mathematics 49, DOI 10.1007/978-3-319-57762-3

The following changes have been made to the text.

In the preface, page xiii, line 23, "than $p(N)$" has been corrected to "then $p(N)$".

The font used in Chapter 1 for ϕ and ψ has been changed to accord with the font used in later chapters.

In Chapter 1, page 12

$$\begin{bmatrix} 0 \\ 0 \end{bmatrix}_0 \text{ has been corrected to } \begin{bmatrix} 0 \\ 0 \end{bmatrix}_q$$

The updated online version of these chapters can be found at
https://doi.org/10.1007/978-3-319-57762-3_1
https://doi.org/10.1007/978-3-319-57762-3_3
https://doi.org/10.1007/978-3-319-57762-3_6
https://doi.org/10.1007/978-3-319-57762-3_10
https://doi.org/10.1007/978-3-319-57762-3_12
https://doi.org/10.1007/978-3-319-57762-3_17
https://doi.org/10.1007/978-3-319-57762-3_30
https://doi.org/10.1007/978-3-319-57762-3_34
https://doi.org/10.1007/978-3-319-57762-3_35
https://doi.org/10.1007/978-3-319-57762-3_36
https://doi.org/10.1007/978-3-319-57762-3_37
https://doi.org/10.1007/978-3-319-57762-3_38

M.D. Hirschhorn, *The Power of q*, Developments in Mathematics 49,
DOI 10.1007/978-3-319-57762-3_44

In Chapter 3, equation (3.4.2),

$$\frac{(q^3;q^3)^3_\infty)^2}{(q;q)^7_\infty} \text{ has been corrected to } \frac{((q;q)^3_\infty)^2}{(q;q)^7_\infty}$$

In Chapter 6, in equations (6.6.6) and (6.6.7) and the Exercise at the top of page 68, the $(-2)^\alpha$ has been corrected to $(-2)^{\alpha-1}$.

In Chapter 10, below the equation (10.7.6), the word "**Exercise**" has been moved to the next line.

In Chapter 10, page 99, "differentin" has been corrected to "different in"

In Chapter 12, page 118,

$$(q^{22},q^{33},q^{44},q^{55},q^{66},q^{77},q^{88},q^{88};q^{121})_\infty$$

has been corrected to

$$(q^{22},q^{33},q^{44},q^{55},q^{66},q^{77},q^{88},q^{99};q^{121})_\infty$$

In Chapter 17, page 161, "equivalent to (17.1.1)" has been corrected to "equivalent to (17.1.1) and (17.1.2)"

In Chapter 30, page 274,

$$= 32\begin{pmatrix} q^3,q^4,q^4,q^8,q^9,q^{12},q^{12},q^{12},q^{15},q^{16},q^{20},q^{20},q^{21},q^{24},q^{24},q^{24} \\ q,q^2,q^2,q^5,q^7,q^{10},q^{10}q^{11},q^{13},q^{14},q^{14},q^{17},q^{19},q^{22},q^{22},q^{23} \end{pmatrix}_\infty ; q^{24}.$$

has been corrected to

$$= 32\begin{pmatrix} q^3,q^4,q^4,q^8,q^9,q^{12},q^{12},q^{12},q^{15},q^{16},q^{20},q^{20},q^{21},q^{24},q^{24},q^{24} \\ q,q^2,q^2,q^5,q^7,q^{10},q^{10},q^{11},q^{13},q^{14},q^{14},q^{17},q^{19},q^{22},q^{22},q^{23} \end{pmatrix}_\infty ; q^{24}.$$

In Chapter 30, page 276, the second occurrence of the line

$$\times((\phi(q^{12})+2q^3\psi(q^{24}))\psi(q^4)+q(\phi(q^4)+2q\psi(q^8))\psi(q^{12}))$$

is incorrect, and has been corrected to

$$\times(\phi(q^{12})\psi(q^4)+2q^2\psi(q^8)\psi(q^{12})+q(\phi(q^4)\psi(q^{12})+2q^2\psi(q^4)\psi(q^{24})))$$

In Chapter 34, page 313,

$$= 4(q,q^2,q^3,q^4,q^4,q^6,q^7,q^8,q^9,q^{10},q^{10}$$

has been corrected to

$$= 4(q, q^2, q^3, q^4, q^4, q^6, q^7, q^8, q^9, q^{10}, q^{10},$$

In Chapter 35, equation (35.1.4), $(1+qt)^{\frac{1}{4}}$ has been corrected to $(1+qt)^{\frac{1}{2}}$.
In equation (35.1.5), $(1+2qt)^{\frac{1}{2}}$ has been corrected to $(1+2qt)^{\frac{1}{4}}$.
In Chapter 36, equation (36.2.6)

$64\dfrac{\psi(q)^7}{\phi(-q^2)^8}$ has been corrected to $64\dfrac{\psi(q)^7}{\phi(-q)^8}$

In Chapter 37, equation (37.3.5)

$\dfrac{abc+q^5c^3}{\phi(q^5)}$ has been corrected to $\dfrac{abc+q^5c^3}{\psi(q^5)}$

$\dfrac{c(ab+q^5c^2)}{\phi(q^5)^2}$ has been corrected to $\dfrac{c(ab+q^5c^2)}{\psi(q^5)}$

In Chapter 38, page 354,

$\dfrac{(q^2;q^2)_\infty^3(q^3;q^3)_\infty^3(q^{12};q^{12})_\infty^3}{(q;q)_\infty(q^4;q^4)_\infty(q^6;q^6)_\infty^8}$ in equation (38.3.2) and below equation (38.3.8)

has been corrected to $\dfrac{(q^2;q^2)_\infty^3(q^3;q^3)_\infty^3(q^{12};q^{12})_\infty^3}{(q;q)_\infty(q^4;q^4)_\infty(q^6;q^6)_\infty^9}$.

In Chapter 38, page 355,

$$8q\frac{(q^3;q^3)_\infty^3(q^4;q^4)_\infty^2(q^{12};lq^{12})_\infty^2}{(q;q)_\infty^7(q^2;q^2)_\infty^3}$$

has been corrected to $8q\dfrac{(q^3;q^3)_\infty^3(q^4;q^4)_\infty^2(q^{12};q^{12})_\infty^2}{(q;q)_\infty^7(q^2;q^2)_\infty^3}$.

In Chapter 38, page 356

$2q\dfrac{(q^2;q^2)_\infty^5(q^{12};q^{12})_\infty^6}{(q^4;q^4)_\infty^2(q^6;q^6)_\infty^9}$ has been corrected to $2q\dfrac{(q^2;q^2)_\infty^3(q^{12};q^{12})_\infty^6}{(q^4;q^4)_\infty^2(q^6;q^6)_\infty^9}$.

The book has been updated with the changes.

Appendix

In this appendix, we prove a small number of somewhat technical identities, some of which are referred to in the text, as well as introduce some more modern notation.

We saw in §1.2 how to prove Euler's identity (1.2.9),

$$(-a;q)_\infty = \sum_{k\geq 0} a^k \frac{q^{(k^2-k)/2}}{(q;q)_k}. \tag{A.1}$$

In a similar manner, we can prove another identity due to Euler, namely

$$\frac{1}{(a;q)_\infty} = \sum_{k\geq 0} \frac{a^k}{(q;q)_k}. \tag{A.2}$$

Both (A.1) and (A.2) can be obtained from the so-called q-binomial theorem,

$$\frac{(at;q)_\infty}{(t;q)_\infty} = \sum_{k\geq 0} \frac{(a;q)_k}{(q;q)_k} t^k,$$

or,

$$\left(\begin{matrix} at \\ t \end{matrix}; q\right)_\infty = \sum_{k\geq 0} \left(\begin{matrix} a \\ q \end{matrix}; q\right)_k t^k. \tag{A.3}$$

If in (A.3) we replace a by $-a/t$, we obtain

$$\left(\begin{matrix} -a \\ t \end{matrix}; q\right)_\infty = \sum_{k\geq 0} \left(\begin{matrix} -a/t \\ q \end{matrix}; q\right)_k t^k. \tag{A.4}$$

Now notice that

M.D. Hirschhorn, *The Power of q*, Developments in Mathematics 49,
DOI 10.1007/978-3-319-57762-3

$$(-a/t;q)_k\,t^k = \left(1+\frac{a}{t}\right)\left(1+\frac{a}{t}q\right)\cdots\left(1+\frac{a}{t}q^{k-1}\right)t^k$$
$$= (t+a)(t+aq)\cdots(t+aq^{k-1})$$

and let $t \to 0$ in (A.4) to obtain

$$(-a;q)_\infty = \sum_{k\geq 0} a^k \frac{q^{(k^2-k)/2}}{(q;q)_k},$$

which is (A.1).

On the other hand, if we set $a = 0$ in (A.3), then replace t by a, we obtain (A.2).

The q-binomial theorem may be proved as follows.

Let

$$f(t) = \left(\frac{at}{t};q\right)_\infty = \frac{(at;q)_\infty}{(t;q)_\infty}.$$

Then

$$f(qt) = \frac{(atq;q)_\infty}{(tq;q)_\infty}$$

and

$$(1-at)f(qt) = (1-t)f(t).$$

If we suppose that

$$f(x) = \sum_{k\geq 0} c_k(a,q)t^k$$

then

$$(1-at)\sum_{k\geq 0} c_k q^k t^k = (1-t)\sum_{k\geq 0} c_k t^k.$$

It follows that for $k \geq 1$,

$$q^k c_k - aq^{k-1}c_{k-1} = c_k - c_{k-1},$$

$$(1-q^k)c_k = (1-aq^{k-1})c_{k-1}$$

and

$$c_k = \frac{1-aq^{k-1}}{1-q^k}c_{k-1}.$$

It follows by induction that

$$c_k = \frac{(a;q)_k}{(q;q)_k} c_0 = \frac{(a;q)_k}{(q;q)_k},$$

and so

$$f(t) = \sum_{k \geq 0} \frac{(a;q)_k}{(q;q)_k} t^k,$$

which is (A.3).

We now introduce some notation.

We define

$$ {}_r\phi_{r-1}\left(\begin{matrix} a_1, & \cdots, & a_r \\ b_1, & \cdots, & b_{r-1} \end{matrix}; q; t\right) = \sum_{k \geq 0} \left(\begin{matrix} a_1, a_2, & \cdots, & a_r \\ q, b_1, & \cdots, & b_{r-1} \end{matrix}; q\right)_k t^k. \tag{A.5}$$

The q-binomial theorem can then be written as

$$ {}_1\phi_0\left(\begin{matrix} a \\ - \end{matrix}; q; t\right) = \left(\begin{matrix} at \\ t \end{matrix}; q\right)_\infty . \tag{A.6}$$

We say “a ${}_1\phi_0$ is summable”. (To a neat product, that is.)

Next we prove an identity known as “q-Gauss”, due to Heine (1847),

$$ {}_2\phi_1\left(\begin{matrix} a, b \\ c \end{matrix}; q; \frac{c}{ab}\right) = \left(\begin{matrix} c/a, c/b \\ c, c/(ab) \end{matrix}; q\right)_\infty . \tag{A.7}$$

In order to do this, we first prove a transformation formula for ${}_2\phi_1$’s, also due to Heine (1847),

$$ {}_2\phi_1\left(\begin{matrix} a, b \\ c \end{matrix}; q; t\right) = \left(\begin{matrix} b, at \\ c, t \end{matrix}; q\right)_\infty {}_2\phi_1\left(\begin{matrix} c/b, t \\ at \end{matrix}; q; b\right). \tag{A.8}$$

We have

$$\begin{aligned} {}_2\phi_1\left(\begin{matrix} a, b \\ c \end{matrix}; q; t\right) &= \sum_{k \geq 0} \frac{(a;q)_k (b;q)_k}{(q;q)_k (c;q)_k} t^k \\ &= \frac{(b;q)_\infty}{(c;q)_\infty} \sum_{k \geq 0} \frac{(a;q)_k (cq^k;q)_\infty}{(q;q)_k (bq^k;q)_\infty} t^k \end{aligned}$$

$$
\begin{aligned}
&= \frac{(b;q)_\infty}{(c;q)_\infty}\sum_{k\geq 0}\frac{(a;q)_k}{(q;q)_k}t^k\sum_{l\geq 0}\frac{(c/b;q)_l}{(q;q)_l}(bq^k)^l \quad \text{by (A.3)}\\
&= \frac{(b;q)_\infty}{(c;q)_\infty}\sum_{l\geq 0}\frac{(c/b;q)_l}{(q;q)_l}b^l\sum_{k\geq 0}\frac{(a;q)_k}{(q;q)_k}(tq^l)^k\\
&= \frac{(b;q)_\infty}{(c;q)_\infty}\sum_{l\geq 0}\frac{(c/b;q)_l}{(q;q)_l}b^l\frac{(atq^l;q)_\infty}{(tq^l;q)_\infty} \quad \text{by (A.3)}\\
&= \frac{(b;q)_\infty(at;q)_\infty}{(c;q)_\infty(t;q)_\infty}\sum_{l\geq 0}\frac{(c/b;q)_l(t;q)_l}{(q;q)_l(at;q)_l}b^l\\
&= \begin{pmatrix} b, at \\ c, t \end{pmatrix}; q\Big)_\infty\ {}_2\phi_1\begin{pmatrix} c/b, t \\ at \end{pmatrix}; q; b\Big),
\end{aligned}
$$

as claimed.

If in (A.8) we set $t = c/(ab)$, we find that

$$
\begin{aligned}
{}_2\phi_1\left(\begin{matrix} a, b \\ c \end{matrix}; q; \frac{c}{ab}\right) &= \left(\begin{matrix} b, c/b \\ c, c/(ab) \end{matrix}; q\right)_\infty {}_1\phi_0\left(\begin{matrix} c/(ab) \\ - \end{matrix}; q; b\right)\\
&= \left(\begin{matrix} b, c/b \\ c, c/(ab) \end{matrix}; q\right)_\infty \frac{(c/a;q)_\infty}{(b;q)_\infty} \quad \text{by (A.6)}\\
&= \left(\begin{matrix} c/a, c/b \\ c, c/(ab) \end{matrix}; q\right)_\infty,
\end{aligned}
$$

which is (A.7).

Here we have just scraped the surface of a vast subject. The interested reader is referred to Andrews (1976), Gasper-Rahman (1990), and Andrews-Askey-Roy (1999).

References

G.E. Andrews, The Theory of Partitions Encyclopedia of Mathematics and its Applications, vol. 2 (Addison-Wesley, Reading, 1976).

G.E. Andrews, R. Askey, R. Roy, *Special Functions*, vol. 71, Encyclopedia of Mathematics and its Applications (Cambridge University Press, Cambridge, 1999).

G. Gasper, M. Rahman, *Basic Hypergeometric Series*, vol. 35, Encyclopedia of Mathematics and its Applications (Cambridge University Press, Cambridge, 1990).

References

1. A. Alaca, S. Alaca, K.S. Williams, On the two-dimensional theta functions of the Borweins. Acta Arith. **124**, 177–195 (2006)
2. K. Alladi, B. Gordon, Vanishing coefficients in the expansion of products of Rogers-Ramanujan type, *The Rademacher Legacy in Mathematics*, vol. 156, Contemporary Mathematics (AMS, Providence, 1994), pp. 129–139
3. G.E. Andrews, A polynomial identity which implies the Rogers-Ramanujan identities. Scripta Math. **28**, 297–305 (1970)
4. G.E. Andrews, Ramanujan's "Lost" notebook III the Rogers-Ramanujan continued fraction. Adv. Math. **41**, 186–208 (1981)
5. G.E. Andrews, *Generalized Frobenius Partitions*, vol. 301, Memoirs of the American Mathematical Society (AMS, Providence, 1984)
6. G.E. Andrews, B.C. Berndt, *Ramanujan's Lost Notebook, Part I* (Springer, Berlin, 2005)
7. G.E. Andrews, B.C. Berndt, *Ramanujan's Lost Notebook, Part III* (Springer, Berlin, 2012)
8. G.E. Andrews, D. Bressoud, Vanishing coefficients in infinite product expansions. J. Aust. Math. Soc. Ser. A **27**, 199–202 (1979)
9. G.E. Andrews, F.G. Garvan, Dyson's crank of a partition. Bull. Am. Math. Soc. **18**, 167–171 (1987)
10. G.E. Andrews, M.D. Hirschhorn, J.A. Sellers, Arithmetic properties of partitions with even parts distinct. Ramanujan J. **23**, 165–181 (2010)
11. A.O.L. Atkin, Proof of a conjecture of Ramanujan. Glasg. Math. J. **8**, 14–32 (1967)
12. A.O.L. Atkin, H.P.F. Swinnerton-Dyer, Some properties of partitions. Proc. Lond. Math. Soc. **4**, 84–106 (1954)
13. R.J. Baxter, Hard hexagons: exact solution. J. Phys. A **13**, L61–L70 (1980)
14. B.C. Berndt, *Ramanujan's Notebooks, Part III* (Springer, Berlin, 1991)
15. B.C. Berndt, *Number Theory in the Spirit of Ramanujan*, vol. 34, Student Mathematical Library (AMS, Providence, 2006)
16. B.C. Berndt, G. Choi, Y.-S. Choi, H. Hahn, B.P. Yeap, A.J. Yee, H. Yesilyurt, J. Yi, *Ramanujan's Forty Identities for the Rogers-Ramanujan Functions*, vol. 188, Memoirs of the American Mathematical Society (AMS, Providence, 2007)
17. B.C. Berndt, R.A. Rankin, *Ramanujan Letters and Commentary*, vol. 9, History of Mathematics (AMS, Providence, 1995)
18. R. Blecksmith, J. Brillhart, I. Gerst, Parity results for certain partition functions and identities similar to theta function identities. Math. Comput. **48**, 29–38 (1987)
19. R. Blecksmith, J. Brillhart, I. Gerst, Some infinite product identities. Math. Comput. **51**, 301–314 (1988)

M.D. Hirschhorn, *The Power of q*, Developments in Mathematics 49,
DOI 10.1007/978-3-319-57762-3

20. J.M. Borwein, P.B. Borwein, A cubic counterpart of Jacobi's identity and the AGM. Trans. Am. Math. Soc. **323**, 691–701 (1991)
21. J.M. Borwein, P.B. Borwein, F. Garvan, Some cubic modular identities of Ramanujan. Trans. Am. Math. Soc. **343**, 35–47 (1994)
22. Z. Cao, Y. Hu, Exact covering systems, quadratic forms, and product identities for theta functions, preprint
23. L. Carlitz, M.V. Subbarao, A simple proof of the quintuple product identity. Proc. Am. Math. Soc. **32**, 42–44 (1972)
24. L. Carlitz, M.V. Subbarao, On a combinatorial identity of Winquist and its generalization. Duke Math. J. **39**, 165–172 (1972)
25. H.-C. Chan, On the Andrews-Schur's proof of the Rogers-Ramanujan identities. Ramanujan J. **23**, 417–431 (2010)
26. H.-C. Chan, Ramanujan's cubic continued fraction and an analog of his "Most Beautiful Identity". Int. J. Number Theory **6**, 673–680 (2010)
27. H.-C. Chan, Ramanujan's cubic continued fraction and Ramanujan-type congruences for a certain partition function. Int. J. Number Theory **6**, 819–831 (2010)
28. H.H. Chan, P.C. Toh, New analogues of Ramanujan's partition identities. J. Number Theory **130**, 1898–1913 (2010)
29. S.H. Chan, H. Yesilyurt, The periodicity of the signs of the coefficients of certain infinite products. Pac. J. Math. **225**(1), 13–32 (2006)
30. W.Y.C. Chen, B.L.S. Lin, Congruences for bipartitions with odd parts distinct. Ramanujan J. **25**, 277–293 (2011)
31. W.Y.C. Chen, E.X.W. Xia, Proof of a conjecture of Hirschhorn and Sellers on overpartitions. Acta Arith. **163**, 59–63 (2014)
32. S. Chowla, Congruence properties of partitions. J. Lond. Math. Soc. **9**, 247 (1934)
33. S. Cooper, The quintuple product identity. Int. J. Number Theory **2**, 115–161 (2006)
34. S. Cooper, On Ramanujan's function $k(q) = r(q)r(q^2)^2$. Ramanujan J. **20**, 311–328 (2009)
35. S. Cooper, *Development of Elliptic Functions According to Ramanujan, Originally by K. Venkatachaliengar*, vol. 8, Monographs in Number Theory (World Scientific, Singapore, 2011)
36. S. Cooper, M.D. Hirschhorn, On some sum-to-product identities. Bull. Aust. Math. Soc. **63**, 353–365 (2001)
37. S. Cooper, M.D. Hirschhorn, On some infinite product identities. Rocky Mountain J. Math. **31**, 131–139 (2001)
38. S. Cooper, M.D. Hirschhorn, Factorizations that involve Ramanujan's function $k(q) = r(q)r(q^2)^2$. Acta Math. Sin. Engl. Ser. **27**(12), 2301–2308 (2011)
39. S. Cooper, M.D. Hirschhorn, R.P. Lewis, Powers of Euler's product and related identities. Ramanujan J. **3**, 153–158 (1999)
40. S. Cooper, M.D. Hirschhorn, J.A. Sellers, Partitions into four squares. Proc. Jangjeon Math. Soc. **8**, 73–94 (2005)
41. P.G.L. Dirichlet, J. Math. **21** (1840) 3–6, Werke, pp. 463, 466
42. F.J. Dyson, Some guesses in the theory of partitions. Eureka (Cambridge) **8**, 10–15 (1944)
43. A.B. Ekin, Inequalities for the crank. J. Comb. Theory Ser. A **83**, 283–289 (1998)
44. H.M. Farkas, I. Kra, Partitions and theta constant identities, *Analysis, Geometry, Number Theory, the Mathematics of Leon Ehrenpreis*, vol. 251, Contemporary Mathematics (AMS, Providence, 2000), pp. 197–203
45. H.M. Farkas, I. Kra, *Theta Constants, Riemann Surfaces and the Modular Group*, vol. 37, Graduate Studies in Mathematics (American Mathematical Society, Providence, 2001)
46. N.J. Fine, *Basic Hypergeometric Series and Applications*, vol. 27, Mathematical Surveys and Monographs (AMS, Providence, 1988)
47. F.G. Garvan, A simple proof of Watson's partition congruences for powers of 7. J. Aust. Math. Soc. Ser. A **36**, 316–334 (1984)
48. F.G. Garvan, New combinatorial interpretations of Ramanujan's partition congruences mod 5, 7 and 11. Trans. Am. Math. Soc. **305**, 47–77 (1988)

49. F. Garvan, D. Stanton, Sieved partition functions and q-binomial coefficients. Math. Comput. **55**, 299–311 (1990)
50. F. Garvan, D. Kim, D. Stanton, Cranks and t-cores. Inventiones Mathematicae **101**, 1–17 (1990)
51. B. Gordon, Some continued fractions of the Rogers-Ramanujan type. Duke Math. J. **32**, 741–748 (1965)
52. R.Wm. Gosper, Some identities for your amusement, in *Ramanujan Revisited, Proceedings of the Centenary Conference, University of Illinois at Urbana-Champaign 1–5 June 1987*, pp. 607–610
53. M.B. Green, J.H. Schwarz, E. Witten, *Superstring Theory I, Introduction, 25th Anniversary Edition* (Cambridge University Press, Cambridge, 2012)
54. C. Gu, L. Wang, E.X.W. Xia, Congruences modulo 81 and 243 for generalized Frobenius partitions with 6 colors. Acta Arith. **175**(3), 291–300 (2016)
55. C. Gugg, A new proof of Ramanujan's modular equation relating $R(q)$ with $R(q^5)$. Ramanujan J. **20**, 163–177 (2009)
56. C. Gugg, Modular equations for cubes of the Rogers-Ramanujan and Ramanujan-Göllnitz functions and their associated continued fractions. J. Number Theory **132**, 1519–1553 (2012)
57. M. Hall Jr., Lectures by Godfrey Harold Hardy on the Mathematical Work of Ramanujan, Institute of Advanced Study, Princeton (1936)
58. G.H. Hardy, *Ramanujan. Twelve Lectures on Subjects Suggested by His Life and Work* (AMS Chelsea, Providence, 1999)
59. G.H. Hardy, P.V. Seshu Aiyar, B.M. Wilson, *Collected Papers of Srinivasa Ramanujan* (AMS Chelsea, Providence, 2000)
60. G.H. Hardy, E.M. Wright, *An Introduction to the Theory of Numbers*, 4th edn. (Oxford University Press, Oxford, 1960)
61. M.D. Hirschhorn, Partitions and Ramanujan's continued fraction. Duke Math. J. **39**, 789–791 (1972)
62. M.D. Hirschhorn, Simple proofs of identities of MacMahon and Jacobi. Discret. Math. **16**, 161–162 (1976)
63. M.D. Hirschhorn, Polynomial identities which imply identities of Euler and Jacobi. Acta Arith. **XXXII**, 73–78 (1977)
64. M.D. Hirschhorn, A simple proof of Jacobi's four-square theorem. J. Aust. Math. Soc. Ser. A **32**, 61–67 (1982)
65. M.D. Hirschhorn, A simple proof of Jacobi's two-square theorem. Am. Math. Mon. **92**, 579–580 (1985)
66. M.D. Hirschhorn, A simple proof of Jacobi's four-square theorem. Proc. Am. Math. Soc. **101**, 436–438 (1987)
67. M.D. Hirschhorn, A generalisation of Winquist's identity and a conjecture of Ramanujan. J. Indian Math. Soc. **51**, 49–55 (1987)
68. M.D. Hirschhorn, A generalisation of the quintuple product identity. J. Aust. Math. Soc. Ser. A **44**, 42–45 (1988)
69. M.D. Hirschhorn, A birthday present for Ramanujan. Am. Math. Mon. **97**, 398–400 (1990)
70. M.D. Hirschhorn, Ramanujan's partition congruences. Discret. Math. **131**, 351–355 (1994)
71. M.D. Hirschhorn, On the expansion of Ramanujan's continued fraction. Ramanujan J. **2**, 521–527 (1998)
72. M.D. Hirschhorn, An identity of Ramanujan, and applications, *in q-Series from a Contemporary Perspective*, vol. 254, Contemporary Mathematics (AMS, Providence, 2000), pp. 229–234
73. M.D. Hirschhorn, Winquist and the Atkin–Swinnerton-Dyer congruences for modulus 11. Australas. J. Comb. **22**, 101–104 (2000)
74. M.D. Hirschhorn, Partial fractions and four classical theorems of number theory. Am. Math. Mon. **107**, 260–264 (2000)
75. M.D. Hirschhorn, Some formulae for partitions into squares. Discret. Math. **211**, 225–228 (2000)

76. M.D. Hirschhorn, On the expansion of a continued fraction of Gordon. Ramanujan J. **5**, 369–375 (2001)
77. M.D. Hirschhorn, Three classical results on representations of a number, in *The Andrews Festschrift, Seventeen Papers on Classical Number Theory and Combinatorics*, ed. by D. Foata, G.-N. Han (Springer, Berlin, 2001), pp. 159–165
78. M.D. Hirschhorn, On partitions into four squares of equal parity. Australas. J. Comb. **24**, 285–291 (2001)
79. M.D. Hirschhorn, The number of representations of a number by various forms. Discret. Math. **298**, 205–211 (2005)
80. M.D. Hirschhorn, The case of the mysterious sevens. Int. J. Number Theory **2**, 213–216 (2006)
81. M.D. Hirschhorn, The case of the mysterious sevens. Aust. Math. Soc. Gaz. **35**, 113–114 (2008)
82. M.D. Hirschhorn, A letter from Fitzroy House. Am. Math. Mon. **115**, 563–566 (2008)
83. M.D. Hirschhorn, The number of representations of a number by various forms involving triangles, squares, pentagons and octagons, in *Ramanujan Rediscovered, Proceedings of a Conference on Elliptic Fuctions, Partitions, and q-Series in Memory of K. Venkatachaliengar, Bangalore, 1–5 June, 2009*. RMS Lecture Notes Series, vol. 14 (2010), pp. 113–124
84. M.D. Hirschhorn, Ramanujan's "most beautiful identity". Am. Math. Mon. **118**(9), 839–845 (2011)
85. M.D. Hirschhorn, On the 2- and 4-dissections of Ramanujan's contiunued fraction and its reciprocal. Ramanujan J. **24**, 85–92 (2011)
86. M.D. Hirschhorn, A note on overcubic partitions. N. Z. J. Math. **42**, 229–234 (2012)
87. M.D. Hirschhorn, My Contact with Ramanujan, Journal of the Indian Mathematical Society New Series, Special Volume to Commemorate the 125th Birth Anniversary of Srinivasa Ramanujan and the National Mathematics Year–2012 (2013), pp. 33–43
88. M.D. Hirschhorn, Factorizations of certain q-series identities of Ramanujan and others. Ramanujan J. **31**, 15–22 (2013)
89. M.D. Hirschhorn, A short and simple proof of Ramanujan's mod 11 partition congruence. J. Number Theory **139**, 205–209 (2014)
90. M.D. Hirschhorn, An identity that may have changed the course of history, *Ramanujan 125*, vol. 627, Contemporary Mathematics (AMS, Providence, 2014), pp. 103–109
91. M.D. Hirschhorn, Some congruences for 6-coloured generalized Frobenius partitions. Ramanujan J. **40**, 463–471 (2016)
92. M.D. Hirschhorn, F. Garvan, J.M. Borwein, Cubic analogues of the Jacobian theta function $\theta(z, q)$. Can. J. Math. **45**, 673–694 (1993)
93. M.D. Hirschhorn, D.C. Hunt, A simple proof of the Ramanujan conjecture for powers of 5. J. Reine Angew. Math. **326**, 1–17 (1981)
94. M.D. Hirschhorn, J.A. Sellers, Some relations for partitions into squares, in *Special Functions 1999, Proceedings of the International Workshop on Special Functions, Hong Kong* (World Scientific, 2000), pp. 118–124
95. M.D. Hirschhorn, J.A. Sellers, On a problem of Lehmer on partitions into squares. Ramanujan J. **8**, 279–287 (2004)
96. M.D. Hirschhorn, J.A. Sellers, Partitions with odd parts distinct. Ramanujan J. **22**, 273–284 (2010)
97. M.D. Hirschhorn, J.A. Sellers, Arithmetic relations for overpartitions. J. Comb. Math. Comb. Comput. **53**, 65–73 (2005)
98. C.G.J. Jacobi, Fundamenta nova theoriae functionum ellipticarum, Königsberg (1829)
99. P. Kesava Menon, On Ramanujan's continued fraction and related identities. J. Lond. Math. Soc. **40**, 49–54 (1965)
100. B. Kim, The overcubic partition function mod 3, *Ramanujan Rediscovered, Proceedings of a Conference on Elliptic Fuctions, Partitions, and q-Series in Memory of K. Venkatachaliengar, Bangalore*, 1–5 June, 2009 Ramanujan Math. Soc. Lect. Notes **14**, 157–163 (2010)
101. O. Kolberg, Some identities involving the partition function. Math. Scand. **5**, 77–92 (1957)
102. D.H. Lehmer, On the partitions of numbers into squares. Am. Math. Mon. **55**, 476–481 (1948)

103. R.P. Lewis, Z.-G. Liu, A conjecture of Hirschhorn on the 4-dissection of Ramanujan's continued fraction. Ramanujan J. **4**, 347–352 (2000)
104. L. Lorenz, Contributions à la theorie des nombres. Tidsskrift for Mathematik **3**, 97–114 (1871)
105. P.A. MacMahon, *Combinatory Analysis* (Cambridge University Press, Cambridge, 1916)
106. R.S. Melham, Analogues of Jacobi's two-square theorem, preprint, 16 August 2007, http://maths.science.uts.edu.au/maths/wiki/RayMelham
107. R.S. Melham, Analogues of Jacobi's two-square theorem: an informal account. Integers **10**, 83–100 (2010)
108. S. Ramanujan, On certain arithmetical functions. Proc. Camb. Philos. Soc. **XXII**, 159–184 (1916)
109. S. Ramanujan, Some properties of $p(n)$, the number of partitions of n. Proc. Camb. Philos. Soc. **XIX**, 207–210 (1919)
110. S. Ramanujan, Proof of certain identities in combinatory analysis. Proc. Camb. Philos. Soc. **XIX**, 214–216 (1919)
111. S. Ramanujan, Notebooks, II, Tata Institute of Fundamental Research, Mumbai (2012)
112. S. Ramanujan, *The Lost Notebook and Other Unpublished Papers* (Springer, Berlin, 1988)
113. L.B. Richmond, G. Szekeres, The Taylor coefficients of certain infinite products. Acta Scientiarum Mathematicarum **40**, 347–369 (1978)
114. L.J. Rogers, Second memoir on the expansion of certain infinite products. Proc. Lond. Math. Soc. (2) **XXV**, 318–343 (1894)
115. C. Sandon, F. Zanello, Warnaar's bijection and colored partition identities I. J. Comb. Theory Ser. A **120**, 28–38 (2013)
116. C. Sandon, F. Zanello, Warnaar's bijection and colored partition identities II. Ramanujan J. **33**, 83–120 (2014)
117. I. Schur, Ein Beitrag zur additiven Zahlentheorie und zur Theorie der Kettenbrüche. Berliner Sitzungsberichte **23**, 301–321 (1917)
118. J.A. Sellers, Elementary proofs of congruences for the cubic and overcubic partition functions. Australas. J. Comb. **66**(2), 191–197 (2014)
119. P.C. Toh, On representations by figurate numbers: a uniform approach to the conjectures of Melham. Int. J. Number Theory **9**, 1055–1071 (2013)
120. H. Valentier (ed.), Oeuvres Scientifique de L. Lorenz, II, Librarie Lehmann & Stage, Copenhagen (1904)
121. S.O. Warnaar, A generalization of the Farkas and Kra partition theorem for modulus 7. J. Comb. Theory Ser. A **110**, 43–52 (2005)
122. G.N. Watson, Theorems stated by Ramanujan (VII): theorems on continued fractions. J. Lond. Math. Soc. **4**, 39–48 (1929)
123. G.N. Watson, Ramanujans Vermutung über Zerfällungsanzahlen. J. Reine Angew. Math. **179**, 97–128 (1938)
124. L. Winquist, An elementary proof of $p(11n+6) \equiv 0 \pmod{11}$. J. Comb. Theory **6**, 56–59 (1969)
125. X. Xiong, The number of cubic partitions modulo powers of 5. Scientia Sinica **41**(1), 1–15 (2011)
126. H. Yesilyurt, A generalization of a modular identity of Ragers. J. Number Theory **129**, 1256–1271 (2009)
127. H. Yesilyurt, Elementary proofs of some identities of Ramanujan for the Rogers-Ramanujan functions. JMAA **388**, 420–434 (2009)
128. B. Yuttanan, New properties for the Ramanujan-Göllnitz-Gordon continued fraction. Acta Arith. **151**, 293–310 (2012)
129. H.S. Zuckerman, Identities analogous to Ramanujan's identities involving the partition function. Duke Math. J. **5**, 88–110 (1939)
130. B. Zwiebach, *A First Course in String Theory*, 2nd edn. (Cambridge University Press, Cambridge, 2009)

Index

M.D. Hirschhorn, *The Power of q*, Developments in Mathematics 49,
DOI 10.1007/978-3-319-57762-3